MEGACITIES AND GLOBAL CHANGE
MEGASTÄDTE UND GLOBALER WANDEL

herausgegeben von

Frauke Kraas, Martin Coy, Peter Herrle und Volker Kreibich

_____

Band 20

Alexander Follmann

Governing Riverscapes

Urban Environmental Change along the River Yamuna
in Delhi, India

Franz Steiner Verlag

Umschlagabbildung:
Old Railway Bridge over the Yamuna in New Delhi, India © Alexander Follmann

Bibliografische Information der Deutschen Nationalbibliothek:
Die Deutsche Nationalbibliothek verzeichnet diese Publikation in der Deutschen
Nationalbibliografie; detaillierte bibliografische Daten sind im Internet über
<http://dnb.d-nb.de> abrufbar.

Dieses Werk einschließlich aller seiner Teile ist urheberrechtlich geschützt.
Jede Verwertung außerhalb der engen Grenzen des Urheberrechtsgesetzes
ist unzulässig und strafbar.
© Franz Steiner Verlag, Stuttgart 2016
Druck: Hubert & Co, Göttingen
Gedruckt auf säurefreiem, alterungsbeständigem Papier.
Printed in Germany.
ISBN 978-3-515-11430-1 (Print)
ISBN 978-3-515-11435-6 (E-Book)

CONTENTS

LIST OF FIGURES 11

LIST OF TABLES 14

LIST OF MAPS 14

GLOSSARY 15

LIST OF ABBREVIATIONS 16

SUMMARY 19

ZUSAMMENFASSUNG 21

PREFACE 23

ACKNOWLEDGEMENTS 25

I   INTRODUCTION AND RESEARCH CONTEXT 27
   1   Introduction 27
       1.1   Delhi: a tale of two cities, but only one river 29
       1.2   Change of perspective: from the riverfront to riverscapes 31
       1.3   Discourse and urban environmental policy-making 35
       1.4   Governance and urban political ecology 36
       1.5   Research agenda and questions 37
       1.6   Outline of the book 40

   2   India's megacities: challenging urban ecologies 41
       2.1   Challenges of urbanization in India 42
       2.2   Post-liberalization urban India 43
       2.3   Making India's cities 'world-class' 46
       2.4   India's urban environmental awakening? 47
       2.5   Reclaiming India's urban rivers 51

II THEORETICAL REFLECTIONS 55
   3   Governing riverscapes: theoretical and conceptual reflections 55
       3.1   Governance research 55

		3.1.1 Good governance – the normative perspective 57
		3.1.2 A (critical) descriptive perspective on governance 58
		3.1.3 'Unpacking' governance processes – the analytical perspective 60
		3.1.4 Urban governance 61
		3.1.5 Environmental governance 62
		3.1.6 Urban environmental governance – an analytical approach 63
		3.1.7 Participation in urban environmental governance 66
	3.2	Urban Political Ecology 67
		3.2.1 Introducing UPE 67
		3.2.2 Deconstructing social constructions 72
		3.2.3 The production of urban environments 75
		3.2.4 The urban environment as a hybrid 77
		3.2.5 From land- and waterscapes to hybrid riverscapes 81
		3.2.6 Riverscapes in neighboring disciplines 84
	3.3	Riverscapes as hybrids – a framework for research 86
		3.3.1 Hybridity of riverscapes and the interplay of materiality and discourse 87
		3.3.2 A UPE-governance approach to study riverscapes 90
		3.3.3 Redefining the research questions 91

III METHODOLOGY AND METHODS 93

4 Methodology 93

	4.1	Research philosophy 93
	4.2	The chosen discourse analytical approach 94
	4.3	The argumentative approach to discourse 96
		4.3.1 Story-lines and discourse-coalitions 96
		4.3.2 Discourse structuration and discourse institutionalization 97
	4.4	The application of the argumentative approach in this study 98

5 Methods and fieldwork 100

	5.1	Data collection and fieldwork 100
		5.1.1 Triangulation of data and analysis 101
		5.1.2 Primary data: multi-temporal mapping and observation 101
		5.1.3 Primary data: interviews 102
		5.1.4 Primary data: Yamuna Future Workshop 107
		5.1.5 Secondary data: the text corpora 108

		5.2	From data processing to analysis	111
		5.3	Reflections on the research process	112

IV SITUATING THE CASE STUDY ... 115

6 The megacity of Delhi ... 115

		6.1	The physical features of Delhi	115
		6.2	The growth of the city	117
		6.3	The socio-spatial mosaic of Delhi	120
		6.4	The 'world-class' agenda – Delhi as a global city	122

7 Introducing Delhi's riverscapes ... 123

		7.1	The river Yamuna	123
		7.2	The river Yamuna in Delhi	125
		7.3	Water abstraction and flow	126
		7.4	Pollution and environmental degradation	129
		7.5	The risk of monsoonal flooding	133
		7.6	Interim conclusion: major challenges	135

8 Urban environmental governance: introducing actors and policy debates ... 137

		8.1	Political and administrative organization in India	137
		8.2	Delhi's administrative history and current set-up	138
		8.3	Urban governance framework: the state actors	139

			8.3.1 Agencies on the national level	141
			8.3.2 Agencies on the regional level	143
			8.3.3 Agencies on the state level	144

		8.4	Environmental governance framework: the state actors	144

			8.4.1 Agencies on the national and interstate level	146
			8.4.2 Agencies on the state level	149

		8.5	Overview of the state agencies involved in governing Delhi's riverscapes	149
		8.6	India's urban rivers: legal provisions and debates	151
		8.7	The courts and the river: overview of selected orders	155
		8.8	Environmental NGOs working on the Yamuna	157
		8.9	Interim conclusion: a complex urban environmental governance set-up	160

V HISTORICAL-GEOGRAPHICAL INSIGHTS INTO THE MATERIAL AND DISCURSIVE REMAKING OF DELHI'S RIVERSCAPES ... 163

9 The early remaking of Delhi's riverscapes ... 163

		9.1	From the Mughal period to colonial times	163
		9.2	A new riverfront for the imperial capital	170

Contents

- 9.3 The riverfront ideas of the early post-colonial period: the 'recreation plan' 171
- 9.4 Taming the river for the development of the city 173
- 9.5 Delhi's riverscapes under the first Master Plan 175
- 9.6 Interim conclusion 176

10 Channelization and riverfront development – a persistent idea 179
- 10.1 The emergence of the channelization idea 179
- 10.2 Learning from the floods of 1978 180
- 10.3 DDA's plans of the 1980s and 1990s 182
- 10.4 Channelization in the second Master Plan (MPD-2001) 186
- 10.5 Land acquisition for channelization 189
- 10.6 Interim conclusion 192

11 Drafting the Zonal Development Plan 194
- 11.1 Change of land use for pocket III: the starting point for channelization 194
- 11.2 Seeking approval for channelization – the draft ZDP of 1998 195
- 11.3 Negotiating scientific 'facts' and recommendations 199
- 11.4 The ZDP draft of 2006: proposing 'partial' channelization 201
- 11.5 'Wasted land' and the 'fear of encroachment' versus 'ecosystem services' 205
- 11.6 Floodplain zoning: institutionalization or co-optation? 208
- 11.7 The ZDP draft of 2008: Seeking post-facto approval for developments 210
- 11.8 Interim conclusion 213

VI THE RECLAMATION OF DELHI'S RIVERSCAPES FOR A WORLD-CLASS CITY IN THE MAKING 215

12 Multi-temporal analysis of land-use change (2001–2014) 215
- 12.1 Selection of research area for land-use change analysis 215
- 12.2 Land-use changes 2001–2014 216
- 12.3 Opening up the River Zone: the urban-mega projects 220

13 'Cleansing' and 'reclaiming' Delhi's riverscapes 223
- 13.1 Agriculture in Delhi's riverscapes 223
  - 13.1.1 The governance of farming Delhi's riverscapes 224
  - 13.1.2 Reclaiming the land for urban development 226
- 13.2 Cleansing Delhi's riverscapes from the slums 230
  - 13.2.1 The slum demolitions (2004–2006) 231
  - 13.2.2 'Green' evictions: discursive reasoning and the role of the courts 232
  - 13.2.3 Fuzzy boundaries, legal categories: 300 meters for the river 236

13.2.4 Demolitions, religious structures and
the channelization scheme 239
13.2.5 Background rationalities:
plans for a riverfront promenade 240

13.3 Remaining residential spaces in Delhi's riverscapes 241
13.4 The effects of simplification:
permitting construction beyond 300 meters 244
13.5 Interim conclusion 245

14 The political ecology of embankments and urban mega-projects 247

14.1 The Akshardham Temple Complex 247

14.1.1 An offspring of the channelization scheme 248
14.1.2 The construction of the Akshardham Bund 250
14.1.3 Religious and political prestige – a powerful coalition 250
14.1.4 Objection against the Akshardham Temple 251

14.2 The Commonwealth Games Village 253

14.2.1 The secret 'approval' of the site 253
14.2.2 Remaking environmental clearances 256
14.2.3 Formation of environmental protest:
Yamuna Jiye Abhiyaan 258
14.2.4 The Games Village in the courts 262
14.2.5 A new boundary of the river?
Multiple readings of the Akshardham Bund 263
14.2.6 The river in between science, politics and
time constraints 266
14.2.7 Agency of the river?
Drainage problem versus floods 268
14.2.8 Economic (ir)rationalities: the bailout package
and corruption 270
14.2.9 Ground realities compared to planning schemes 271

14.3 'Development must take place' and
'leave the river to the experts' 272
14.4 Interim conclusion 274

VII AFTER THE URBAN MEGA-PROJECTS: PURIFYING DELHI'S RIVERSCAPES FOR RESTORATION, CONSERVATION AND BEAUTIFICATION 277

15 'Managing dissent': signs of change 277

15.1 A new authority and a moratorium for the river? 278
15.2 A plan for the river 284
15.3 Interim conclusion 287

16 Redefining the river's boundaries 289

16.1	The model: the Yamuna Biodiversity Park	289
16.2	Scaling up: from the Biodiversity Park to the Biodiversity Zone	290
16.3	Actions on the ground: the Golden Jubilee Park	293
16.4	Evicting the 'green keepers'? An attempt to develop a counter-story line	298
16.5	Remaking the plan: the influence of the NGOs and the courts	301

    16.5.1 The evolvement of the Debris Case at the NGT 301
    16.5.2 Rezoning the river – DDA's move to redefine the boundaries of the river 307
    16.5.3 The action plan for 'Restoration and Conservation' 310
    16.5.4 The orders of the NGT: 'Maily se Nirmal Yamuna' 311

  16.6   Interim conclusions 315

VIII GOVERNING DELHI'S RIVERSCAPES: A SYNTHESIS 317

17 Reflection and discussion 317

  17.1   Revisiting the main planning discourses and story-lines 317

    17.1.1 The main discourses and story-lines 318
    17.1.2 The persistent story-lines 321
    17.1.3 'Planned' versus 'unplanned' encroachment 322
    17.1.4 DDA's reluctant planning for the River Zone 324

  17.2   The hybridization and purification of Delhi's riverscapes 326
  17.3   Urban environmental governance: prevailing insights 330

18 Conclusion 335

  18.1   Riverscapes: (mega-)urban challenges and the question of governance 337
  18.2   Evaluation of the conceptual approach: do riverscapes make sense? 339

REFERENCES 341

APPENDIX 377

MAPS 385

COLOR PHOTOGRAPHS 392

LIST OF FIGURES

Figure 1: River Zone (Zone O) and research area 30
Figure 2: Ghats along the Yamuna in Delhi (Yamuna Bazaar) 52
Figure 3: Ghats along the Ganga in Benares (Varanasi), Uttar Pradesh 52
Figure 4: Demolished slums, Sabarmati riverfront in Ahmedabad, Gujarat 52
Figure 5: Riverfront slums along the Yamuna in Delhi 52
Figure 6: Sabarmati Riverfront Development Project in Ahmedabad, Gujarat 53
Figure 7: Advertisement for the Sabarmati riverfront project, Ahmedabad 53
Figure 8: Rivers and the spheres of governance 64
Figure 9: The practices of 'hybridization' and 'purification' 79
Figure 10: Hybridity of riverscapes 87
Figure 11: The material and discursive production of hybrid riverscapes 88
Figure 12: The embeddedness of urban environmental governance 90
Figure 13: Overview of the physical features of Delhi 116
Figure 14: The growth of Delhi 119
Figure 15: Schematic map of the Yamuna up to Delhi 128
Figure 16: Discharge of polluted water at Okhla Barrage, South Delhi 131
Figure 17: Polluted water for irrigation, Agra Canal headworks, South Delhi .. 131
Figure 18: Annual maximum water level at the Old Railway Bridge 133
Figure 19: Government structure of the NCT of Delhi 139
Figure 20: Urban governance framework 140
Figure 21: Environmental governance framework 145
Figure 22: Governance framework for Delhi's riverscapes 161
Figure 23: Delhi's riverscapes ca. 1807 164
Figure 24: Guide bund, Geeta Colony Bridge 188
Figure 25: ITO Barrage 188
Figure 26: DTTDC's plan for channelization and riverfront development (2006) 204
Figure 27: Floodplain Zoning (O-Zone) prepared by DDA 209
Figure 28: Land-use change in the research area by main categories (2001–2014) 219

List of figures

Figure 29: Metro depot (DMRC) at Shastri Park, East Delhi 221
Figure 30: Information and Technology Park (DMRC) at Shastri Park, East Delhi 221
Figure 31: Planning schemes for 'pocket III' 249
Figure 32: Cross section Akshardham Bund 250
Figure 33: Protest camp 'Yamuna Satyagraha' 259
Figure 34: Tents along the Noida Link Road during flood 269
Figure 35: Flooded construction worker camps next to CWGs-Village 269
Figure 36: Standing water behind the Akshardham Bund 269
Figure 37: Standing water in front of the Akshardham Bund 269
Figure 38: Signage erected during the construction period of the CWGs-Village 272
Figure 39: Construction of Ring Road By-Pass Road near Old Railway Bridge 295
Figure 40: Ring Road By-Pass Road at Geeta Colony Bridge intersection 295
Figure 41: Debris along the Yamuna Pushta Embankment near Geeta Colony .. 302
Figure 42: Debris dumping along guide bund near Yamuna Bank Metro 302
Figure 43: Solid waste recycling near ISBT Bridge, Shastri Park, East Delhi 302
Figure 44: Slums on debris material in front of Shastri Park Metro Depot 302
Figure 45: Public notice boards posted along the embankments 304
Figure 46: Rezoning proposal for River Zone (Zone O) by DDA 307
Figure 47: DTC Millennium Bus Depot 309
Figure 48: 'Land bank' protected by Ring Road By-Pass Road 309
Figure 49: Main discourses and story-lines of Delhi's riverscapes 317
Figure 50: Pollution of the river Yamuna, eastern bank, near CWGs-Village 392
Figure 51: Pollution river Yamuna near ghats, Yamuna Bazar area 392
Figure 52: Riverfront promenade along the river Musi in Hyderabad, Telangana 392
Figure 53: Golden Jubilee Park along the Yamuna 392
Figure 54: Agriculture and temporary settlements of farmers along the Yamuna 393
Figure 55: Plant nursery, Yamuna Bank area 393
Figure 56: Fencing off the river, Kalindi Kunj, South Delhi 393
Figure 57: Okhla Bird Sanctuary, Kalindi Kunj, South Delhi 393
Figure 58: Unauthorized Colony, Jogabai Extension, Kalindi Kunj area 393

List of figures

Figure 59: Unauthorized Colony, Majnu-Ka-Tila (New Aruna Nagar) 394
Figure 60: Akshardham Temple 394
Figure 61: CWGs-Village Entry Gate 394
Figure 62: Construction work on the Akshardham Bund towards the river 394
Figure 63: Akshardham Bund and CWGs-Village 394
Figure 64: Farming Bela Estate near Golden Jubilee Park 395
Figure 65: Former construction workers camp near CWGs-Village 395
Figure 66: Yamuna Biodiversity Park Phase I 395
Figure 67: Yamuna Biodiversity Park Phase II 395
Figure 68: Yamuna Biodiversity Park Phase I Amphitheatre 395
Figure 69: Yamuna Biodiversity Park Phase I Butterfly Park 395
Figure 70: Golden Jubilee Park during flood of 2010 396
Figure 71: Golden Jubilee Park 396
Figure 72: Model for the development of Golden Jubilee Park (phase I) 396

LIST OF TABLES

Table 1: Terminology used for discourse analysis 99

Table 2: Overview and types of conducted in-depth interviews 106

Table 3: Population growth NCT of Delhi 1951–2011 118

Table 4: Area and population of the river basin 124

Table 5: Flood discharge and maximum water level
during major flood events..135

Table 6: Agencies involved in governing Delhi's riverscapes 150

Table 7: Environmental NGOs working on the Yamuna in Delhi 158

Table 8: Land use in the different stretches of the Yamuna in Delhi 217

Table 9: Land-use change in the research area 2001–2014 218

Table 10: Existing and proposed land use for River Zone (Zone O)...................285

LIST OF MAPS

Map 1: Delhi and the river Yamuna 385

Map 2: Land-use change in the research area (north) 2001–2014......................386

Map 3: Land-use change in the research area (south) 2001–2014 387

Map 4: Time series (research area)...388

Map 5: The development of the interrelated mega-projects
(focus area I)...390

Map 6: Changing land uses: slums and the Golden Jubilee Park
(focus area II)...391

GLOSSARY

akhada	a traditional wrestling ground and training center on the riverbank
bāngar	older alluvial soils along rivers, generally safe from floods
bhag	garden, orchard
crore	ten million
dhobi	washerman
ghat	riverside area with a series of steps leading down to the river
headworks	barrage
jamuna par	Trans-Yamuna, East Delhi
jheel	lake
jhuggi jhopdi	huts of the poor, slum (also JJ cluster)
jhuggies	huts of the poor in a slum
khadar / khādir	fertile soils of the low-lying, flood-prone areas along the river
kuccha / kacha houses	houses made of wood, mud and other organic materials
lakh	one hundred thousand
mahābhārata	major Sanskrit epics of ancient India
nadī	river
nallah / nala	stream or drain
Panchayat	local (village) governing body, elected council in rural areas
pandit / panda	Hindu religious scholar and teacher (pundit)
pooja	prayer, worship
pucca houses	solid and permanent houses made of brick, stone and concrete
pushta	bund or embankment, also referred to as riverbank
samādhi	shrine, memorial site

LIST OF ABBREVIATIONS

ANT	actor-network theory
ASL	above sea level
BAPS	Bochasanwasi Shri Akshar Purushottam Swaminarayan
BBMB	Bhakra-Beas Management Board
BCM	billion cubic meter
CGWA	Central Groundwater Authority
CGWB	Central Groundwater Board
CNG	compressed natural gas
CPCB	Central Pollution Control Board
CPWD	Central Public Works Department
CRZ	Coastal Regulation Zone
CSE	Centre for Science and Environment
cusec	cubic feet per second
cumec	cubic meter per second
CWC	Central Water Commission
CWGs	Commonwealth Games
DBU	designated best use
DCB	Delhi Cantonment Board
DDA	Delhi Development Authority
DIT	Delhi Improvement Trust
DJB	Delhi Jal Board
DMRC	Delhi Metro Rail Corporation (in short: Delhi Metro)
DND	Delhi-Noida-Delhi
DPCC	Delhi Pollution Control Committee
DTTDC	Delhi Tourism and Transportation Development Corporation
DUAC	Delhi Urban Arts Commission
DUSIB	Delhi Urban Shelter Improvement Board
EAC	Environmental Appraisal Committee
EIA	Environmental Impact Assessment
FDI	foreign direct investment
GAP	Ganga Action Plan
GDP	Gross Domestic Product
GNCTD	Government of the National Capital Territory
GoI	Government of India
HFL	High Flood Level

HLC	High Level Committee
IAS	Indian Administrative Service
ICT	information communication technology
IFCD	Irrigation and Flood Control Department
INTACH	Indian National Trust for Art and Cultural Heritage
IPCC	Intergovernmental Panel on Climate Change
ITO	Income Tax Office
JJ	Jhuggi jhopdi (JJ-cluster = 'slum')
JNNURM	Jawaharlal Nehru National Urban Renewal Mission
JNU	Jawaharlal Nehru University
L&DO	Land and Development Office
LG	Lieutenant Governor
LIFE	Legal Initiative for Forests and Environment
MCD	Municipal Corporation of Delhi
MoEF	Ministry of Environment and Forests
MoHuPA	Ministry of Housing and Urban Poverty Alleviation
MoU	Memorandum of Understanding
MoUD	Ministry of Urban Development
MoUEPA	Ministry of Urban Employment and Poverty Allevation
MoWR	Ministry of Water Resources
MPD	Master Plan for Delhi
NCRPB	National Capital Region Planning Board
NCT	National Capital Territory
NDA	National Democratic Alliance
NDMC	New Delhi Municipal Corporation
NEERI	National Environmental Engineering Research Institute
NGO	Non-Government Organisation
NGT	National Green Tribunal
NH	National highway
NHF	Natural Heritage First
NOIDA	New Okhla Industrial Development Authority
NRCD	National River Conservation Directory
PIL	Public Interest Litigations
RRZ	River Regulation Zone
PWD	Public Works Department
RTI	Right to Information
SANDRP	South Asia Network for Dams, Rivers and People
SKAD	Sociology of Knowledge Approach to Discourse
SPA	School of Planning and Architecture

STP	Sewage Treatment Plan
TAG	Technical Advisory Group
UN	United Nations
U. P.	Uttar Pradesh
UPE	Urban Political Ecology
UYRB	Upper Yamuna River Board
Rs.	Indian Rupees
WWF	World Wide Fund for Nature
WYC	Western Yamuna Canal
YAP	Yamuna Action Plan
YJA	Yamuna Jiye Abhiyaan
YRDA	Yamuna River Development Authority
YREMC	Yamuna Removal of Encroachments Monitoring Committee
YSC	Yamuna Standing Committee
ZDP	Zonal Development Plan

SUMMARY

The inherent complexity of environmental change along urban rivers in the megacities of the Global South requires a change of perspective going beyond the riverfront. In order to overcome the binary conceptualizations of nature/culture and river/city, this study uses the notion of *riverscapes* as a single terminology referring to the riverine landscape formed by the natural forces of the river and human interventions. By linking a discourse analytical approach with theoretical concepts from governance research and urban political ecology, this study develops the theoretical framework of *riverscapes* to study environmental change along urban rivers.

Interlinked with the opening and liberalization of the Indian economy, the vision to make Delhi a 'world-class' city has transformed the urban landscape of the megacity in manifold ways. The river Yamuna, which divides the city into two parts, was historically degraded to a foul-smelling drain by the city's untreated sewage and was the neglected 'backyard' of the megacity for a long time. The reclamation of the floodplain areas for the planned development of the city has been discussed in Delhi since the late 1970s. For decades, models of European riverfronts dominated the discourses around the river and the urban imaginaries of the planners.

Ecological risks and opposition from environmental groups have prevented large-scale channelization of the river. However, the perception of the river's floodplain as 'wasted land' has transformed it into a pivotal space in the remaking of the city in the twenty-first century. The large-scale slum demolitions and development of urban mega-projects along the banks of the Yamuna are characteristic for dynamic land-use changes in post-liberalization urban India.

The research presented in this book focuses on the multiple city-river relationships and current processes of urban environmental change. The results highlight that dynamic land-use changes and the reclamation of ecologically sensitive spaces are deeply connected to changing discursive framings of the role and function of these socio-ecological hybrids in the remaking of cities. Through analysis of the discourses surrounding Delhi's riverscapes, the study shows how dominant discourses and their associated story-lines have remained persistent over long periods of time and how these discourses have influenced the current processes of urban environmental change and governance.

ZUSAMMENFASSUNG

Für ein besseres Verständnis der Komplexität von urbanen Landnutzungsveränderungen und Umweltproblemen entlang von Flüssen in den Megastädten des Globalen Südens bedarf es empirischer Untersuchungen, die über eine enge räumliche Fokussierung auf die *riverfront* hinausgehen. Hierzu ist es notwendig, die moderne Dichotomie von Natur und Kultur sowie Fluss und Stadt aufzubrechen, um das Fluss-Stadt-Verhältnis neu zu definieren. Diese Studie entwickelt in diesem Kontext ein neues Verständnis für die Erforschung von urbanen Flusslandschaften als sogenannte *riverscapes*. Das theoretische Konzept der *riverscapes* basiert auf einer Verknüpfung von Governance-Forschung und Ansätzen der Urban Political Ecology mit diskursanalytischen Ansätzen.

Im Zuge wirtschaftlicher Liberalisierung sowie fortschreitender Globalisierungsprozesse ist es das Ziel der Stadtentwicklungspolitik, die indische Hauptstadt Delhi in eine „Weltklasse-Stadt" zu verwandeln. Diese ambitionierte Zielsetzung der Stadtentwicklungspolitik verschärft die urbanen Landnutzungskonflikte innerhalb der Megastadt. Ein räumlicher Fokus der Stadterneuerungsmaßnahmen liegt hierbei insbesondere auf der Flussaue der Yamuna, die in der Vergangenheit auf Grund der monsunalen Überschwemmungen sowie der starken Verschmutzung des Flusses städtebaulich unberücksichtigt blieb und im Zuge dessen Raum für Marginalsiedlungen bot. Eine städtebauliche Entwicklung der Uferbereiche nach dem Vorbild westlicher Flüsse wurde bereits seit den späten 1970er Jahren diskutiert, jedoch auf Grund ökologischer Risiken nicht umgesetzt. Eine Eindeichung und Entwicklung von ausgewählten Arealen der Flussaue erfolgte erst im Zuge städtebaulicher Großprojekte nach der Jahrtausendwende. Zusammen mit großflächigen Slumräumungen sind diese Entwicklungen charakteristisch für aktuelle Landnutzungsveränderungen und -konflikte in den indischen Megastädten.

Der Fokus der Untersuchungen liegt auf den vielfältigen Fluss-Stadt Beziehungen und dynamischen Veränderungen in der Flussaue der Yamuna. Die Studie zeigt, dass aktuelle Stadtentwicklungsprojekte eng verknüpft sind mit sich verändernden Stadtentwicklungs- und Umweltschutzdiskursen.

PREFACE

This study was submitted as a doctoral thesis to the Faculty of Mathematics and Natural Sciences of the University of Cologne and was defended on 17th June 2015. Prof. Dr. Boris Braun and Prof. Dr. Frauke Kraas were the reviewers.

ACKNOWLEDGEMENTS

Although it seems impossible to acknowledge everybody who supported me in Delhi and Cologne during the last few years, I would like to thank some key individuals here.

Very special thanks go to my supervisor, Boris Braun, who always encouraged me in my work and gave me all the support a PhD student needs, including a great office environment and plenty of freedom to do my research. I would like to especially thank Boris for the confidence and motivation he gave me through reading and critiquing my drafts. I thank my Indian 'supervisor' N. Sridharan from the School of Planning and Architecture (SPA) in Delhi for the great support provided throughout the years.

I am especially grateful to my interview partners in Delhi. Without their patience and explanations this study would not have been possible. My special thanks goes to Manoj Misra, who welcomed me so many times, introduced me to the farmers along the Yamuna, invited me to court proceedings and conferences, and shared his knowledge on the river with me during multiple hours of discussions. Further, I would like to thank Sudha Mohan for providing additional help and always arranging snacks and drinks for the long talks on the river.

My thanks also goes also to Ravi Agarwal, Manu Bhatnagar, Anupam Mishra, Suresh Rohilla, Diwan Singh, Vikram Soni, and Himanshu Thakkar. All of them have shared their own understandings of the river with me, which have had a lasting effect on my understanding of Delhi's riverscapes. I am grateful for the wonderful talks we had. I further need to thank Rahul Choudhary and Rittwick Dutta from the Legal Initiative for the Environment (LIFE) for digging out all the old court files.

Special reference needs to be given to the late Ramesh C. Trivedi, who explained to me almost everything related to the pollution of the Yamuna. Also to the late Ramaswamy R. Iyer, who helped me enormously to better understand the governance of water and rivers in India. With their deaths, India's rivers have lost important spokesmen. I am grateful for the opportunity to learn from them.

I am further very thankful for the hospitality of my interview partners within the different government departments. A special thank goes to the members of the DDA Landscape Unit, who openly discussed with me their planning schemes for Delhi's riverscapes. I gained the most impressive insights into the everyday changes and governance of Delhi's riverscapes from talks with the farmers (who for their own security shall remain anonymous). I have rarely met friendlier people. Talking to many of them would not have been possible without my translators in the field. I thank Sarandha Jain and Kavita Ramakrishnan for accompanying me on some of the visits and providing this translation

A word of appreciation goes to the German Academic Exchange Service (DAAD) providing the initial scholarship for my time at the SPA in 2009–2010 and the German Academic Scholarship Foundation (Studienstiftung des deutschen

Volkes) for funding my studies for two and a half years (2011–2013). Thanks for institutional support go to the team of Max Mueller Bhavan, and Robin Mallick in particular, who helped to organize the Yamuna Future Workshop in March 2013.

My personal research greatly benefited from the affiliation with the 'Chance2Sustain - Urban Chances, City Growth and the Sustainability Challenge' project. Sincere thanks to Loraine Kennedy and Isa Baud for taking me on-board and inviting me to workshops and conferences and to Véronique Dupont for her inspiring work on Delhi and her thoughtful advice. In Delhi, discussions with Amita Baviskar, Bérénice Bon, Timothy Karpouzoglou, and Awadhendra Sharan have been a source of inspiration. As part of the urban workshop series, organized by the Centre de Sciences Humaines (CSH) and the Centre for Policy Research (CPR), I was able to discuss my work on the river in November 2011 with a larger audience in Delhi. I would like to thank Marie-Hélène Zerah and Partha Mukhopadhyay for inviting me, and the participants for their comments.

A series of discussions with Anna Zimmer over the last few years has influenced my growing interest in Urban Political Ecology and socio-ecological hybrids. Her insightful remarks and sharp feedback helped me enormously. A thousand thanks for the great support. The theoretical concept made major progress based on the valuable comments by Erik Swyngedouw and the other participants during a PhD workshop at the Politecnico di Milano in May 2014. I am grateful to Gloria Pessina for the opportunity to participate.

Heartfelt thanks for all the support also go to the whole team at the Institute of Geography. Special thanks go to Frauke Kraas for her valuable comments on the conceptual framework and the research design, to Regine Spohner and Ulrike Schwedler for preparing the excellent maps, Veronika Selbach for her advice, and Peter Dannenberg for giving me support and time to finish this book. My wonderful colleagues at the institute made long office days enjoyable: thank you Amelie, Annika, Benjamin, Birte, Carsten, Fabian, Franziska, Gerrit, Harald, Holger, Katharina, Madlen, Mareike, Marie, Pamela, Petra, Sebastian, Steffi, Tibor, Tine, and Valerie.

In the final phase, Anna Zimmer, Jürgen Schiemann, Tine Trumpp, Veronika Selbach, Valerie Viehoff, Megha Sud, Lisa-Michéle Bott, Birte Rafflenbeul, and Marie Pahl provided important support in proof reading, as well as Regine Spohner, Benjamin Casper and Benedict Vierneisel in cartography and graphics. Additionally, I thank Hannah Stanley for final proofreading. Special thanks for the wonderful time in Sheikh Sarai B-24 go to my wonderful flat mates and friends Adil, Anna, Dawa, Imran, Karen and Nanu, as well as the Barista group from SDA market, New Delhi.

Finally, the great support of my family deserves my heartfelt thanks. My parents supported me wholeheartedly in all respects. I am also grateful to my siblings and grandparents who were always interested in my work. Lastly, my deepest gratitude to Natalie for all her support and love during the stressful years and long stays abroad. I owe you so much.

I INTRODUCTION AND RESEARCH CONTEXT

1 INTRODUCTION

The megacities of the Global South have emerged as hot spots of global environmental change; both as drivers of this change, as well as experiencing the intense adverse effects (KRAAS 2003, 2007, KRAAS & MERTINS 2014, PARNELL et al. 2007, SINGH 2015, SORENSEN & OKATA 2011). Due to their scale, dynamics and complexity, the largest cities of the world face multiple socio-environmental challenges and their future development is at the center of public debate and scientific research (DAVIS 2006, KRAAS et al. 2014, PARNELL & OLDFIELD 2014, SORENSEN 2011, UN HABITAT 2010).

The dynamic processes of urbanization and associated land-use changes in the megacities of the Global South are driven by a multiplicity of actors embedded in complex global-local relations (HEINRICHS et al. 2012, HOMM 2014). In many cases the governance of megacities in the Global South is characterized by sectoral approaches lacking integrated planning and inter-sectoral coordination (BAUD & DHANALAKSHMI 2007, FARIA et al. 2009, KRAAS & MERTINS 2014, MITTAL et al. 2015). An omnipresent urban informality adds to the multiple challenges for urban governance in many cities of the Global South (ALSAYYAD & ROY 2004, MCFARLANE 2012, MCFARLANE & WAIBEL 2012, ROY 2005, 2009b).

In light of climate change, urban transformations in the megacities of the Global South are intimately linked to the challenge of making cities less vulnerable and more resilient (AßHEUER 2014, AßHEUER & BRAUN 2011, BIRKMANN et al. 2010, GARSCHAGEN 2014, HANSJÜRGENS & HEINRICHS 2014, HORDIJK & BAUD 2011, OTTO-ZIMMERMANN 2011, 2012). Along with specific local challenges of resource overexploitation, environmental degradation and associated health problems ongoing processes of mega-urbanization raise multiple questions of sustainability and socio-environmental justice (AGGARWAL & BUTSCH 2011, RADEMACHER & SIVARAMAKRISHNAN 2013b).

The environmental question is "generally often circumscribed to either rural or threatened 'natural' environments or to 'global' problems", but the central role of the global urbanization process is still under-represented in the environmental debate (SWYNGEDOUW 2004: 9). This neglect of urban nature has been connected to the modern separation of nature and society through which 'the city' has for a long time been considered to be the very antithesis to nature (see among others ANGELO & WACHSMUTH 2014, CHILLA 2005b, HARVEY 1996b, HEYNEN et al. 2006b, KEIL 2003, KEIL & GRAHAM 1998, TREPL 1996). The city, seemingly entirely created by humans, was not considered as a natural ecosystem. As a result, environmental problems in the cities of the Global South have long been largely ignored (HARDOY & SATTERTHWAITE 1991). Only since the turn of the millennium have urban environmental problems in the Global South, especially air and water pollution, come into increasing focus of both state and non-state actors. As a result, a growing 'urban environmentalism' has resulted in new forms of urban environmental governance (BENTON-SHORT & SHORT 2013, BRAND & THOMAS 2005, FARIA et al. 2009, SHUTKIN 2001, VÉRON 2006, WHITEHEAD 2013).

Within this evolution, the modern city/nature dichotomy is challenged. 'Nature' is to some extent brought back into the city, yet this 'reintroduction of nature' into the urban realm does not follow any consistent narrative, but is rather fragmented in space and time, and often emerges as contradictory and highly politicized. Furthermore, despite a growing recognition of the importance of the urban environment, "a paradoxical form of inaction is the norm when it comes to implementing urban environmental solutions" (SHEPPARD 2006: 299). With regard to environmental conflicts in the cities of the Global South, numerous authors have argued that priority has often been given to ecological issues linked to larger questions of intergenerational equity, climate change and natural resource depletion ('green agenda') over the basic needs of the poor and the multiple challenges of the poverty-environment nexus ('brown agenda') (BARTONE et al. 1994, MCGRANAHAN & SATTERTHWAITE 2000, WATSON 2009).

In this context, the current urban environmental politics and socioecological transformations in India's megacities appear as an especially interesting case. Due to continuing population growth and the multiple effects of economic liberalization, India's megacities have been facing dynamic transformations raising manifold questions of urban sustainability and socio-environmental justice. India's current urbanization process poses multiple challenges for urban environmental governance to balance environmental protection and economic development, and the desires of a growing and increasingly assertive middle class[1] and the basic needs of the urban poor (BAVISKAR 2003, 2011a, DE MELLO-THÉRY et al. 2013, MAWDSLEY 2009, RADEMACHER & SIVARAMAKRISHNAN 2013a, TRUELOVE & MAWDSLEY 2011, VÉRON 2006).

Taking the case study of the river Yamuna in India's capital city Delhi, this study seeks to study these multiple governance challenges by linking questions of urban land-use change and urban redevelopment strategies to questions of river pollution and environmental degradation. City-river relationships reflect larger changes in socio-natural configurations and socioecological transformations (HOLIFIELD & SCHUELKE 2015, RADEMACHER 2011). Or more broadly, as HEIKKILA (2011: 33) frames it: "The manner in which societies interact with 'their' rivers tells us as much or more about themselves as it does about the rivers per se." By analyzing the river-city nexus, this study aims to shed light on the socioecological transformation in urban India beyond the physical space of the river Yamuna in Delhi.

---

[1] Writings on the role of India's (emerging) urban middle class(es) tend to use a vague definition (BROSIUS 2010, ELLIS 2011, FERNANDES 2006, GHERTNER 2011d, MAWDSLEY 2004, SRIVASTAVA 2009). For a more detailed discussion see among others SRIDHARAN (2011).

1.1 Delhi: a tale of two cities, but only one river

Delhi has experienced rapid urbanization since India's Independence in 1947. With a population of approximately 17 million India's capital is today one of the largest megacities in the world. The long history of the city has always been closely connected to the river Yamuna, which is often referred to as the lifeline and the green lung of the city (DDA 2007). The ecologically sensitive river zone is the largest remaining natural feature and a crucial life supporting ecosystem of the megacity. The river Yamuna divides the city of Delhi into two parts, referred to as West Delhi and East Delhi (see Map 1, page 385).

The two parts of the megacity are characterized by distinctly different urban morphologies. The historic cores of the city and all major institutional areas are located in West Delhi. West Delhi is a comparatively 'green' city; especially the central, planned areas of the city which feature large green and recreational spaces. In contrast, East Delhi has largely grown informally and unplanned. The area is characterized by higher densities and generally poorer residential areas (CENSUS OF INDIA 2011a, MISTELBACHER 2005: 25). East Delhi is considerably lacking in terms of infrastructure provisions and adequate open and green spaces. An increasing number of bridges and new metro lines connects both parts of the city today, but the river's remaining 'undeveloped' floodplain is between one and three kilometers wide and still forms a major physical barrier separating the 'two cities'. This dichotomy of West Delhi and East Delhi needs to be taken into account because it influences the city-river relationship and the discourses associated with the river.

The growth of the megacity on both sides of the river has come at a large social and environmental cost. The extraction of the river's freshwater for agriculture and drinking water purposes, and increasing quantities of sewage released by the ever growing city have turned the sacred river, worshipped by Hindus since time immemorial, into a "sewage canal" (CSE 2007). The degradation of the riparian zone to a foul-smelling drain expresses a state of neglect regarding its protection and socio-ecological importance (see Figure 50 and Figure 51, page 392). A World Bank funded study in 2003 suggested that the Yamuna in Delhi "is perhaps the most threatened riverine ecosystem in the world because of the immense anthropogenic pressures on this riparian habitat" (BABU et al. 2003: 1).

The river's ecological importance for the city is acknowledged by several environmental policies and legislations. In the city's Master Plan, the city's central planning body, the Delhi Development Authority (DDA), has defined the River Zone (Zone O) as a special planning zone (see Figure 1). By using capital letters for 'River Zone' the author intends to highlight that this spatial demarcation and its associated planning regulations are defined by the DDA. The demarcation of the 'River Zone' itself is problematic and the policy-making process surrounding it is outlined in this study.

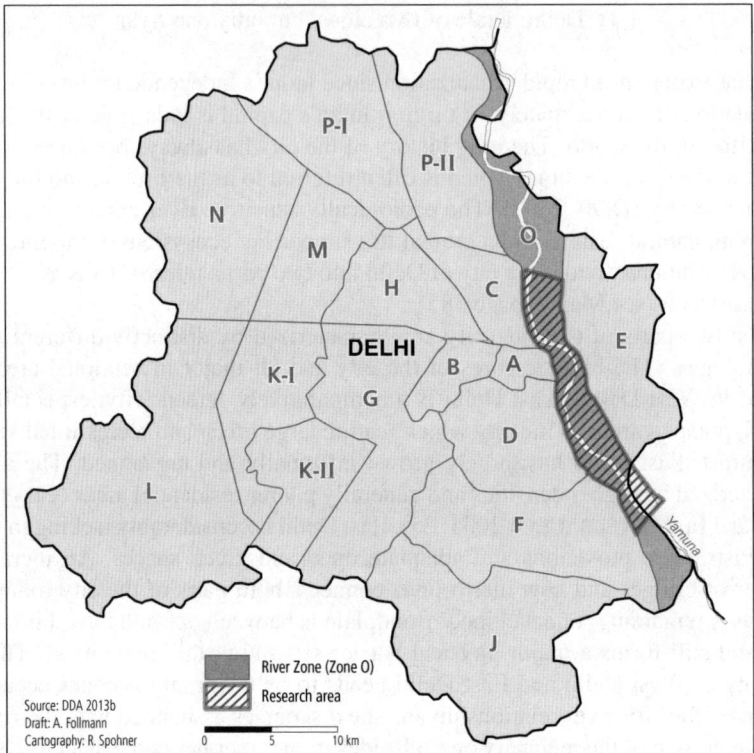

Figure 1: River Zone (Zone O) and research area

The ecologically sensitive River Zone covers about 97 square kilometers and stretches through the heart of the National Capital Territory (NCT) of Delhi for which the DDA envisaged "strict pollution control measures and eco-sensitive land use controls" (DDA 2007). The River Zone has been further earmarked as a groundwater recharge zone (CGWB 2012). Its fertile land is still widely used for agriculture and horticulture. While the area along the river has faced considerable development pressures in the last decades, it has not been decisively protected by law.

For decades the flood-prone and vulnerable riparian zone was a place where the marginalized 'urban poor' were forced to settle in default of sufficient residential provisions for the poorest sections of the population (DUPONT 2008). The slums used to be home to a population of several hundred thousand before massive slum demolitions in 2003 to 2006 'reclaimed' the riverfront areas for the 'planned' redevelopment of the city in the twenty-first century. Since the opening and liberation of the Indian economy in the early 1990s, the development of the city experienced a massive boost which found expression in the idea of transforming Delhi 'from walled city to world city' (BAVISKAR 2007a, DUPONT 2011).[2]

Along with a major city beautification drive, the riverfront was envisioned to be transformed into a 'world-class' space through the development of urban mega-projects, and the creation of extensive green and recreational areas. The riverfront emerged as a pivotal space in the remaking of the city on its way to become a global metropolis, especially in the course of the preparation for the Commonwealth Games held in Delhi in 2010 (BATRA & MEHRA 2008, BAVISKAR 2011c, BHAN 2009, FOLLMANN 2014, 2015, FOLLMANN & TRUMPP 2013, SHARAN 2015).

The river Yamuna in Delhi tells *a tale of two cities, but only one river*: West Delhi versus East Delhi; the informal, unplanned city versus the formal, 'planned' city; the needs of the urban poor versus the desires of the emerging urban middle class; the environmental challenges of a highly polluted river versus the dream of a beautified riverfront; the old, traditional versus the new 'world-class' urban India. These multiple framings of the changing river-city discourses are the focus of this book.

Analyzing urban environmental change along the river Yamuna in Delhi, its underlying governance structures, and the associated discourses aims to engage simultaneously with the multiple, and often contradictory, material and discursive realities of the complex river-city nexus. The overarching research motivation of this study is a better understanding of urban environmental change and governance in the megacities of the Global South. Delhi and the river Yamuna are an especially instructive case study for similar processes in other megacities in India and beyond.

1.2   Change of perspective: from the riverfront to riverscapes

Recent riverfront developments from around the world show that the river-city interface is of particular importance for the long term economic, social and environmental sustainability of cities (SANDERCOCK & DOVEY 2002: 151). Pushed by real estate interests and an ever more recognized aesthetic and recreational value of the urban waterfront, cities around the globe have redeveloped their riverfronts (see among others CHANG & HUANG 2011, HEIKKILA 2011). In order to transform the cities' economic landscapes and enhance their competitiveness, in many cases urban mega-projects have played a key role in riverfront revitalization (cf. DEL CERRO SANTAMARÍA 2013, FAINSTEIN 2008, REN & WEINSTEIN 2013).

Urban geographers, architects and planners have typically viewed urban rivers as elements of the urban landscape and, thus, the lens which they deployed was generally calibrated from a city-centric point of view. This nexus between the river

---

[2] The slogan 'from walled city to world city' originates from a campaign of India's largest English-language daily newspaper 'The Times of India' in 2004. The campaign targeted at rebranding the city's image.

and the city is commonly referred to as *the riverfront* or *waterfront*[3], demarcating the actual material space where the city and the river meet. On a global scale, the literature on urban rivers has focused on riverfront redevelopments, and the analysis most often sticks to the seemingly clear boundary demarcating the aquatic realm of the river from the terrestrial realm of the city (see among others DESAI 2012b, SANDERCOCK & DOVEY 2002, SAVAGE et al. 2004).

Concentrating on the riverfront, the city-river nexus in urban planning, architecture and urban studies is generally reduced spatially and functionally to the redevelopment of the riverfront for either recreational uses or real estate development. In many cities around the globe riverfronts further hold key transportation infrastructure; especially roads and highways. The term 'river*front*' points to a liminal space at the edge of the water, and thereby implicitly refers to fixed spatial boundaries. Furthermore, the term reflects a certain notion of urban (real estate) development. Despite these common associations, the city-river nexus is much more complex and stretches far beyond the water's edge.

An analysis focusing on the riverfront without taking into account the multiple connections between the city and the river at large is therefore inappropriate for a better understanding of urban environmental change. Holistic approaches to urban rivers as constitutive elements of the urban landscape, which highlight their importance for urban development and their multiple connections to urban sustainability are rare. Overall, little research has been done on the governance and socionature of urban rivers (CASTONGUAY & EVENDEN 2012b, DESFOR & KEIL 2000, 2004, GANDY 2006a, HAGERMAN 2007, HOLIFIELD & SCHUELKE 2015, LEIGH BONNELL 2010, RADEMACHER 2007, 2008, 2011).

The problem of the 'riverfront approach' becomes clear when trying to demarcate the spheres of the river and the city. Rivers and their internal landscapes are complex. They comprise of the (active) river channel(s), where the water flows, and the land along the river, which is often made up of alluvial materials transported and deposited by the river. Together with other fluvial features such as meanders and ox-bow lakes, the water channel and the river banks form what is generally referred to as the floodplain or riverbed.[4] Floodplains are the result of complex interaction of fluvial processes. The geomorphological characters of a floodplain are essentially influenced by the stream power (e.g. velocity of the river) and the sediment character of the river course. As the name implies, a floodplain is inundated by the river's water during floods. Despite the fact that these definitions are simple, a diverse and ambiguous scientific terminology of floodplain exists. Hydrologists and engineers generally consider the floodplain to be the surface next to the river channel, which is underwater during the floods of a given time period. Such a definition disregards the geomorphological process responsible for the creation of any

---

3   Waterfront also refers to seafronts or lakefronts. While these *-fronts* share certain characteristics, they are not directly comparable. Therefore the discussion here sticks to the term riverfront referring to the waterfront along the river.
4   The term riverbed, generally referring to the bottom of the river, is often used synonymously to floodplain.

floodplain (geomorphic history). Therefore focusing on a genetic definition of floodplain, NANSON & CROKE (1992: 460) define floodplain "as the largely horizontally-bedded alluvial landform adjacent to a channel, separated from the channel by banks, and built of sediment transported by the present flow-regime" under the present hydro climatic conditions.

In summary, the short discussion presented here reveals that if one engages with different terminologies denoting certain elements of (urban) rivers, competing definitions exist and difficulties arise defining their spatial extent. Multiple questions arise: Where to draw the boundaries between the floodplain of the river and adjacent land? How to demarcate the riverfront? Where to draw the boundary between the river and the city?

These kinds of questions reveal the problematic modern dichotomy of nature and culture, non-human and human. Critical thinkers consequently hold that it is time, especially for geographers, to think beyond these binaries, since

> "the stuff of the world is made up of both things and relationships that simply cannot be separated into boxes labelled 'nature' and 'society'. [And] ontologically, the society–nature distinction makes no sense as the basis for an understanding of our world" (LOFTUS 2012: 2–3).

Similar to how NEIL SMITH (2006: xi) has framed it: "[…] few ingrained assumptions will look so wrongheaded or so globally destructive as the common-sense separation of society and nature". Yet, the "distinction between society and nature is so familiar and fundamental as to seem unquestionable" (CASTREE & MACMILLAN 2001: 208) and geographical research has often been unable to overcome this dichotomy; especially with regard to urban rivers.

The literature going beyond the riverfront perspective and aiming to develop a more holistic understanding of the city-river relationship concentrates to a large extent on urban rivers in Western, industrialized countries and often focuses on questions of river restoration (see among others CASTONGUAY & EVENDEN 2012b, DESFOR & KEIL 2000, 2004, EDEN & TUNSTALL 2006, EDEN et al. 2000, GANDY 2006a, HOLIFIELD & SCHUELKE 2015, KELMAN 2003, LEIGH BONNELL 2010, WHITE 1995, WORSTER [1985] 1992). The character of these urban rivers has often been determined by an urban-industrial riverine landscape including large areas of land occupied by industries, shipyards and ports. Studies focusing on urban rivers beyond the 'West' have concentrated on, for example, the pollution of urban rivers (cf. for China ECONOMY 2004, HEIKKILA 2011, for Nepal RADEMACHER 2011) or the challenges and opportunities of delta regions (cf. for the Mekong RENAUD & KUENZER 2012).

It is important to acknowledge that multiple non-human and human processes from the flow and meandering of the river itself (including all physical, chemical and biological processes) to the use and disposal of water and the development of its banks (including human interventions) have formed urban rivers. As such, a 'natural' or 'fixed' boundary separating the river (nature) from the city (culture) consequently does not make sense, but is rather a social construction applied especially by urban and environmental planners.

The inherent complexity of urban environmental change along urban rivers in the megacities of the Global South – as indicated for the case of Delhi above – requires a change of perspective looking beyond the riverfront. Therefore, in order to overcome the binary conceptualizations of nature/culture and river/city, this study uses the notion of *riverscapes* as a single terminology referring to the riverine landscape coined by the natural forces of the river and numerous human interventions. Thus, first of all the notion of riverscapes is to be understood to refer to a physical space and an observable environment: a river, its riverbed, its floodplain and its surrounding environment. This study, however, aims to develop riverscapes into a spatial and analytical conceptual framework.

The idea of conceptualizing riverscapes as a theoretical framework for research has evolved from ERIK SWYNGEDOUW's writings on the waterscape (SWYNGEDOUW 1999, 2004, 2009) and the progression of his thoughts by others (BAGHEL 2014, BAVISKAR 2007b, DOVEY 2005, KAÏKA 2005, MOLLE 2009, NÜSSER 2014, NÜSSER & BAGHEL 2010, RADEMACHER 2011, ZIMMER 2012b).

ARJUN APPADURAI's (1990, 1996) concept of *scapes* is also important, since the development of urban rivers has certainly been shaped by technology and engineering expertise as well as the globalized, modern ideas of controlling rivers in the name of development and progress (cf. BAGHEL 2014: 17, NÜSSER & BAGHEL 2010: 231). According to APPADURAI, globalization has resulted in "new cosmopolitanisms" constituted by "transnational cultural flows" (APPADURAI 1996: 49). While being influenced by these deterritorialized flows, Delhi's riverscapes are not deterritorialized spaces. The suffix 'scapes' "allows [...] to point to the fluid, irregular shapes" (APPADURAI 1990: 297) which are characteristic both for the physical landscape coined by a river, as well as the multiple discursive representations of a river. Therefore, like landscapes (COSGROVE 1984, COSGROVE & PETTS 1990), riverscapes need to be understood as something socially constructed and discursively produced over time. Riverscapes are cultural landscapes being shaped by human and nature. Thus, besides their physical materiality they are loaded with cultural meanings and discursive representations (WINCHESTER et al. 2003: 4).

With regard to APPADURAI's understanding of *scapes*, and in order to emphasize the multiple nexuses between the river and the city, the inherent spatial heterogeneity of an urbanized river, and the temporal dynamics of its making and remaking both physically and discursively, the term riverscapes is intentionally used in plural in this book.

1.3 Discourse and urban environmental policy-making

This study deals with Delhi's riverscapes as a matter of discourse being shaped by different narratives and knowledges[5] of a multiplicity of actors. A discourse, as DRYZEK (2005: 9) frames it, "is a shared way of apprehending the world" and through the use of language in its multiple forms (spoken and written) a discourse is formed and eventually "enables those who subscribe to it to interpret bits of information and put them together into coherent stories". Thus, the discursive representation is a process which involves multiple actors and multiple acts of interpretation of different narratives. A discourse can therefore be defined as "a specific ensemble of ideas, concepts, and categorizations that are produced, reproduced, and transformed in a particular set of practices and through which meaning is given to physical and social realities" (HAJER 1995: 44). It is assumed that dominant, and especially hegemonic, perceptions and ideas shaping discourses find their way into environmental and spatial policies and eventually (if implemented) affect the physical ground realities.

In addition to analyzing urban environmental change in terms of its material transformations in space and time, this research is dedicated to trace, interpret and deconstruct the history and development of discursive representation of Delhi riverscapes in order to be able to understand the different actors' perceptions and ideas. Bringing these two perspectives (material and discursive) together, allows a holistic perspective on the making of urban environmental policies (cf. among others DESFOR & KEIL 2004, FISCHER & FORESTER 1993, HAJER 1995).

In this context, urban environmental polices need to be understood to be "co-constructed" through networks of different actors from the local to the global level (KEELEY & SCOONES 2003: 3, cf. LATOUR 1993, LATOUR 2004). Environmental science plays a central role in this process since scientific facts create authority and legitimacy for other actors (especially the state) to press for an institutionalization of policies based on these 'facts' (FORSYTH 2003, STOTT & SULLIVAN 2000). As outlined by KEELEY & SCOONES (2003: 26–27), science aims at "generating universalizable statements". Furthermore, the making of policies is not a straight process from problem-setting to policy decisions and their implementation, but is rather a contested field. Scientific results "create facts by closing controversies" and are often used by actors to legitimize a certain framing of a policy in order to "shortcut" political bargaining processes on otherwise contested governance terrains (LATOUR 1993).

KEELEY & SCOONES (2003: 26–27) further highlight that "policy-makers delimit areas for scientific enquiry" and thereby determine to some extent the production of specific 'facts' to please certain interests. Moreover, the 'facts' found by scientists are often negotiated with the sponsors and commissioners of the studies before the scientific 'facts' are published. Accordingly, there is skepticism surrounding facts and policies as they are co-constructed and co-produced through an

---

5 The term 'knowledges' is intentionally used in plural in this book to highlight different forms and sources of knowledge.

argumentative process (HAJER 1995). This study will outline this co-production of science and policy with regard to Delhi's riverscapes.

1.4 Governance and urban political ecology

The theoretical and conceptual roots of this study are grounded in governance research and urban political ecology (UPE). Governance in this study is understood as the processes of interactions, negotiations and bargaining among a multiplicity of state and non-state actors (KOOIMAN 2003, RHODES 1996). Governance researchers have intensively studied urban and environmental policy-making (cf. DANIELL & BARRETEAU 2014, DAVIDSON & FRICKEL 2004, HOHN & NEUER 2006, REED & BRUYNEEL 2010). Nevertheless, the theoretical concepts and explanation value of governance research are limited and the governance approach emerges to be of limited epistemological value to explain the underlying motives, driving forces and powers of the involved actors. This study aims to widen the analytical perspective in governance research on the urban environment by combining it with theoretical perspectives originating from the heterogeneous field of UPE. Such an approach seems to offer the opportunity to develop a deeper understanding of urban environmental change and urban environmental policy-making.

While governance research has often focused on the 'management' and 'coordination' of urban environmental challenges (UNESCAP 2005), UPE scholarship has often been marked by a radical critique of both the triggers (e.g. capitalism, domination of nature) and the solutions (e.g. new technologies) offered by technocratic governance approaches (HEYNEN et al. 2006a, KEIL 2003, LAWHON et al. 2014, LOFTUS 2012). Drawing on political ecology's long established "central theme" of the "politicized environment" (BRYANT 1998: 79, cf. BRYANT & BAILEY 1997: 27), UPE scholarship has highlighted that urban environmental governance "is not a mere matter of proper management but equally one of power and politics" (VÉRON 2006: 2093).

Recent post-structuralist research within political ecology[6] (ESCOBAR 1996b, FORSYTH 2003, NATTER & ZIERHOFER 2002, PEET & WATTS 1996) has highlighted the importance of ideas and discourses influencing environmental policy-making, rather than drawing on structuralist approaches often prevailing in governance research. While more traditional approaches in political ecology have also tended to focus on structural explanations, a second generation of political ecology has been highly influenced by post-modernist, post-colonialist and post-structuralist thoughts (cf. BENTON-SHORT & SHORT 2013, ESCOBAR 1996b, 2010, FORSYTH 2003, 2008, NATTER & ZIERHOFER 2002, PEET & WATTS 1996), as well as feminist approaches, actor-network theory (ANT), and other concepts originating from the

---

6   The term post-structuralist political ecology is used in the following as a broad headline to subsume newer strands in political ecology, which have considerably refrained from deploying structural explanations (see for a similar classification ESCOBAR 1996a, NÜSSER & BAGHEL 2010, ROCHELEAU 2008).

field of science studies (cf. HARAWAY 1991, LATOUR 1993, 2004, ROCHELEAU 2008, ROCHELEAU et al. 1996, WHATMORE 2002).

Furthermore, UPE scholarship has been motivated by writings highlighting that the city is not to be understood as the antithesis to nature (see among others HARVEY 1996b, HEYNEN et al. 2006b, KEIL 2003, TREPL 1996). This school of thought has further aimed to bring together the binaries of nature and society by understanding the city as a hybrid (SWYNGEDOUW 1996) and "retheorize urbanization itself as a process of socionatural and not only social transformation" (ANGELO & WACHSMUTH 2014: 3).

> "[…] poststructuralist thought offers us not only a different way of thinking about nature, but a different way of understanding and doing environmental politics. […] the domain of politics is no longer understood as limited to political institutions – parliament, law, etc. – but enlarged to include concepts and knowledges that inform debates within these arenas" (BRAUN & WAINWRIGHT 2001: 60).

The aim of adopting these thoughts in this study is to develop a holistic understanding of the driving actors and discourses of urban environmental change by deconstructing the binary thinking of nature and society – the river and the city.

1.5 Research agenda and questions

As outlined above, there is a lot written about environmental change from the global to the local scale. Despite this, scholars have only recently started to engage with the complexity of urban environmental change in fast-growing megacities of the Global South. Many studies on urban environmental governance in the Global South focus on sectoral approaches. In the case of Delhi, studies have been dedicated to topics including, but not limited to, water supply (MARIA 2008, SELBACH 2009, ZÉRAH 2000), wastewater (SINGH 2009, ZIMMER 2012b, 2015a), air pollution (ESCUDERO 2001, GHOSH 2007, KUMAR 2012, SHARAN 2013, VÉRON 2006) or solid waste management (GIDWANI 2013).

This study provides an alternative perspective to existing research by focusing on Delhi's riverscapes. The aim of this is to gain insights into urban environmental change starting from a spatial rather than a sectoral approach. Therefore, this study focuses explicitly on an ecologically sensitive space of the urban landscape – the riverscapes – in order to draw a holistic picture of how and why urban environmental change occurs, which spatial and ecological impacts are connected to this, and how and why urban environmental change might have accelerated in recent decades. In doing so, this study focuses on the river Yamuna in Delhi with regard to both the river's role in the process of mega-urbanization and the impacts of mega-urbanization on the river. Such an approach follows CASTONGUAY & EVENDEN's (2012a) appeal to study urban rivers with regard to their associated risks as well as opportunities. The multiple city-river relationships are examined from a geographical perspective. This perspective, while emphasizing the historical spatial changes, relates those to current socio-environmental challenges and governance processes.

When applying such a spatial approach it is essential to take into account different scales. Within this research, scales are understood as descriptive and analytical, since both the administrative limits of cities and the boundaries of the river basins are based on social constructions rather than being naturally given (COLTEN 2012, MARSTON 2000). While using different analytical scales to fully understand the relationship between the city and the river, from a spatial perspective this study focuses on Delhi's riverscapes in the very heart of the megacity from Wazirabad in the north, to Okhla in the south (see Map 1, page 385). This segment of the river has been facing the highest development pressures and is characterized by the highest levels of environmental degradation and dynamic land-use changes. Together these two aspects pose multiple challenges for the urban environmental governance of Delhi's riverscapes, which are exemplary for similar processes in other megacities of the Global South.

The ensuing four key research questions follow from the described approach. The first question of this study addresses the process of urban environmental change:

What are the urban environmental changes in Delhi's riverscapes?

This first research question is addressed in this study through a detailed analysis of the dynamic land-use changes in Delhi's riverscapes and the mutual relations between land-use change and environmental degradation.

Urban environmental change is caused by multiple actions from a range of different actors. In the context of urban environmental governance the different actors' actions and interests are understood to be controlled, coordinated, planned and synchronized – in short governed – through urban environmental policies and regulations (e.g. land-use plans, pollution policies). Thus, the first key question needs to be specified through a second question engaging with the ways environmentally sensitive urban spaces like riverscapes are governed:

How are Delhi's riverscapes governed?

This question explicitly includes an exploration of the roles and responsibilities of the different institutions involved in governing environmentally sensitive urban areas. In this context, DE MELLO-THÉRY et al. (2013: 214) argue that urban and environmental policies "comprise of a set of laws, norms, rules and institutions, but they are also anchored in a set of cultural and social representations". Traditionally, policies have been understood as approaches to govern problems, and policy research has focused on the implementation of these laws, regulations and institutions (ibid.). This perspective has evolved, and social science research has shown that the *process* of problem framing (the social construction of problems and seeing the problem first of all) and the making of public policies, is at least as important as questions of implementation (see among others DESFOR & KEIL 2004, FISCHER 2000, 2003, FISCHER & FORESTER 1993, HAJER 1993, 1995, KEELEY & SCOONES 1999, 2003, LASCOUMES & LE GALES 2007, LEES 2004, MELS 2009). In addition, often the (re)making of policies and their failed implementation are closely linked. Therefore besides exploring the kind of policies and institutions which exist to govern Delhi's riverscapes, this study aims to critically analyze the contemporary and historic processes of urban environmental policy-making for Delhi's riverscapes.

The first two key questions already indicate that the making of urban environmental policies needs to be understood as a process. Therefore, besides reviewing and analyzing existing policies and institutions this study is interested in exploring the different policies and institutions which are discussed among the actors in different environments. A discourse analytical approach emerges in this context as especially insightful, since it offers the methodological tools to explore the (re)making of urban environmental policies (BRAND & THOMAS 2005, DESFOR & KEIL 2004, HAJER 1995, JACOBS 2006, LEES 2004, WHATMORE & BOUCHER 1993). The third overarching research question therefore aims to include the discursive framing of environmental knowledges in the policy-making process:

How are environmental knowledges discursively co-produced by different actors, and how are these discourses reflected in the urban environmental governance of Delhi's riverscapes?

Engaging with this key research question, the research in this study also addresses the multiple challenges of public participation in urban environmental policy-making in the mega-urban context of the Global South. For example, how are non-state actors involved in governing Delhi's riverscapes? How do they interact with the multiple agencies of the state? Who is able to participate in the discursive co-construction of urban environmental policies and who is excluded? Whose interests and discursive framings of environmental problems are reflected in urban environmental policies and whose are marginalized? With regard to these questions, this study focuses on the role of middle-class dominated, environmental non-governmental organizations (NGOs) and the farmers cultivating large parts of Delhi's riverscapes. The author focuses on these two groups of non-state actors, which are themselves heterogeneous, since they are assumed to have played an important role in the remaking of Delhi's riverscapes either materially or discursively.

Finally, it is assumed that policies and institutions governing environmentally sensitive urban areas are bound to change over time, especially when confronted with increasing development pressures of dynamic urban transformations. In the case of Delhi, the realization of urban mega-projects in the course of the preparation for the Commonwealth Games in the first decade of the 21st century marks a time frame of profound urban restructuring, which triggered urban environmental change in Delhi's riverscapes. Consequently, a final key question needs to be answered:

How has urban environmental policy-making for Delhi's riverscapes responded to recent development pressures of dynamic urban transformation and restructuring?

As the riverscapes concept is developed by linking UPE and governance research, the case study will further reveal to what extent the concept emerges as a useful and insightful theoretical grounding for analyzing urban environmental change in a mega-urban context of the Global South.

1.6 Outline of the book

The book is structured into eight parts, including this introduction and the following outline of the research context highlighting the challenges of urban environmental change in India's megacities. In Part II of this book the theoretical groundings are laid out including a discussion of governance and insights from UPE as the conceptual roots of this study. This theoretical chapter sets out a conceptual framework for the analysis of hybrid riverscapes.

The methodological thoughts and applied methods are outlined in Part III. Part IV situates the case study by introducing the megacity of Delhi, the river Yamuna and the underlying governance structures. Based on these insights Part V then outlines historical-geographical insights into Delhi's riverscapes by engaging with the ever-changing material and discursive realities of Delhi's riverscapes. This Part lays the foundation for a deeper understanding of the more recent changes in Delhi's riverscapes.

Focusing on the period from 2001 to 2014, Part VI then analyzes the reclamation of Delhi's riverscapes for a world-class city in the making. Part VII presents the attempts to remake the urban environmental policies after the Commonwealth Games in order to 'protect' and 'conserve' the remaining green and open spaces along the river. Part VIII then merges the insights of the empirical chapters, draws the conclusions and evaluates the conceptual approach.

2 INDIA'S MEGACITIES: CHALLENGING URBAN ECOLOGIES

India is undergoing dynamic and profound inter-related transformations. In addition to an increase in population, the country is simultaneously experiencing inter alia: sustained economic growth, continued economic liberalization, global integration, a transition from an aid recipient to a donor nation in the development context, and a shift from a predominantly rural country to an increasingly urbanized society. India, once known as *the land of villages*, is being transformed into an urbanized nation (SHAW 2007, SIVARAMAKRISHNAN et al. 2007).

India's cities continue to grow both in population and territory, resulting in massive changes in the urban peripheries (DUPONT 2005a, GURURANI 2013, HOMM 2014, JAIN et al. 2013, NARAIN & NISCHAL 2007, SAMI 2013). In addition, the country is experiencing increased densification as a result of redevelopment projects in the central parts of the cities (BANERJEE-GUHA 2011, COELHO & RAMAN 2013, DOSHI 2013, KENNEDY 2009, WEINSTEIN 2009). Concurrently, India's megacities are aiming to move up the ladder in the global urban system by initiating ambitious urban development projects aiming to boost the cities' economic base by making them competitively viable and investment friendly especially for foreign capital (BAVISKAR 2007a, DUPONT 2011, GOLDMAN 2011b, McKINSEY 2003, NISSEL 2001). This has resulted in increasing development pressure, especially on land but also on other already scarce natural resources such as drinking water.

Likewise, efforts for city beautification and urban ecological restoration (including riverfronts and water bodies) emerge as "idioms through which the cities position themselves in the global arena as well as direct strategies for capital accumulation through real estate value" (COELHO & RAMAN 2013: 147). Furthermore, these urban (re)development initiatives and their associated policies increasingly aim to satisfy the desires of a growing urban middle class for green and clean cities (ANJARIA 2009, ARABINDOO 2012, BAVISKAR 2002, 2011a, CHATTERJEE 2004, ELLIS 2012, FERNANDES 2006, GHERTNER 2011d). In this context, the contemporary Indian city is

> "neither a deferred Western city nor the anticipated East Asian metropolis. It is a different configuration of things and discourses that merits its own accounts. It is an account that must, for this very reason, be a historically situated one" (SHARAN 2014: 2).

The historic and current relationships of the megacity Delhi and the river Yamuna are emblematic of these larger development trends representing India's challenging urban ecologies. This chapter outlines that transforming India's cities, and the capital city in particular, into sustainable urban environments and socio-ecologically just metropolises "is the single greatest opportunity that governments, entrepreneurs, and innovators face in the coming decades" (RADEMACHER & SIVARAMAKRISHNAN 2013b: 1). Yet, existing socio-environmental problems and the changing urban ecologies accelerated by economic liberalization and global integration pose also profound challenges for urban environmental governance.

2.1 Challenges of urbanization in India

India's population growth is becoming more and more an urban phenomenon. In the period from 1951 to 2011, India's urban population grew from approximately 62 million to approximately 377 million (CENSUS OF INDIA 2011c). Nevertheless, today only around a third of India's population (31%) lives in urban agglomerations and in comparison to the global level of urbanization, India's level of urbanization remains quite low (UN 2011). It is however projected that by 2050, half of India's population – more than 800 million – will live in cities (UN 2014: 22).

A large portion of this urban growth will occur in the country's million-plus cities and megacities, even when decadal growth rates of the country's megacities are decreasing considerably (CENSUS OF INDIA 2011b). Recent projections for the urban agglomeration of Delhi forecast a growth of another 10 million from 25 million in 2014 to 36 million by 2030; Mumbai is expected to grow in the same period by about 7 million, Bangalore by 5 million, Kolkata, Chennai and Hyderabad by more than 4 million, and Ahmedabad and Surat by more than 3 million (UN 2014: 26–27).

India's urban agglomerations have emerged as the economic drivers of the country. In 2008, they accounted for more than half (58%) of India's Gross Domestic Product (GDP) and it is projected that the urban share will increase to about 70% by 2030 (MCKINSEY GLOBAL INSTITUTE 2010: 16). Furthermore, India's cities make up about 80–85% of the country's total tax revenues and, thus, contribute the lion's share to funding development and wealth redistribution (MCKINSEY GLOBAL INSTITUTE 2010: 18). As *motors of development* the country's large agglomerations are key regions for India's future, but the pace of large-scale urbanization "poses huge problems for city governments" (BAUD & DEWIT 2008b: 2).

India's urbanization takes place very unevenly, creating fragmented spaces of hope and despair, wealth and poverty, environmental beautification and degradation. Overall, India's megacities "pose especially threatening scenarios of deterioration of human quality of life and environmental degradation" (HUST 2005: 1). Elements causing this degradation include, among others, high levels of air pollution, uncontrolled dumping of solid waste (including toxics and e-waste), the degradation of urban water bodies by untreated sewage as well as its reclamation for urban development leading in turn to rapidly declining groundwater aquifers, an intensification of water shortages and a loss of biodiversity. In particular, the urban poor are forced to live in highly precarious urban environments in once marginal spaces – formerly seemingly forgotten urban spaces along rivers, canals, lakes, railroad tracks or highways – which have concurrently come increasingly into the focus of urban development projects.

India's urban environmental problems have remained largely neglected by national policies as environmental policy making has concentrated largely on the management of natural resources, and in turn predominantly on rural areas (TIWARI et al. 2015: 23). It appeared that urban India was seen as the antithesis to nature and was not in the focus of environmental conservation. In addition, Mahatma Gandhi's ideal image of a pleasant, autonomous and self-sufficient village life and his negative attitude towards the big cities for a long time remained quite powerful in the Indian political scene after independence (JODHKA 2002: 3351). The burgeoning demand of the cities for resources was typically perceived as a major threat, which was encroaching on the rural, natural environment. 'Nature' was not to be found in the Indian cities; in contrast, it was in need of protection against urbanization and industrialization.

Urban development was "rarely at the centre of developing planning" (SIVARAMAKRISHNAN 2011: 1). This was evident also in India's five-year plans. The first two five-year plans (1951 and 1956) prepared by the Planning Commission almost completely neglected urban policies with the exception of regulations for housing and urban land prices (cf. KENNEDY & ZÉRAH 2008, MAHADEVIA 2003, RAMACHANDRAN 1989). In the fifth five-year plan of 1974 the term 'Urban Development' was included for the first time as a separate chapter (see for details SIVARAMAKRISHNAN 2011, SIVARAMAKRISHNAN 2015). Since 1991, economic liberalization has reshaped India's economic base and thereby redefined the role of the cities. This has led to a new focus on urban development resulting in dynamic spatial transformations and new forms of urban governance (cf. BAUD & DEWIT 2008a, HARRIS 2007, HUST & MANN 2005, KUNDU 2003, SHAW 2007). The eleventh five-year plan finally acknowledged the important role of the megacities and requested metropolitan planning beyond the municipal limits (SIVARAMAKRISHNAN 2015: xxix). Yet, until today, "the fear of the big city persists" (ibid.) and, thus, the megacities continue to have an "ambivalent status in India's overall development strategy" (KENNEDY 2014: 115).

2.2 Post-liberalization urban India

The post-liberalization phase in India is characterized by economic and urban growth accompanied by profound spatial and structural changes (see among others BAUD & DEWIT 2008b, HARRIS 2003, HUST 2005, JENKINS 1999, KENNEDY 2014, RUET & TAWA LAMA-REWAL 2009, SRIDHARAN 2008). In order to dismantle a system of state control, the 1991 adopted structural adjustment program has led to a wide-ranging liberalization of India's economic policies. Far-reaching reforms in trade and industrial policies were needed to overcome the doctrine of import substitution, opening the Indian economy for private capital and the global market (JENKINS 1999: 16–17). This global economic integration has been accompanied and facilitated by a process of "state restructuring" in India, during which the state's role has been redefined in order to foster economic growth (KENNEDY 2014). According to JENKINS (1999: 18), these policy changes and initiatives on the subnational scale are a logical consequence, since the economic reforms forced state governments "to compete with one another to create investor-friendly climates".

Economic liberalization and the opening up of India's national economy to the global market left its mark on India's cities. The capital city of Delhi has certainly been at the center of these debates and experienced a major transformation and face-lifting to become a world-class city (BAVISKAR 2007a, 2011b, DUPONT 2011). In this context urban scholarship has highlighted the multi-layered challenges of changing urban governance, focusing especially on the integration of public-private partnerships and private actors in former tasks of the state as well as the devolution of powers from the regional states and its urban development authorities to the local municipal bodies (BAUD & DEWIT 2008a, HUST & MANN 2005, RUET & TAWA LAMA-REWAL 2009, SIVARAMAKRISHNAN 2011, 2015). A series of authors outline that the global economic integration of India's metropolises and changing patterns of urban governance have led to an emergence of new political alliances and growth coalitions to transform the Indian city (DUPONT 2011, GHOSH et al. 2009, KENNEDY 2014, KENNEDY & ZÉRAH 2008, SHATKIN 2014). Besides the growing influence of the private sector in urban governance, a series of authors also emphasize the increasing role of India's new middle classes and elite groups in urban policy-making (ANJARIA 2009, ARABINDOO 2012, BAVISKAR 2003, CHATTERJEE 2004, ELLIS 2012, FERNANDES 2006, GHERTNER 2011a, ROY 2004, ZÉRAH 2009). Highlighting questions of equal citizenship and socio-environmental justice, PARTHA CHATTERJEE (2004: 131) raised the question "if Indian cities are becoming bourgeois at last?"

Spatially, these developments have manifested in the form of urban mega-projects and a further informalization, illegalization and marginalization of the urban poor leading to substantial criminalization of their activities and often large-scale displacement (see among others BATRA 2010, DUPONT 2011, MAHADEVIA & NARAYANAN 2008b). Despite this transition, large parts of the Indian cities are still dominated by informal processes and planned urban development often concentrates on certain, highly-visible and high-profile areas; e.g. airports, office parks, malls, upmarket residential areas (see among others DATTA 2012, KUDVA 2009, NISSEL 2009). Thus, "elite-driven visions of urban change" compete with "the daily incursion on the planned city by the poor" (SHATKIN 2014: 1).

Overall, the dynamic transformations in Indian cities meet with a multiplicity of agencies characterized by often overlapping jurisdictions and ill-defined responsibilities. This has led to the situation that the Indian judiciary has played a tremendous role in governing the urban environment – both with regard to urban environmental problems and the displacement of the urban poor (DEMBOWSKI 1999, 2001, DUPONT & RAMANATHAN 2008, MEHTA 2006, RAMANATHAN 2006, RUBIN 2013).

Remarkably, the transformation of India's urban agglomeration, including the country's largest megacities, has not been driven and regulated by the urban local bodies. Due to an inadequate provision for local self-governance in the Indian Constitution, municipalities generally have a rather limited mandate and financial budget (KENNEDY 2014: 112, SHATKIN & VIDYARTHI 2013: 10). Traditionally, municipalities are in charge of basic service delivery (water supply, sewage disposal, street lightning, solid waste management, etc.). Larger questions of socio-economic development are not in their direct purview as the authorities in charge of urban development and the reframing of urban policies in India have always been located on the state or central level (BAUD & DEWIT 2008a, HUST & MANN 2005, KENNEDY 2014, RUET & TAWA LAMA-REWAL 2009).

This dominance of the state governments, and in Delhi especially the central government, in urban governance was meant to change with the passing of the 74th Constitutional Amendment Act in 1992, which aimed at the devolution of governing powers (including e.g. financial responsibilities and sources of revenue, district and metropolitan planning committees) to the urban local bodies (KENNEDY 2014: 116, NANDI & GAMKHAR 2013, SHATKIN & VIDYARTHI 2013: 10, SIVARAMAKRISHNAN 2011, VAIDYA 2009: 14). Though, the devolution of governance powers to the local level did not meet the ambitious targets, largely due to political considerations on the state level, and resulted in "large gaps between policy intent and implementation" (NANDI & GAMKHAR 2013: 59). Accordingly, in the majority of the states, "a power vacuum" still exists on the metropolitan level (REN & WEINSTEIN 2013: 108) and the reforms for decentralization have had only limited impact on the actual urban governance structures. Due to the status of Delhi as the National Capital Territory, main functions of urban governance (e.g. land-use planning) have not come under control of the city's municipalities, but largely remain with agencies under purview of the central government.

The economic liberalization on the city level most notably included the encouragement of privatization of basic services through public-private partnerships (including introduction of user fees), the liberalization of land and real estate market through the repeal of Urban Land Ceiling Acts and change in Rent Control Acts, the opening of real estate sector for 100% Foreign Direct Investment (FDI) and the easement of land use conversions (BANERJEE-GUHA 2009: 98, cf. NANDI & GAMKHAR 2013, SHATKIN 2014). India's urban reforms have therefore been predominantly neoliberal focusing on a market-driven urban restructuring as "capital is flooding India's cities like never before" (GOLDMAN 2011a: 229). Nonetheless, an increasing shortage of land and rising land values coupled with various claims to land and often ambitious land records are characteristic for Indian megacities. Therefore "land has become viewed by planners and investors as the key obstacle" (GOLDMAN 2011a: 231). Urban development policies have thus aimed for a larger flexibility to facilitate fast-track planned urban development (ROY 2009b, SHATKIN 2011, 2014).

Drawing on the neoliberal agenda of economic liberalization, the Indian state itself, as ROY (2005, 2009b) argues, makes use of a "calculated informality" in order to ensure its own "territorialized flexibility" in the future development process of the cities. Following ROY's conception, informality is not invariably "the object of state regulation" but rather often "produced by the state itself" (ROY 2005: 149). Therefore it is the (Indian) state deciding "what is informal and what is not, and to determine which forms of informality will thrive and which will disappear" (ROY 2005: 149). ROY's interpretation of informality as "a mode of urbanization" (ROY 2005: 148) highlights that what is commonly viewed as a loss of formal governability and control (see among others KRAAS 2007) is in the case of the Indian cities in fact partly a deliberately produced urban informality by the state itself to decide case by-case whether for example certain land uses are desired or not (ROY 2005: 148).

Based on her work on the East Kolkata Wetlands, ROY (2003, 2004) argued that state agencies use the governing tool of "unmapping" in order to allow for a "considerable territorialized flexibility to alter land use, deploy eminent domain, and to acquire land" (ROY 2009b: 81). Unmapping is described as a consecutive process through which the state agencies intentionally disregard current land uses (e.g. informal settlements, agricultural uses) as well as ecological sensitive areas (e.g. urban water bodies) in the plan making process. Based on this vagueness of the land-use plans and planning policies, the state agencies aim to remain flexible in replanning and remapping of the urban landscape in order to 'future-proof' and accommodate the new elements of the 'world-class' city (ROY 2009b).

2.3 Making India's cities 'world-class'

Since the turn of the century, the ambition to become 'world-class' has led to a massive urban restructuring of India's megacities (HARRIS 2007, KENNEDY 2014, KENNEDY & ZÉRAH 2008, MAHADEVIA & NARAYANAN 2008b, SHATKIN 2014, SHAW & SATISH 2007), particularly in the city of Delhi (BATRA 2010, BAVISKAR 2007a, DUPONT 2011). This vision of the political elites and state authorities in charge of urban development essentially refers to converting the country's megacities into competitive cities which attract global capital investments. Modern transportation infrastructure, high-end residential projects, and exclusive shopping and entertainment multiplexes have been put on the agenda of urban development and a new image-building process has promoted the development of clean and green cities (DUPONT 2011, RADEMACHER & SIVARAMAKRISHNAN 2013a, SHATKIN 2014).

With a larger focus on the cities of the Global South, ROBINSON (2002, 2006) outlined how formerly analytical concepts like the "world city hypothesis" (FRIEDMANN 1986) and the "global city" (SASSEN 1991) have over time become "the aspiration of many cities around the world" (2002: 548) determining urban development policies, which in turn has often led to negative consequences – principally for the urban poor. In the Indian context, BATRA (2010: 17) constitutes that "the transformation of major metropolitan cities into 'world-class' cities has increasingly come to be the very raison d'être of urban development policy". A series of authors have shown that the 'world-class' agenda has led to intensified socio-spatial polarization, slum demolitions and overwhelming negative consequences for the urban poor (DUPONT 2011). GOLDMAN (2011a) speaks in this context of a new form of "speculative urbanism", which follows its own imperatives and creates new mechanisms in urban governance in order to suit the interests of (global) real estate investors. The role model cities in this process are typically cities such as Hong Kong, Singapore or Shanghai (BENG HUAT 2011, GOLDMAN 2011b, MAHADEVIA & NARAYANAN 2008a, ONG 2011).

Against this background, the central government has made significant funds available for large-scale urban redevelopment, especially through the Jawaharlal Nehru National Urban Renewal Mission (JNNURM) launched in 2005

(MoUEPA & MoUD 2005, SIVARAMAKRISHNAN 2011). JNNURM must be interpreted as a response to the existing infrastructural deficits and aims at improving the cities' basic infrastructure systems including transportation networks, water supply, sewerage and drainage, solid waste management, but also includes ecological restoration of water bodies (SIVARAMAKRISHNAN 2011). The scheme has induced and funded a large number of infrastructure projects. Furthermore, the new focus on public-private partnerships has introduced major shifts in urban governance towards a growing influence of private actors (real estate developers, domestic and foreign investors, landowners and other private businesses). The reform-linked urban development program JNNURM is to be interpreted as a logical result of India's post-liberalization politics. In short, JNNURM aims at transforming India's metropolises into world-class cities and this means making them more investment friendly. In this context, urban poverty "was identified as the prime cause for unregulated urban growth, environmental damage and growing crime rates" in India's cities (BANERJEE-GUHA 2009: 97).

The multinational consulting firm McKinsey has framed the world-class agenda in post-liberalization urban India as 'India's urban awakening' (MCKINSEY GLOBAL INSTITUTE 2010). This urban awakening has promoted Delhi (BAVISKAR 2007a, DUPONT 2011) and Mumbai (MCKINSEY 2003, NIJMAN 2007) to become global metropolises and world-class cities. But also the emerging megacities Ahmedabad (CHATTERJEE 2011a, DESAI 2012a), Bangalore (DITTRICH 2003, 2007, GOLDMAN 2011b, SHAW & SATISH 2007), Chennai (HOMM 2014, HOMM & BOHLE 2012) and Hyderabad (DAS 2015, KENNEDY 2007, KENNEDY & ZÉRAH 2008) have tried to position themselves as regional growth engines competing for (foreign) investments and businesses. As a result, the 'cleansing' and 'greening' of India's cities emerged as a key strategy in the overall transformation of the Indian city with urban waterfronts as an important locale in 'world-class' making. Overall, an environmental discourse for cleaning and greening the Indian cities is clearly apparent in India's urban governance. The Delhi's Government's campaign "Clean Delhi, Green Delhi" launched in September 2003 is emblematic for the way this discourse has been pushed to generate public awareness.

The next section outlines to what extent these developments might be interpreted as an 'urban environmental awakening'. These aspects are of major significance to understand why the river-city discourses in Delhi and other Indian cities have shifted in recent times.

2.4 India's urban environmental awakening?

As MAWDSLEY (2004: 79) outlines, "[t]he environment in colonial and postcolonial India has been widely explored as a site of both material and discursive conflict, often emblematic of broader social and political struggles." From forests to rivers, and from the Himalayas to the coastline, the relationship between nature and society has been a highly controversial and political topic almost all over India. Resource conflicts, environmental degradation and pollution have led to the creation and proliferation of many different forms of environmentalism and organized environmental movements in India (for an overview see MAWDSLEY 2004, WILLIAMS & MAWDSLEY 2006a, 2006b). Environmental non-governmental organizations (NGOs) have emerged in India as important agents since the 1980s and their number is increasing. In 2007, the directory of environmental NGOs compiled by WWF-India encompassed about 2,300 environmental NGOs. This number increased to about 3,500 environmental NGOs by the end of 2013 (ENVIS CENTRE ON NGO AND PARLIAMENT 2014). These groups are engaged in environmental research, public education, information dissemination and training in a large number of fields. The debate on environmental activism is too wide a field to be discussed in a chapter like this and, thus, shall only be shortly sketched here. It is important to understand the larger canvas of environmental activism in India, including its historical roots and discourses, in order to analyze the tools, strategies and the overall role of environmental activism in current processes of urban transformation.

The literature dealing with environmental activism in India can generally be divided into two separate strands. With a strong focus on rural areas, many studies have focused on the nexus of environmental change and livelihood issues regarding natural resources like forests or water (see for example AGRAWAL & SIVARAMAKRISHNAN 2001, BAVISKAR 1995, 1997, GADGIL & GUHA 1994, GUHA 1989, GUHA & MARTINEZ-ALIER 1997, MCCULLY 1996). The research on environmental conflicts in India after Independence has focused to a great extent on some historically influential, predominantly rural, social-environmental movements (e.g. Chipko Andolan and Narmada Bachao Andolan). Only since the beginning of the twenty first century, has a shift towards urban environmental change and problems been evident in the literature (e.g. BAVISKAR 2002, CHAPLIN 1999, DEMBOWSKI 2001, SHARAN 2002). Even more recently, urban issues have been discussed in a rapidly growing body of literature (cf. BAVISKAR 2010, 2011a, KARPOUZOGLOU & ZIMMER 2012, KUMAR et al. 2012, MAWDSLEY 2009, NEGI 2011, RADEMACHER & SIVARAMAKRISHNAN 2013b, TRUELOVE & MAWDSLEY 2011, VÉRON 2006, WILLIAMS & MAWDSLEY 2006a, 2006b, ZÉRAH & LANDY 2013, ZÉRAH 2007, ZIMMER 2012b, 2015a, 2015b).

From environmental tragedies (e.g. the Bhopal gas leak disaster) to the current ecological challenges of environmental degradation in Indian cities, environmental activism has in many cases triggered legislative and regulatory action; especially through the courts (DIAS 1994, DIVAN & ROSENCRANZ 2001, GONSALVES 2010). On the governmental side, the integration of the environmental issues in urban governance has, however, been rather reluctant and a recrafting of the city-nature relationship remains a major challenge in urban India (TIWARI et al. 2015: 175).

A multi-layered, yet often contradictory, 'urban environmental awakening' in India is nevertheless observable. Urban environmental issues are talked about today in India more widely and more frequently than ever before. India's mass media (both print and TV) covers environmental violations on a daily basis. The dissemination of new information and communication technology (ICT) has further made information more easily accessible, and has accelerated information exchange between different environmental groups (e.g. via e-mail and social networks). Most importantly, the Right to Information (RTI) Act of 2005 granted the citizens the right to access a range of (environmental) information from state authorities, which had earlier remained largely outside the reach of civil society organizations (NAIB 2013).

In line with the new opportunities to access knowledge about environmental as well as the underlying governances processes and reasons, numerous environmentally-motivated Public Interest Litigations[7] (PILs) have been filed challenging environmental degradation and depleting natural resources in India since the 1980s (DIVAN & ROSENCRANZ 2001, MEHTA 2009a, 2009b). WILLIAMS & MAWDSLEY (2006b: 665) even speak of an "explosion of environmentally-oriented public interest litigation". Therefore the judiciary, being "a spectator to environmental despoliation for more than two decades" in the last decades, by reacting to PILs as well as acting *suo motu* entirely at the courts' own initiative, has "assumed a pro-active role of public educator, policy maker, super-administrator, and more generally, amicus environment" (DIVAN & ROSENCRANZ 2001: 1). Through PILs the government and its authorities have often been forced to comply with its own environmental regulations (WILLIAMS & MAWDSLEY 2006b: 666) and the courts have often taken a pro-active role setting up court-appointed committees for the enquiry of facts and figures as well as to oversee the implementation of the orders (see among others DEMBOWSKI 2001, DIVAN & ROSENCRANZ 2001, GILL 2012, 2013, MEHTA 2009a, 2009b, RAJAMANI 2007, RAZZAQUE 2004).

Delhi has undoubtedly been a hot spot of enviro-legal activism (judicial activism through the courts). The courts played an important role in struggles for better air quality, the rejuvenation of urban water bodies, the struggles of solid waste disposal, as well as the protection of the Delhi Ridge. They have equally played a crucial role in the efforts to clean the Yamuna. All these environmental concerns and associated policy-making have been influenced, if not driven, by PILs filed by environmental NGOs and the courts' directions to these PILs, or *suo motu* action of the courts (FOLLMANN 2014, 2015, GILL 2014, MEHTA 2009a, 2009b, SHARAN 2002, 2013, 2014, VÉRON 2006).

The literature on India's urban environmental awaking is by no means entirely positive, but rather emerges as highly critical of the urban socio-environmental changes associated with this recent wave of urban environmentalism. Focusing on a wide variety of urban environmental issues from air quality to urban green space

---

7   PILs are "not for the purpose of enforcing the right of one individual against another" but rather PILs must "promote and vindicate public interest" (CHAKRABARTY & PANDEY 2008: 147). Thus, the *locus standi* which traditionally allows only affected persons to file litigation in the court based on sufficient connection to and harm from the existing law or action challenged, was deliberately bypassed to allow well-intentioned petitions in the public interest to be dealt with by the courts. PILs were intended to be helpful for demanding the rights of people, who are socially or economically disadvantaged and who are not able to demand their rights on their own. PILs offer the opportunity that a 'third' party can seek justice at the court for these groups and in public interest (ibid.).

and wastewater management, many of the recent writings critically engage with the role and impact of the growing urban middle class (see among others BAVISKAR 2002, 2007a, 2011a, GANDY 2008, GHERTNER 2011b, MAWDSLEY 2004, 2009, RADEMACHER & SIVARAMAKRISHNAN 2013a, SHARAN 2002, 2013, TRUELOVE & MAWDSLEY 2011, VÉRON 2006, ZÉRAH & LANDY 2013, ZÉRAH 2007). Principally due to the middle classes' "strong representation in the [English] media, politics, scientific establishment, NGOs, bureaucracy, environmental institutions and the legal system" (MAWDSLEY 2004: 81), the middle classes are often able to mobilize the public discourse in such a manner that their environmental ideologies of a clean and green city are equated with 'the larger public interest' (BAVISKAR 2011a). Similar, GANDY (2008: 121) highlights that "the political strength of the urban middle classes in India appears to be growing as part of a new discourse of 'environmental improvement'." CHATTERJEE (2004) evokes this process:

> "In metropolis after Indian metropolis, organized civic groups have come forward to demand from the administration and the judiciary that laws and regulations for the proper use of land, public spaces and thoroughfares be formulated and strictly adhered to in order to improve the quality of life of citizens. Everywhere the dominant cry seems to be to rid the city of encroachers and polluters and, as it were, to give the city back to its proper citizens" (CHATTERJEE 2004: 140).

Thus, in many cases the discourse of the 'proper citizens' has been deployed by the middle classes through middle class activism, especially against the urban poor; or as CHATTERJEE (2004) has termed it the 'civil society' against the 'political society' – the 'citizens' against 'population'. While the civil society has managed to organize and articulate their demands, the political society has largely failed to do so or has not been heard by neither politics, state authorities nor the civil society in questions of urban environmental change (CHATTERJEE 2004). Underlying middle-class values of an idealized green and clean urban environment coupled with the existence of an already relatively exclusive hegemonic public sphere dominated by the urban middle class have produced a powerful "elite-led NIMBYism" (WILLIAMS & MAWDSLEY 2006b: 668). In Delhi this has resulted in urban beautification, the demolition of slums (BHAN 2009, DUPONT 2007, DUPONT & RAMANATHAN 2008) and the banning of polluting industries (BAVISKAR et al. 2006, KUMAR 2012, NEGI 2011, SHARAN 2002, 2013, VÉRON 2006).

In this context, the conflicting outcomes of a middle-class driven and centered environmentalism have provocatively been termed as "bourgeois environmentalism" by BAVISKAR (2003, 2011a). Bourgeois environmentalism refers to an underlying "double-think" of middle-class environmentalism: on the one hand they are the leading voices for cleaning and greening the cities or creating protected parks and sanctuaries to conserve wildlife and biodiversity; on the other hand their resource-intensive and affluent lifestyles especially result in environmental destruction (BAVISKAR 2002, 2003, 2011a). However, environmental values, beliefs and behaviors of India's middle class are heterogeneous and, consequently, it is too simplistic to "make any sweeping generalizations about the nature of middle class environmental activism" in India (MAWDSLEY 2004: 90). Nevertheless, the middle-class dominated 'civil society' (CHATTERJEE 2004) and its resulting bourgeois environmentalism appear as major constrains for requesting an intensified participation of the CHATTERJEE's 'civil society' in urban environmental governance in India (MAWDSLEY 2009, TRUELOVE & MAWDSLEY 2011).

While a lot has been written about the middle class in India, environmental NGOs – often middle-class dominated – have been acknowledged for their critical role as 'watchdogs' in the context of environmental governance (MAWDSLEY 2004). Studies focusing on their role in urban environmental governance are still rare (for notable exceptions see DEMBOWSKI 2001, VÉRON 2006). Therefore by examining the role of environmental NGOs in the governance of Delhi's riverscapes, this study wants to contribute to an empirically-informed understanding of the diversity, challenges and contradictions of urban environmental activism in India. In contrast, the farmers from Delhi's riverscapes are, according to CHATTERJEE, part of the 'political society'. It is assumed that they therefore have faced major difficulties to participate in the urban environmental policy-making and spatial remaking of Delhi's riverscapes.

2.5 Reclaiming India's urban rivers

India's rivers have long been critically discussed with regard to their high levels of pollution (ALLEY 2002, HABERMAN 2006, MISRA 2010a, TRIVEDI 2010) and ongoing conflicts around large-scale developments of dams (BAVISKAR 1995, ERLEWEIN 2013, NÜSSER 2014, NÜSSER & BAGHEL 2010). One of the most controversial debates is the idea of 'inter-linking of rivers' (CHATTOPADHYAY 2005, IYER 2003a, SINGH 2003). The new focus on large-scale urban riverfront projects has extended this debate on rivers into the urban context. Studying India's riverscapes with a focus on the interplay of environmental discourses and practices in a rapidly changing urban context like Delhi is therefore highly relevant to better understand the changing relationships between (urban) nature and current processes of urbanization in India.

Several peculiarities emerge when analyzing India's urban rivers in comparison to those in other cities across the world. Due to their religious significance, Indian rivers and their riverfronts have been of great importance since ancient times. Yet in modern times, India's urban rivers are generally degraded and contaminated by the cities' untreated wastewater (ALLEY 2002, COLOPY 2012, HABERMAN 2006, SINGH 2007, SINHA & RUGGLES 2004). For India's sacred rivers like the Yamuna or the Ganga (Ganges) there remains more than one understanding of 'pollution'. River pollution scientifically measured as a contamination by fecal coliforms or chemical pollutants etc. emerges as largely detached from, and irrelevant to, the cultural-religious belief of purity associated with the holiness of the water by Hindu devotees (HABERMAN 2006, 2011: 4, SHARAN 2014: 29). An ecologically 'dead' river in the scientific understanding therefore remains a holy river in the cultural-religious sense, even when drinking its water or bathing along its banks seriously endangers human health. Nonetheless, the riverfronts and the once glorious ghats have become neglected backyards along many, but not all urban rivers in India (see Figure 2 and Figure 3). Furthermore, slums dominate the rivers' edge in many cities (see Figure 4 and Figure 5).

Figure 2: Ghats along the Yamuna in Delhi (Yamuna Bazaar)
Source: Follmann, November 2011

Figure 3: Ghats along the Ganga in Benares (Varanasi), Uttar Pradesh
Source: Follmann, January 2010

Figure 4: Demolished slums, Sabarmati riverfront in Ahmedabad, Gujarat
Source: Follmann, November 2011

Figure 5: Riverfront slums along the Yamuna in Delhi
Source: Follmann, September 2014

Overall, the rejuvenation of India's rivers is a pressing challenge for sustainable urban development, particularly as India's urban agglomerations also source a large share of their drinking water from rivers. At the same time, major monsoonal rivers, like the Ganga and its tributary the Yamuna, have preserved their comparatively wide floodplains even in urban areas. This is due to their very nature as monsoonal rivers and the major risks of flooding that goes along with it. Although 'Western models' of river channelization and riverfront development have been discussed in India since the late 1970s, the country's limited financial resources for large-scale channelization and riverfront development helped to preserve much of the lands in the floodplains compared to rivers in other parts of the world.

In recent years, a combination of factors has led to a new emphasis on urban riverfront redevelopment in India. With India's urban environmental awakening and the cities' ambitious world-class agendas riverfront developments have emerged as a major spatial focus in urban beautification initiatives in the cities of Delhi (BAVISKAR 2011c, FOLLMANN 2014), Ahmedabad (DESAI 2012b) and Chennai (COELHO & RAMAN 2010, 2013). Large-scale riverfront development projects have further been planned, or have already been implemented, in the cities across India including Agra, Guwahati, Hyderabad, Kolkata, Lucknow, Mumbai, Pune, and Surat.[8] New financial opportunities arising from the economic growth, high pressures on urban land markets, and innovations in river engineering provide the economic and technological basis for a new focus on the urban riverfronts in India's cities. Politicians, urban planners and (foreign) real estate companies therefore argue for the reclamation of the urban riverfronts from slums and other unwanted uses. The Sabarmati Riverfront Development Project in Ahmedabad is considered to be the largest project under construction and most widely discussed in the literature (CHATTERJEE 2009, 2011b, DESAI 2012b, MATHUR 2012, PESSINA 2012, VARMA PAKALAPATI 2010). The project highlights that channelization and riverfront development are connected to massive construction activities in order to control flooding and reclaim land for development (see Figure 6 and Figure 7).

Figure 6: Sabarmati Riverfront Development Project in Ahmedabad, Gujarat
Source: Follmann, November 2011

Figure 7: Advertisement for the Sabarmati riverfront project, Ahmedabad
Source: Follmann, November 2011

Without exception urban riverfront projects in India aim to reclaim the riverfront from slums and the urban poor in order to develop green and recreational infrastructures like parks and riverfront promenades (see Figure 52 and Figure 53, page 392).

---

8 Studies focusing on urban riverfronts in India are available for the river Yamuna in Agra (HARKNESS & SINHA 2004, SINHA & HARKNESS 2009), Mathura (BHARGAVA 2006), the Hoogly River in Kolkata (BEAR 2011), the Mulla and Mutha in Pune (BARVE & SEN 2011), the Musi in Hyderabad (COHEN 2011) and the Gomti River in Lucknow (NAGPAL & SINHA 2009). On the other projects information are provided by feasibility studies or reports published by environmental NGOs and social activists.

High-class residential projects and massive transport infrastructure developments (e.g. highways, flyover, metro lines, metro stations) are further characteristic for India's new riverfronts. The riverfront development projects are advertised as environmental upgradation and river rejuvenation, which besides cleaning up the water of the river are also supposed to bring the riverfront back to life. These projects have more often than not been carried out in the form of urban mega-projects (ALTSHULER & LUBEROFF 2003, GELLERT & LYNCH 2003), which have significant socio-ecological impacts on the urban riverscapes.

Contradictorily, while many observers have called attention to violations of environmental laws and regulations, urban mega-projects are discursively framed as environmental upgradation (COELHO & RAMAN 2013, DESAI 2012b, FOLLMANN 2015, GHERTNER 2011b, GILL 2014, MATHUR 2012, PESSINA 2012). These projects further underpin that procedures for environmental clearance in India are often inadequate (see among others KOHLI & MANJU 2005, LELE et al. 2010, MENON & KOHLI 2009, PANIGRAHI & AMIRAPU 2012). While some of these projects were for some time delayed by court cases, they could not be stopped by environmental activism and often received final approval from the courts.

The desire to develop the urban riverfronts encounters lacking environmental regulations for the protection of urban rivers. India's urban environmental governance is generally characterized by a multiplicity of agencies, overlapping jurisdictions as well as fragmented and ill-defined responsibilities. The long-lasting debate for the rejuvenation of India's rivers as well as recent debates for large-scale riverfront developments are showcase examples for the mutual recrimination within India's (urban) environmental governance framework. Conflicts are evident between different levels of government as well as between different sectoral departments. Thus, large-scale riverfront development projects often go ahead largely unchallenged and without in-depth environmental impact assessment (EIA), monitoring and control.

This study aims to highlight the challenging urban ecologies connected to the remaking of India's urban riverscapes. The case study of the river Yamuna in Delhi emerges in this context as an extremely relevant and insightful example to highlight the multiple debates, challenges and contradictions associated with 'riverfront' development in India. By engaging with both strands of debate, ecological rejuvenation as well as riverfront development, the theoretical reflections outlined in the following chapter aim to provide a conceptual framework which allows analysis of the multiple city-river linkages in a more holistic way, beyond the Indian context.

II THEORETICAL REFLECTIONS

3 GOVERNING RIVERSCAPES: THEORETICAL AND CONCEPTUAL REFLECTIONS

"[…] nature is too important to be left to natural scientists; we need writers, critics, and poets to describe much of the inherent and apparent beauty in terms that do not neutralize or objectify nature to a simple subject of science" (MOSELEY et al. 2014: 291).

This chapter aims to develop an understanding of urban environmental policy-making and governance from a political ecology viewpoint by reviewing selected literature on governance research and urban political ecology. Based on this, the second part of the chapter introduces the theoretical and conceptual framework for the analysis of the governance of hybrid riverscapes, which will be applied throughout this study to analyze urban environmental change along the river Yamuna in Delhi.

3.1 Governance research

Governance research is concerned with the different forms of cooperation and negotiation between the state, private actors and the civil society. New partnerships and arrangements between the state and other non-state actors on different scales have become an important research focus of an ever growing body of literature from various fields including geography. Consequently, the concept of governance gained wide popularity.

The international discourse on, and theorization of, the concept of governance has especially been coined within the political sciences. Different authors trace back the introduction of the term governance within scientific discourses to the work of ROSENAU & CZEMPIEL (1992) and their book 'Governance Without Government: Order and Change in World Politics'. In Germany, it has been the work of RENATE MAYNTZ at the Max-Planck-Institute for the Study of Societies which had, and continues to have, a lasting effect on the understanding of governance in the social science including human geography (see e.g. MAYNTZ 2003, 2004, 2005, 2009a, 2009b, MAYNTZ & SCHARPF 1995). Traditionally, the term 'governance', which derived from Latin and Greek words for the *steering of boats,* was used synonymously to describe the actions and affairs of the government or the process of governing involving multiple stakeholders (JESSOP 1998, MAYNTZ 2003, RHODES 1996, STOKER 1998). Newer understandings of governance generally refer to certain behaviors and specific activities, as well as underlying systems of rule (ROSENAU & CZEMPIEL 1992: 4). Further, the term governance points explicitly to "a shifting pattern in styles of governing" (STOKER 1998: 17). More precisely "governance signifies a change in the meaning of government" including "a *new* process of governing; or a *changed* condition of ordered rule; or the *new* method by which society is governed" (RHODES 1996: 652–653 [emphasis in original]). Governance theory therefore draws on institutions, both structures and procedures, and the theoretical ideas have derived from questions of alternative ways in the coordination of transactions and interaction between different actors embedded in certain networks (KOOIMAN 2003, LE GALÈS 1998: 492f.). Governance, as an umbrella concept, gained great popularity because it is generally understood to be much wider than when using the term 'government' and has the characteristic "to cover a whole range of institutions and relationships involved in the process of governing" (PIERRE & PETERS 2000: 1, cf. WILLIAMS 2009).

The literature further highlights the differentiation between government ("backed by formal authority") and governance ("backed by shared goals that may or may not derive from legal and formally prescribed responsibilities") (ROSENAU & CZEMPIEL 1992: 4). The term 'government' is generally either understood as the system by which a country is governed (e.g. democratic government) or defined as the group of people governing a country, or a part of a country, e.g. state, provinces, municipalities, etc., at a certain point of time. Discussing the term government, STOKER (1998: 17) refers to the "monopoly of legitimate coercive power" of the "formal institutions of the state" in decision-making and enforcing these decisions. In such a definition, government includes politics as well as the state's bureaucracy covering the legislative, executive and judiciary powers of the state. Following a broader consensus (see e.g. JESSOP 1998: 30), in this study, 'government' explicitly refers to state actors. In contrast, governance refers to the coexistence of collective regulations, negotiations and the cooperation between public and private actors, as well as sovereign government decision making (JESSOP 1997, KOOIMAN 2003, LE GALÈS 1998, MAYNTZ 2005). It thereby includes the "evolving forms of government and regulation that transcend those based on traditional state hierarchies and market systems" (CASTRO 2007: 102) and points towards the more or less voluntary co-operation of state and non-state actors within different governance processes (ROSENAU & CZEMPIEL 1992). In other words, while the classical understanding of 'to govern' was defined by the sovereign power of the state to exert power over its territory (national to local level), governance theory acknowledges the role of the state in "leading, steering and directing" public policies, "but without the [unabated] primacy of the sovereign state" (LE GALÈS 1998: 495).

Governance research reveals that it is too complex to follow the simplistic classic Weberian division of state organization in decision-making (politics) as well as implementation (bureaucracy), since both the state and the civil society are not "monolithic bounded entities" (LEACH et al. 2007: 7). Furthermore, in both spheres of the state – politics and bureaucracy – the different agencies of the state debate, negotiate and bargain with a multiplicity of other state actors as well as a wide range of non-state-actors "creating networks and blurred boundaries" (ibid.). Governance research aims to analyze how these networks and blurred boundaries influence decision-making processes and their implementation under the involvement of multiple state and the civil society actors. Or to put it differently, the key interest of governance research is to deconstruct decision-making and implementation processes taking into account existing institutions and actor networks.

There emerges "a tendency to confuse governance as an empirical phenomenon with theories about how this phenomenon operates and can be understood" (PIERRE & PETERS 2000: 14). Consequently, governance is discussed by PIERRE & PETERS (2000) as both *structure* and *process*. Similar MAYNTZ (2005) underlines the *dual nature* of the term governance which refers both to the structure or the regulating framework for action ("Handeln regelnde Struktur") and the process of regulation itself ("Prozess der Regelung"). Both the structure and process dimension of governance are determined by existing institutions. Through the actions and interactions the different actors involved in governance "structure and restructure the systems they are part of" (KOOIMAN 2003: 16). Such an understanding of governance reflects the concept's connection to the larger debates on structure and agency
(GIDDENS 1984). MAYNTZ (2003: 27) therefore distinguishes two understandings of governance "both distinct from political guidance and steering". Firstly, governance denotes "a new mode of governing that is distinct from the hierarchical control model" and stands for "a more cooperative mode where state and non-state actors participate in mixed public/private networks" (MAYNTZ 2003: 27). The second new understanding of governance is much broader and includes "different modes of co-ordinating individual actions, or basic forms of social order" and has derived from empirical findings showing that co-ordination between different actors typically works differently to pure hierarchical or pure market logics (MAYNTZ 2003: 28).

This study applies the concept of governance to an analysis of urban environmental change from a geographical perspective. In order to do this, a framework is needed to conceptualize governance from an analytical perspective. In this context, it is important to distinguish such an analytical perspective from a normative perspective, focusing on good governance and a critical descriptive perspective on governance focusing on the shift from government to governance (cf. ZIMMER 2012b). After briefly presenting these different perspectives on governance in the following sections, the different conceptualizations are related to both urban governance and environmental governance. The two spheres are then merged in an urban environmental governance framework developed for this study.

3.1.1 Good governance – the normative perspective

Good governance generally refers to efficient, citizen-oriented political and administrative principles following the rule of law and creating supportive conditions for economic development (MAYNTZ 2005). In doing so, good governance has inherently become a buzzword in the development context (ZIMMER 2012b: 19) and has served as a "mantra for donor agencies as well as donor countries" (NANDA 2006: 269). It has been used both as a guideline for administrative reforms and an evaluation standard for the creditworthiness of developing countries like India.

The normative perspective of good governance is closely connected to the neoliberal economic reforms of the 1990s. The neoliberal reforms associated with privatization and liberalization rested on a the strong belief by international development agencies – especially the World Bank and the International Monetary Fund – that "governance structures of the market would be an improvement over the governance structures of the Keynesian state" (LEACH et al. 2007: 5). These international actors "aimed to produce a degree of global consensus over what constitutes 'good governance', and to make governance reform a legitimate and central issue within their development agendas" (WILLIAMS 2009: 606).). When reviewing the development discourse since the 1980s, it is evident that these international agencies were first to press for a 'rolling back' of the state, which transformed to more "sophisticated forms of public sector reform conducted in partnership with both the private sector and the civil society" (WILLIAMS 2009: 609). This resulted in "structural adjustment programs", which have been the precondition for bailing countries (like India) out of debt.

The theoretical underpinnings of the good governance agenda are contested. In particular, their neoliberal character has been widely criticized. MAYNTZ (2005) highlights that the normative concept of good governance causes difficulties when used as an analytical instrument. Speaking of good governance and setting indicators for measuring always assumes the existence also of *bad governance*. From a postcolonial perspective the concept of good governance needs to be further viewed critically in terms of the transferability of Western concepts into states and cities in completely different contexts. The use of the term good governance suggests that finding a solution for a socio-environmental problem like river pollution or the conservation of floodplains is mainly an organizational problem. This implies that it is a problem which can be fixed if a common goal is formulated and a set of adequate policies is implemented (as per good governance principles). The normative governance rhetoric suggests a *one-size-fits-all* approach, but in different cultural, social and political settings governance plays out in very different ways (CORNWALL 2004: 9). Therefore reducing governance to rational and technical management makes it an apolitical concept (cf. FREY 2008, HEWITT DE ALCÁNTARA 1998, LEFTWICH 1993, ZIMMER & SAKDAPOLRAK 2012).

3.1.2 A (critical) descriptive perspective on governance

From a descriptive perspective of governance, the traditional role of government "as the appropriate, legitimate and unchallenged vehicle for social change, equality and economic development" (PIERRE & PETERS 2000: 2) was superseded when its leitmotif, the Keynesian welfare system, entered in crisis in the late 20th century. After the Second World War scholars drew a largely prescriptive picture of the state's capacity to steer (MAYNTZ 2003: 28–29). After the crisis, in which state-failure was attested, the attention turned to the reasons of this failure (ibid.). As a result, governance as a theoretical concept gained popularity and scholars started to analyze the shifting pattern from government to governance in more depth. Subsequently, the dominant role of the state was largely perceived as problematic and, thus, privatization and decentralization made way for the market to become the dominant force for change (MAYNTZ 2003, PIERRE 1998, PIERRE & PETERS 2000, RHODES 1996, ROSENAU & CZEMPIEL 1992, STOKER 1998). From the early 1990s a discourse emerged within the political science focusing on the question of "what is, and what should be the role of government in society" (PIERRE & PETERS 2000: 3). In addition, the emergence and mainstreaming of the term *governance* is closely related to the transformational processes caused by globalization and liberalization at the global level as well as the decentralization and communalization efforts on the local level (cf. KERSTING et al. 2009).

Broadly defined, the descriptive perspective on governance views new forms of governance as deriving from the integration of non-state actors into former activities of the state. This perspective empirically studies the rise of "new formal or informal institutional arrangements that engage in the act of governing outside and beyond-the-state" (SWYNGEDOUW 2005b: 1991). This governance-beyond-the-state, as SWYNGEDOUW (2005b: 1992) argues, gives "a much greater role in policymaking, administration and implementation to private economic actors on the one hand and to parts of civil society on the other". In other words, governing is no longer the exclusive task of the government, but rather a joined action by the state, private actors and the civil society. In this transition, former core government duties and functions have been transferred to non-state actors under the neoliberal agenda.

In this context, authors criticize that the neoliberal reforms focusing on deregulation and privatization have resulted in a hollowing-out of the state (BRENNER & THEODORE 2002, HARVEY 1989, JESSOP 2002, PECK & TICKELL 2002, STOKER 1998, SWYNGEDOUW 2005b). Other authors interpret the hollowing-out or downscaling of the state as a more strategic form of outsourcing (or devolving) of certain tasks to non-state actors in order to achieve a slimmed-down, but more capable state (JESSOP 1998, LE GALÈS 1998, LEACH et al. 2007, MAYNTZ 2003, RHODES 1997). They view this process to be controlled by the state itself and making it more capable for the state's main functions such as policy making and strategic development. MAYNTZ (2003: 31) argues that it "is not so much a loss of state control as a change in its form", since these new forms of governance take place under the legitimization, oversight and control of the state and the power still rests with the state. The state is still able to "intervene by legislative or executive action" when outsourcing and devolution fails to meet the state's expectations (MAYNTZ 2003: 32). Supporting this argument, JESSOP (1998: 39) describes the role and the power of the state in governance:

> "[…] although governance mechanisms may require specific techno-economic, political, and/or ideological functions, the state typically monitors their effects on its own capacity to secure social cohesion in divided societies. The state reserves to itself the right to open, close, juggle, and re-articulate governance […]."

Therefore hierarchical control by the government and new forms of governance are not "mutually exclusive", but are rather intertwined as "different ordering principles" (MAYNTZ 2003: 32) and the role of the state is rather "shifting than shrinking" (KOOIMAN 2003: 3).

In contrast, critical voices point to the contradictory nature of the changing modes of governance. According to SWYNGEDOUW (2005b: 1993–1994), these changes result in the dangerous tendency that reforms are celebrated as resulting in "increasing democracy and citizen's empowerment" overlooking "their often undemocratic and authoritarian character" and are perceived as a precondition for good governance. Furthermore, as JESSOP (1998: 41) argues, there is doubt that complex objects of governance (e.g. urban environmental change, environmental degradation of rivers, loss of biodiversity) "could ever be manageable" even with the *best* policies, institutions and governance set-up.

Governance research interested in urban environmental change therefore needs to break with the traditional perspectives of both normative conceptions of good governance as well as the (critical) descriptive comparison of traditional modes of government with new forms of government. Rather, it should focus on empirical insights from actually occurring processes of governance. The analytical value of the concept of governance needs to be highlighted in this context.

3.1.3 'Unpacking' governance processes – the analytical perspective

The analytical perspective of governance looks into the actual existing processes of governance with the goal to better understand the interactions among the different actors involved. The ultimate goal of this perspective is consequently to analyze the networks of actors, multi-actor interactions, processes of negotiation and bargaining as well as decision-making. From an analytical perspective governance is thus viewed as both an activity (DEAN 2010: 18) and a process (LEACH et al. 2007: 1), which needs to be better understood by focusing on the practices of the different actors (ZIMMER 2012b: 22).

Following such an analytical approach implies, as HEINELT (2002: 23) argues, that "governance is primarily seen as a 'fact' and is not normatively perceived as good or bad". Following these general principles for analytical governance research, this study understands governance more as an analytical tool than an outcome. In this context, LE GALÈS (1998: 496) proposed an analytical framework for urban (or regional) governance which should focus on "the type of actors and their resources" and "the linkages of different regulations in a territory".

The analysis of urban environmental governance from a geographical perspective therefore needs to examine the actors and their interaction in order to unveil the interests and strategies of the involved actors as well as the underlying power structures (HOHN & NEUER 2006: 293). In doing so, the interactions between different levels of government, political parties, private businesses, special interest groups (e.g. unions, peasant movements or religious groups), international agencies (e.g. international financial institutions, multilateral development banks) and a variety of civil society organizations (including NGOs) have to be analyzed (CASTRO 2007: 107). Such an analytical approach acknowledges the many interrelations and inter-dependencies among the different actors and adequately takes into account that no individual actor "has the knowledge and information required to solve complex, dynamic and diversified societal [and environmental] challenges" (KOOIMAN 2003: 11). Rather, as outlined by KOOIMAN (2003: 4), "various interacting factors [...] are rarely wholly known and not caused by a single factor; technical and political knowledge is dispersed over many actors; and governing objectives are not easy to define and are often submitted to revision".

From a geographical perspective, the decision-making process for, and implementation of, spatial policies emerge here as key analytical components in understanding urban environmental change. By focusing on both "consensus-finding processes" and "causes of tension and conflict" between the different actors (HOHN & NEUER 2006: 293) an analytical approach to governance is able to overcome the risk of reducing governance to an apolitical organizational problem.

3.1.4 Urban governance

New forms of governance have emerged on all scales, from global to local. Without doubt, the urban scale has emerged as "a pivotal terrain where these new arrangements of governance have materialised" (SWYNGEDOUW 2005b: 1992–1993) and *urban governance* has become a major focus in governance research (in all three perspectives). A good starting point for explaining some theoretical perspectives on urban governance is the definition provided by the United Nations Human Settlements Programme (UN HABITAT 2002: 9):

> "Urban governance is the sum of the many ways individuals and institutions, public and private, plan and manage the common affairs of the city. It is a continuing process through which conflicting or diverse interests may be accommodated and cooperative action can be taken. It includes formal institutions as well as informal arrangements and the social capital of citizens."

The above definition suggests that research on urban governance should consequently focus on the processes of planning and management of the city. Governance research must engage with the diverse, often conflicting, interests of multiple actors aiming to understand why certain decisions are taken. In doing so, governance research needs to make explicit the formal rules and informal arrangements influencing decision making. Further, as the shift from government to governance is accepted as a fact, analytical governance research on the city must pay special attention to the role of private actors and the civil society engaging with the state actors in multiple ways. Analytical research on urban governance is therefore complex and engages with multiple questions. These include questions regarding the involvement of all relevant actors, the nature of interaction between the state and the non-state actors as well as the outcomes of these interactions (BAUD & DEWIT 2008b: 7).

Urban governance is mostly framed around the scale of the city and is in many cases understood as a form of 'local' governance on the city level. Nevertheless, the literature shows that the actual urban governance set-up encompasses different levels which go beyond the actual institutions on the city level and include for example regional or national institutions involved in urban issues (e.g. ministries, regional planning bodies, etc.). The focus of the policy-making in urban governance concentrates on the scale of the city. Studies of urban governance have further concentrated on the above described new forms of governance-beyond-the-state or looking at the different arrangements between the state and non-state actors in delivering collective services for the city. In this regard, providers and users of environmental services, as well as their direct links in decision-making have been analyzed (see among others BAUD & DHANALAKSHMI 2007, VAN DIJK 2011).

3.1.5 Environmental governance

In contrast to urban governance, environmental governance is a multi-level approach reflecting various scales (BRAUN et al. 2003, BULKELEY 2005, BULKELEY & BETSILL 2003). Despite this, the local scale has often been black boxed paying too little attention to the environmental governance in, and of, the city (HODSON & MARVIN 2009).

From a (critical) descriptive perspective environmental governance refers to a "qualitative shift" in the way decision-making takes place with regard to natural resources and the utilization of nature (BRIDGE & PERREAULT 2009: 475). In the environmental sector these shifts are significant, since environmental regulation traditionally used to be a key element of the tasks of sovereign states, but today a range of non-state actors on different scales are involved. In this context, environmental degradation is often reasoned to be an outcome of state failure and/or market failure (BAKKER 2003b, LEMOS & AGRAWAL 2006, MOL 2002). Following the neoliberal logic, traditional forms of command and control regulation by the state with regard to the environment are reduced, and market-based legal instruments aiming for an internalization of environmental externalities (e.g. emission rights trading, taxes or subsidies) and voluntary approaches (e.g. voluntary commitments of industries to reduce pollution levels) are introduced (cf. RAZZAQUE 2013).

International institutions (e.g. UNESCAP 2005) have argued that markets and an improvement of existing (state) institutions are central to curb negative environmental impacts in the cities of the Global South (FARIA et al. 2009: 640). In this process, numerous globally accepted environmental doctrines have emerged including the precautionary principle, the polluter-pays principle, and intergenerational equity, which thereof derived the concept of sustainable development. All of these have been highly influenced by institutional and legislative developments on an international scale, including the Rio Declaration on Environment and Development in 1992 (cf. GINTHER et al. 1995, RAZZAQUE 2013) which eventually resulted in certain agendas on the local level (e.g. Agenda 21).

The discussion surrounding environmental governance is generally supportive of the increasing role and participation of non-state actors. Simultaneously, these developments are seen as an inevitable result of the increasing pressure from the civil society, environmental movements, the courts and the media. This view if reflected in the Global South (cf. DEMBOWSKI 1999, 2001, DWIVEDI 2001, ECONOMY 2004, GILL 2012, 2013, GUPTA 2012, HOCHSTETLER 2002, JASANOFF 1997, RAZZAQUE 2004, SONNENFELD & MOL 2006, YANG & CALHOUN 2007), yet in the Indian context, as mentioned earlier, there is critique surrounding the increasing role of the middle class and the courts.

From a normative governance perspective environmental degradation of urban rivers and the encroachment on floodplains in Indian cities could be designated as cases of governance failure; both in terms of state failure (policy and implementation failures, e.g. due to wrong assumptions or corruption) and market failure (e.g. polluter pays principle not implemented). A critical descriptive perspective would focus in this context on the reasons of shifting from regulations by the state (e.g. pollution levels) to the dissemination of market-based legal instruments (e.g. polluter-pays principle). However, both perspectives generally fail to theorize why environmental problems, such as river pollution or the shrinking of floodplains, remain largely persistent even when there is mutual agreement on the need for action.

Here the analytical perspective aims to highlight the complexity of environmental problems including questions of political will, the role of historical developments, and interconnectedness of multiple scales. In doing so, the analytical perspective explicitly engages with the question of *why* certain environmental problems are persistent. It acknowledges that both the classical regulations by the state and market-based legal instruments might fail since environmental problems are complex, and demand for action across scales, as well the interests of the different involved actors vary greatly. From an analytical perspective multiple questions therefore arise going beyond the binary of market or state failure. The research focus thereby shifts to the reasons for the limitations of the existing policies and the ways these policies are framed and implemented (for a more detailed discussion of different framings of environmental governance as an analytical approach see BRIDGE & PERREAULT 2009).

3.1.6 Urban environmental governance – an analytical approach

As the above introduction of urban and environmental governance suggests, urban governance and environmental governance overlap. The term urban environmental governance accordingly attempts to bring these two spheres together. The essence of the argument in this study is to highlight that urban environmental governance stands for more than just a fusion of urban and environmental governance, but rather raises a range of specific questions and problems with regard to urban environmental issues and their governance.

More specifically, analyzing the governance of urban rivers requires an understanding of both urban governance as well as environmental governance, since both spheres (while often evolving more or less autonomously) are involved in governing the river in the city and beyond (see Figure 8).

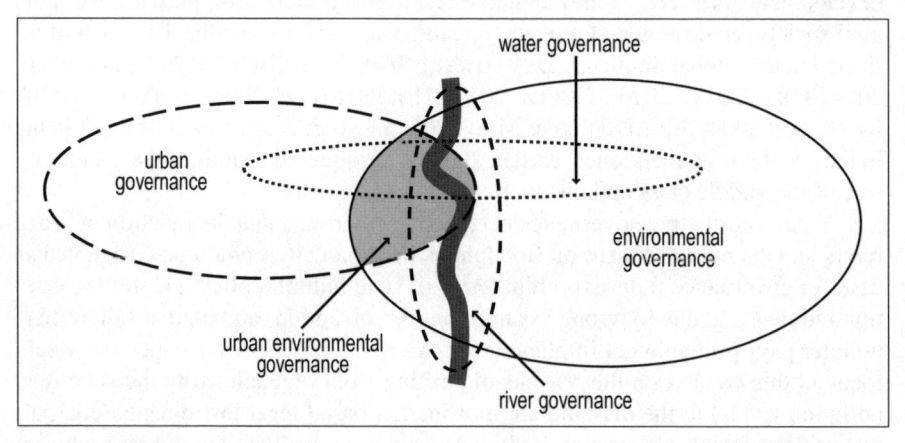

Figure 8: Rivers and the spheres of governance
Source: Graphic by author

Conceptualizing urban environmental governance as a distinct field of policy making emerges here as insightful, since it combines the two governance spheres at the local level while simultaneously taking into account multiple scales. Additionally, sectorial approaches, like water governance, are present both in urban governance and environmental governance. As Figure 8 further indicates, all five approaches come together in governing urban rivers.

The governance of urban rivers is often characterized by sectoral fragmentation and lacks coordinated action across scales. The fragmentation of governance set-ups emerges as both a cause and effect of sectoral approaches in governing the urban environment. The fragmentation is identifiable when analyzing urban environmental policies and the processes of their making as well as the day-to-day processes of their implementation.

In the interests of this study, and following KEIL & DESFOR (2003: 29), two categories of urban environmental policies need to be distinguished which have often been discussed separately in the literature on urban environmental governance. First, *pollution policies* including all legislation, practices, procedures and other institutional arrangements for regulating pollution of land, water (in general) and rivers in particular. Secondly, *land-use policies* including all spatial policies and planning documents which (re)structure or (re)make restrictions on the use of urban space. While the former is generally dealt with in the sphere of environmental governance, the latter is predominantly the matter of urban governance. An integration of both spheres in policy-making is needed since they are closely interrelated functionally and spatially (DESFOR & KEIL 1999, 2000, 2004, KEIL & DESFOR 2003).

An analytical perspective on urban environmental governance must thus question how and why this fragmentation in urban environmental policies occurs. Existing research on urban environmental governance has often remained focused on sectoral approaches (e.g. studying water governance) and as such has often failed to holistically conceptualize the socio-environmental processes constituting urban environmental change. Furthermore, the governance approach emerges to be of limited epistemological value to explain the underlying motives, driving forces and powers of the involved actors. Based on this, it can be speculated that coordinated and collaborative action might somehow not be the main interest of all actors.

As outlined above, governance research (especially the normative perspective) tends to concentrate to a large extent on coordination and collaboration rather than focusing on actual conflicts and different interests of the involved actors. HIRSCH (2006: 190) argues that "the often-apolitical discourses around governance tend to hide the tension between governance as a software techno-fix and governance as an inherently political, context-bound process." In doing so, the normative perspective tends to mask existing tensions between the different actors or the dominance of certain discourses, which often find their way into policies. In this respect, questions of power remain often inadequately addressed (TAYLOR 2007: 300) and the definition of important actors remains blurred.

An analytical perspective, as proposed in this study, has a strong focus on the conflicting issues. The understanding of governance in this study is therefore not reduced to questions of coordination and collaboration among the actors (which largely would draw on the normative categories of 'good' or 'bad' governance), but rather focuses on the interaction, negotiation and bargaining processes among the different actors that have a stake in Delhi's riverscapes. In particular, the negotiation and bargaining processes within contentious issues can provide information on how the different actors define and explain the reasons for environmental problems such as river pollution or the encroachment on urban floodplains. These negotiation and bargaining processes may also lead to improved urban environmental policies. Eventually these may result in coordinated, collaborative actions by the involved actors if agreement is achieved among the actors. Continuing environmental problems like river pollution might be valuable as an indicator for non-coordinated land-use and environmental policies; indicating likewise that the involved actors have not come to an agreement. While a normative perspective on governance would term such a situation as bad governance, an analytical perspective on governance searches for the underlying reasons for the failure of governance.

Urban environmental change is complex, as indicated in the preceding paragraphs. Different actors might perceive and evaluate these changes differently. They might draw on diverse types of knowledge and might further use different language to describe environmental change. Therefore analytical governance research should analyze the way environmental conditions are described and measures are proposed by the involved actors, in order to be able to assess how environmental problems are framed by the different actors, and how they think solutions could be achieved. In doing so, it might also be possible to detect the influence of different actors on policy outputs (DESFOR & KEIL 2004: 12).

3.1.7 Participation in urban environmental governance

Multiple questions arise with regard to participation of the civil society and their impact on policy making in urban environmental governance (DESFOR & KEIL 2000, 2004, HIRSCH 2006). Civil society actors "vary in their capacity to participate" (ACHARYA et al. 2004: 46) and participation in form of public hearings etc. is often ritualized, resulting in only very limited effect on decision-making (ALONSO & COSTA 2004, CORNWALL 2004). In the context of urban governance and direct citizen participation in cities of the Global South, CORNWALL (2002, 2004) has argued that the state has created various forms of "invited spaces" for public participation, but these have only been created for certain sections of the population (the 'proper citizens' of the middle class). She states that the poorer sections of the population have had to create "their own opportunities and terms for engagement" (CORNWALL 2002). Drawing on this work, MIRAFTAB (2004, 2009) distinguishes "invited" and "invented spaces" of participation.

In the setting of Indian cities, research in urban governance has shown that it is always a question of which sections of the population participate; and this question has been discussed beyond CHATTERJEE's (2004) differentiation of the 'political' and 'civil society' outlined earlier (BAUD & NAINAN 2008, ZÉRAH 2009).

BAUD & NAINAN (2008) argue that the poor depend on "negotiated space" to participate, while for the middle class an "executive space" is opening up through new forms of urban governance. When combining these thoughts on participation with KOOIMAN's (2003: 7) argument that new forms of governance should require that involved actors "are willing and able to interact", an analytical approach to urban environmental governance needs to take this willingness of the state actors to interact with the 'political' *and* 'civil society' into account.

The analytical perspective on urban environmental governance can draw on existing conceptualizations by adopting the terminologies of invited/invented (MIRAFTAB 2004, 2009) and negotiated/executive spaces (BAUD & NAINAN 2008) in order to analyze the various forms of interaction, negotiation and bargaining of civil society (e.g. environmental NGOs) but also private actors. Urban environmental governance is thus "clearly not a 'neutral' or apolitical term" (RACO 2009: 623). Yet in the context of the cities in the Global South the discussion has remained "dominated by managerial, technocratic, and generally apolitical approaches" (LAWHON et al. 2014: 1, based on MYERS 2008). Scientific and technical issues tend to dominate urban environmental governance in particular, and the making of urban environmental policies often includes only some perspectives (predominantly expert knowledge or actors which have a strong voice) at the expense of others; especially the poor and marginalized groups (KARPOUZOGLOU & ZIMMER 2012, KEELEY & SCOONES 1999, 2003, WHATMORE 2009).

> "Policy is set out as objective, neutral, value-free, and is often termed in legal or scientific language, which emphasises its rationality. In this way, the political nature of the policy is hidden by the use of technical language, which emphasises rationality and objectivity. But the technical is always in some way political" (IDS 2006: 13).

It is important to recognize that any urban environmental policy reflects political interests, is based on the negotiation process and actions of a multiplicity of actors, and is therefore a product of discourse. Therefore, research on urban environmental governance needs to address the broader influencing factors on urban environmental policy making by focusing on the *process* of policy-making (DESFOR & KEIL 2004, HAJER 1995, 2008, IDS 2006, KEELEY & SCOONES 1999, 2003, LEACH et al. 2007). In order to do this, it is necessary to link knowledge and discourse, actors and networks, as well as politics and interests, to be able to thoroughly understand the process of making environmental policies.

Connecting governance research with key concepts from UPE offers the opportunity to introduce new critical theoretical thoughts into governance research. This is elaborated on in the following sections. The concepts of UPE, as this study will show, empower governance research in various ways. Namely:

- To connect the material outcomes of environmental change to underlying discourses;
- To link the analysis of governance to question of knowledge and power;
- To (re)conceptualize the role of the non-human actors in governance settings;
- To deconstruct sectoral approaches and supposedly 'natural' scales of governance.

UPE therefore rejects the apolitical tendency of governance research by highlighting that the urban environment is highly politicized.

3.2 Urban Political Ecology

A UPE-perspective offers an insightful theoretical framing to widen the analytical perspective in governance research. The following sections outline key concepts of UPE providing an understanding of the urban environment as a socio-natural hybrid. This is the theoretical baseline for the introduction of the analytical framework of this study. In doing so, a post-structuralist UPE perspective is chosen, which emphasizes the social construction of nature and urban environmental change. This offers an insightful starting point to develop a deeper understanding of urban environmental governance.

3.2.1 Introducing UPE

In their introduction to the edited volume *In the Nature of Cities* HEYNEN et al. (2006b: 2) argue that while the world is rapidly urbanizing, both scholarship on ecological modernization and urban studies for a long time have "unjustifiably largely ignored the urbanization process as both one of the driving forces behind many environmental issues and as the place where socio-environmental problems are experienced most acutely". Political ecology had been no exception until the rather recent emergence of UPE.

As a relatively young interdisciplinary field, political ecology has developed since the 1980s. It was predominantly formed based on the thoughts of cultural ecology and Marxist political economy with findings from developing studies. UPE has considerably evolved over time (see among others BLAIKIE 1995, 1999, BRYANT 1992, 1998, FORSYTH 2003, KRINGS 2008, NEUMANN 2008, 2009a, 2009b, 2011, ROBBINS 2012, WALKER 2005, 2006, 2007, WATTS 2008, ZIMMERER & BASSETT 2003). A detailed review of the different theoretical roots and concepts of political ecology and the series of strands within the field would go beyond the scope of this study.

Due to the complexities of environmental change and the wide range of theoretical influences within the interdisciplinary approach of political ecology, political ecology has not developed a single methodology or a clearly defined set of theoretical concepts. Political ecology is rather characterized by a heterogeneity of approaches and a mobilizing of different concepts from a broader school in order to critically explain socio-environmental change (cf. FORSYTH 2003, LOFTUS 2012, NEUMANN 2009a, ROBBINS 2012, ROCHELEAU 2008, WALKER 2006). As a result, political ecologists often start from, or end in, socio-environmental contradictions (ROBBINS 2012: 87) and their findings show the complexity of socio-environmental interactions rather than "producing concise and compelling narratives" (WALKER 2006: 384).

Political ecology is thus "not a theory or a method", but rather "a specific sort of critical attitude" which is shared by "a community of practice" in engaging with socio-environmental questions (ROBBINS 2012: 84-85). Political ecology is known for challenging taken-for-granted narratives about environmental change, deconstructing socio-environmental myths and asking for socio-environmental justice by opposing deterministic, neo-Malthusian, apolitical explanations of environmental change (cf. among others HINCHLIFFE 2008, KEELEY & SCOONES 2003, NEUMANN 2009a, STOTT & SULLIVAN 2000).

Classical approaches in political ecology have tended to focus on structural explanations, however a 'second generation' of political ecology has adopted postmodernist, post-colonialist and post-structuralist thoughts (cf. BENTON-SHORT & SHORT 2013, ESCOBAR 1996b, 2010, FORSYTH 2003, 2008, NATTER & ZIERHOFER 2002, PEET & WATTS 1996), feminist approaches, actor–network theory (ANT), and other concepts originating from the field of science studies (cf. HARAWAY 1991, LATOUR 1993, 2004, ROCHELEAU 2008, ROCHELEAU et al. 1996, WHATMORE 2002). The post-structuralist scholarship within political ecology has drawn on concepts like social construction (see 3.2.2) and hybridity (see 3.2.4). This work has challenged "the stable category of 'nature' that had underlain mainstream political ecology from the beginning" and emerges today to have "much in common with actor–network theory" (ANGELO & WACHSMUTH 2014: 3).

Political ecologists, as ROBBINS (2012: 87) frames it, thus "simultaneously make claims about the state of nature and claims about claims about the state of nature." To unravel these claims, political ecologists often study environmental knowledges and socio-environmental practices of a range of different actors with special interest in socio-environmental change. In doing so, they aim to *denaturalize* seemingly *natural* social and environmental condition by outlining underlying politics and power structures. Questions of knowledge, power and their interaction are therefore central to research in political ecology, because "[k]nowledge and practice are always situated in the web of social power relations that define and produce socio-nature" (SWYNGEDOUW 2004: 22). Critical research in this vein has opened up multiple relationships between environmental change and patterns of political decision-making highlighting often technology-determined, expert-led decision making processes (BAGHEL 2014, FORSYTH 2003, KARPOUZOGLOU & ZIMMER 2012, LAWHON & MURPHY 2012). In this way, political ecology has also incorporated discourse analytical work.

In sum, by combining historical-materialist analysis of the production of nature with representational analysis, political ecology has emerged as an important school of thought in showing how representations of nature are socially constructed and nature is materially and discursively produced (BUNCE & DESFOR 2007: 253–254).

Political ecology has also broadened its scope spatially. Early political ecology studies concentrated on environmental change and conflicts over access to resources (especially land) in rural areas in parts of the Global South (cf. BLAIKIE 1995, 1999, BLAIKIE & BROOKFIELD 1987, BRYANT 1992). Acknowledging the dynamic development of cities on a global scale and the multiple environmental challenges as a consequence thereof, *urban* political ecology has explicitly shifted the focus of critical research to the ecology of the city. UPE is therefore considered as an approach bridging the divide between society and nature in the context of the city.

Inspired by a range of critical social theories, UPE links the analysis of urban environmental change to overarching socio-political systems (KEIL 2003: 724). UPE research highlights that the uneven power relations in the contemporary cities constitute highly unjust socio environmental relations loading much of the burden of environmental degradation on marginal groups. Therefore, UPE research is engaged among others with the following questions: who produces what kind of city for whom? who gains and who suffers from socio-environmental change? Thus, linking back these broader UPE questions to governance and policy-making processes with regard to Delhi's riverscapes, the above questions can also be framed as: who produces what kind of riverscapes for whom? who gains and who suffers from (new and existing) policies governing Delhi's riverscapes? Or even more precisely: who governs Delhi's riverscapes – and for whom?

Engaging with these kinds of questions, UPE focuses on "what is considered 'normal' under current political, economic, and social conditions and thus goes noticed" (ZIMMER 2010: 345 [emphasis in original]). Reconstructing these normalizations and deconstructing hegemonic beliefs, UPE scholars aim to show that urban environmental settings and processes are not natural but rather the outcomes of political, social and environmental struggles. Their work has shown that dominant discourses about the city and the environment overshadow subaltern discourses, resulting in many voices going unheard.

In addition to this explicit environmental justice perspective, UPE research is engaged with questions regarding the agency of nature, drawing on different conceptualizations of the non-human/human relationships from NEIL SMITH's (1984) production of nature thesis, to LATOUR's quasi-objects and socionature (among others LATOUR 1993, 2004)[9], the hybrid geographies described by WHATMORE (2002) or the making of cyborgs (HARAWAY 1991). NEIL SMITH's (1984) provocative thesis that nature is produced has been inspiring for the work of several urban political ecologists, since it is a very fruitful starting point to challenge, deconstruct, politicize and rethink the contemporary (re)making of cities across the world (cf. LOFTUS 2012).[10]

Another root of UPE's approach towards urbanized nature has been studies in the field of environmental history (see among others CRONON 1992, WHITE 1995, and in regard to rivers WORSTER [1985] 1992). Several authors point here to WILLIAM CRONON's book *Nature's Metropolis* (1992) on the rise of the city of Chicago based on its resource-providing, industrialized hinterland (cf. ROBBINS 2012: 73). Out of these rural-urban studies, the concept of urban metabolism emerged as a central metaphor in UPE describing the flow of goods into the city and through the city's infrastructure (cf. GANDY 2004, HEYNEN 2014, HEYNEN et al. 2006b, MONSTADT 2009, PINCETL 2012, ROBBINS 2012). In contrast to its rural counterpart, UPE research initially concentrated on case studies in the Global North, and only recently UPE has turned to its roots in the Global South (LAWHON et al. 2014, ZIMMER 2015b). Research has focused on urban environmental problems like air pollution (VÉRON 2006), water supply (BAKKER 2003a, BUDDS 2009, LOFTUS 2009, SWYNGEDOUW 2004), wastewater management (ZIMMER 2012b, 2015a) or waste management (NJERU 2006, YATES & GUTBERLET 2011), among others.

Informed by postcolonial approaches in urban studies, which aim to reframe urban theoretical concepts away from their general points of reference in the Global North (e.g. EDENSOR & JAYNE 2012, PARNELL & ROBINSON 2012, ROBINSON 2006, ROY 2009a, SIMONE 2010), UPE is critiqued for conceptualizing the urban ecology of the cities in the Global South as "inherently problematic" (ZIMMER 2015b) and thereby often understanding urban environmental governance in these cities as

---

9 LATOUR speaks in this context from *actants* emphasising that both humans and non-humans can act and participate in networks (LATOUR 1996). While this term has stimulated geographical research (cf. BLOK & JENSEN 2011, MURDOCH 1997, 1998, 2006) it is not used in this study. That the river is an actor in the production of Delhi's riverscapes is without any doubt; and this study also aims to show this.

10 The theoretical roots of this UPE work are closely connected to a (neo)Marxist logic. See especially ERIK SWYNGEDOUW's article *'The City as a hybrid: On nature, society and cyborg urbanization'* (1996) and a series of related publications by him and others (HEYNEN et al. 2006a, SWYNGEDOUW 1997, 1999, 2004, 2006a, 2006b, SWYNGEDOUW & KAÏKA 2000, 2013). This work has been further inspired by DAVID HARVEY (1989, 1996a, 1996b).

"failed" (LAWHON et al. 2014: 9).[11] Nevertheless, both LAWHON et al. (2014) and ZIMMER (2015b) acknowledge recent research within UPE focusing on the everyday environmental practices and governance in the cities of the Global South as a way forward. They call for a reconceptualization of power in UPE towards a more diffuse and relational understanding rather than drawing on the ideologically pre-defined evils of capitalism, neoliberalism and class differences. In this context, ZIMMER (2010: 351) highlights that actor-oriented approaches are "only rarely taken into account" in UPE.

Nevertheless, UPE has made important contribution to a (re)conceptualization of the human-environment relationships in the city, provided critical insights into urban environmental change, highlighted unequal access to urban resources and thereby raised many question related to urban environmental governance.

Analyzing the governance of Delhi's riverscapes with an explicitly actor and policy oriented approach bridges UPE and governance research. Additionally, a reconfiguration of political ecology thinking based on governance research emerges as an insightful starting point to empower actor and policy oriented research on the urban environment. UPE offers a wide range of theoretical thoughts to conceptualize the nature-society relationship (e.g. urban metabolism, hybrids) as well as the knowledge-power nexus within society (e.g. post-structuralism, feminism, science studies) to challenge apolitical myths in urban environmental governance.

Moreover, aligning UPE research with actor-oriented approaches from governance research opens up new opportunities for a better understanding of who *shapes* (in the sense of material production as well as policy making) the urban environment in certain ways and not others; and who is involved in these processes and who is not. A UPE based approach to urban environmental governance, as GABRIEL (2014: 43) argues, needs to reveal the "diverse and overlapping governmental rationalities" constituting urban environmental change.

This study, through focusing on the river-city nexus and its governance, responds further to the critique by ANGELO & WACHSMUTH (2014) that UPE shows a "methodological cityism" focusing too much on "the politics of nature within cities" and, thereby, paying less attention to the urbanization of nature beyond the administrative limits of the city. While the megacity of Delhi is chosen here as the site (and to some extent level) of analysis, the study takes into account the environmental challenges which are reaching far beyond the city's boundary. In doing so, this study focuses less on Delhi than on environmental change along the Yamuna. Yet, it is apparent that the riverscapes of the Yamuna are most dynamically transformed – or urbanized – within the megacity of Delhi, having tremendous impacts on the upstream and downstream. Therefore this study contributes to the very recent debates in UPE going beyond the scale of the city. The deconstruction of the social

---

11 In her book 'Indian Cities' ANNAPURNA SHAW summarizes this attitude of urban studies coined by *the West* towards the cities of the Global South: "For decades, cities in poor countries have been regarded as a kind of misfit representation of the real city and their modernity, a kind of inauthentic, borrowed modernity. [...] their existence and future have always been questioned by experts and contrasted to the proper cities or the well-planned and managed cities of the West" (SHAW 2012: xix).

construction of nature emerges as a starting point to develop an analytical framework for this study.

3.2.2 Deconstructing social constructions

An increasing number of human geographers – and political ecologists in general – have been working on the deconstruction of traditional representations and conceptualizations of *nature* and *the environment* (see among others CASTREE & BRAUN 2001, CHILLA 2005b, DEMERITT 2002, GEBHARDT et al. 2007, HEYNEN et al. 2006a, SWYNGEDOUW 1996). Especially the often taken for granted dichotomy of nature and culture has been rethought. Naturalized and hegemonial social and environmental 'facts' or 'truths' have long been questioned (cf. BIRD 1987, FITZSIMMONS 1989). Underlying social norms and conventions, narratives and discourses, are scrutinized and power structures are unveiled. Research interests, as BRAUN & WAINWRIGHT (2001: 41) outline, have thereby shifted "from questions of nature's material existence and transformations to questions of epistemology" to some extent. This means that questions of knowledge of, and/or about, nature have become of central interest in empirical research, because "knowledge is socially and historically produced rather than found" (BRAUN & WAINWRIGHT 2001: 46, cf. DEMERITT 1998, HARAWAY 1991).

In other words, assumed ways of understanding the world are actually constructed and ultimate 'truth' about nature does not exist, but is always "a *relative* truth" (BIRD 1987: 258). Scholarship on the social construction of nature draws chiefly on post-structural thinking influenced by Foucaultian thoughts (cf. MURDOCH 2006). These writings acknowledge that "truth is an effect of power, one that is formed through language and enforces social order by seeming intuitive or taken for granted" (ROBBINS 2012: 70). By focusing on divergent perceptions, interpretative frames and communication patterns within a society (CHILLA 2005a: 185), research in this vein does not aim to find the final *truth*, "but instead opens up ways of seeing […] differently" (BRAUN & WAINWRIGHT 2001: 61). Social construction, consequently, is formed through discourse. As FOUCAULT (1984: 73) termed it, "regimes of truth" make these social constructions appear as true.[12]

Drawing on a Foucaultian and Deleuzian poststructural perspective ESCOBAR (1996b: 326) argues that "different discursive regimes" have dominated the way the relationship between nature and society has been articulated, framed and defined. From this perspective, a materialist analysis of (urban) environmental change cannot be undertaken without taking into account discursive regimes and the dialectic of material change and discursive regimes (ibid.).

---

12 While FOUCAULT has referred to the methodology of unveiling these regimes of truth as archaeology, he has not outlined any clear methodology to do so. Nevertheless, today a wide variety of discourse analytical approaches draw on his work in order to study discourses. For a more detailed account of the methodological challenges see Part IV.

"The post-structuralist analysis of discourse is not only a linguistic theory; it is a social theory, a theory of the production of social reality which includes the analysis of representations as social facts, inseparable from what is commonly thought of as 'material reality'. Post-structuralism focuses on the role of language in the construction of social reality; it treats language not as the reflection of 'reality' but as constitutive of it" (ibid.).

Discourse is thus defined as "the articulation of knowledge and power" in the "process through which social reality inevitably comes into being" (ibid.). However, this constructivist approach is "not uncontroversial" (ROBBINS 2012: 124). DEMERITT (1998) gives an overview of the multiple understandings of social construction, which ROBBINS (2012: 127–131) summarizes as "radical" and "soft" constructivism. Radical constructivism sees the environment as a complete intervention of human imagination suggesting "that things are true because they are held to be true by the socially powerful and influential" (ibid.: 127). This approach denies non-human actors and 'natural' processes and, thus, makes it "unattractive" for most political ecologists (ibid.: 128). A soft constructivist position, in contrast, exposes that "our concepts of realty are real and have force in the world, but that they reflect incomplete, incorrect, biased, and false understandings of an empirical reality" (ibid.). This study adopts the 'soft' understanding of social construction most common in political ecology, since "nature cannot be (re)produced outside social relations, neither is it reducible to them" (WHATMORE & BOUCHER 1993: 167).

The debate within human geography discourse about the social construction of nature is multifaceted. *Nature* is already a complex word with at least three different but closely intertwined meanings (CASTREE 2005: 8–9, DEMERITT 2002: 777–778, based on WILLIAMS 1983):

1) Nature = the external, material, non-human world (in the meaning of wilderness, pristine nature untouched by humans);
2) Nature = the essential quality or character of something being natural (the essence of something being of natural origin); and
3) Nature = an inherent force (directing/ordering the actions of both humans and non-humans alike).

Based on this trilogy of meanings of nature, CASTREE (2005: 8–9) emphasizes that only nature in the first meaning is connected to the separation of humans and non-humans, while the second and third meanings of nature include both humans and non-humans. A constructivist approach to nature therefore assumes that nature, or certain elements of nature, only become relevant through social discourses about/of nature, in which nature is given certain meaning and valuation. Such an understanding claims that "nature is not at all – or not simply – 'natural' but in fact a *human construction*" (CASTREE & MACMILLAN 2001: 209 [emphasis in original]). In other words, nature as it is commonly understood, does not exist but is rather discursively defined or constructed by humans in order to denote *the things out there*. Yet, this does not mean that nature does not exist. In contrast, as CASTREE (1995: 35) argues, nature "is both ontologically real and active". ESCOBAR (1996b: 325) might have best condensed this understanding by claiming that the social construction of nature is "entirely different from saying 'there is no real nature out there'". This constructivist perspective to (urban) nature is thus explicitly critical of objective facts and/or accepted truths (CHILLA 2005a: 187, HAJER & VERSTEEG 2005: 176).

In summary, empirical research on the social construction of nature is interested in the process of how and why certain assumed beliefs about nature are constituted (CASTREE 2005: 18). Thus, by adopting insights from post-structuralist debates, this study links environmental discourses and narratives with material environmental change. It rejects a priori explanation for environmental change and problems (e.g. capitalism leads to environmental degradation) and rather understands environmental discourses and narratives "as convenient yet simplistic beliefs about the nature, causes and impacts of environmental problems" (FORSYTH 2008: 758). These environmental narratives are also referred to as "story-lines" (HAJER 1993, 1995, 2006) and uncritical, highly simplistic, apolitical environmental concepts (e.g. desertification, deforestation, soil erosion) are further referred to as "environmental orthodoxies" (FORSYTH 2003, LEACH & MEARNS 1996, NEUMANN 2009a).

The characteristics of environmental orthodoxies and story-lines are that they "persist despite the accumulation of empirical evidence that would overturn or modify them" (NEUMANN 2009a: 228). Political ecology research has therefore been engaged since the 1980s with the questions of how and why certain "institutionalized beliefs about environmental change come into place" as well as how and why "alternative, more inclusive, ways of addressing environmental problems" are not mainstreamed (FORSYTH 2008: 758). A post-structuralist perspective combined with insights from governance research helps here to overcome the "usual positionality" of state actors and private industry in opposition to environmental NGOs and activists; a common limitation of early work in political ecology focusing on structural political analysis (ibid.).

As this study will highlight, a simplistic drawing of boundaries between the different actors is certainly not helpful in understanding the multiple processes of negotiation and bargaining between the different actors with regard to urban environmental change. The position of the different actors is rather fluid and varying. For the study of Delhi's riverscapes a set of questions are of specific interest:

– How do relative truths of Delhi's riverscapes (e.g. environmental problems, extent of the floodplain) develop into widely accepted, taken-for-granted truths and by whom?
– How do story-lines and their hegemonic truths dominate discourses with regard to Delhi's riverscapes?
– How do these narratives and discourses influence policy-making for Delhi's riverscapes?
– How are these policies implemented and by whom in governing Delhi's riverscapes?

3.2.3 The production of urban environments

In contrast to the social construction of nature, NEIL SMITH (1984), based on KARL MARX's historical materialism, provocatively claimed that nature is *socially produced* rather than *socially constructed*. While SMITH's argument has much in common with the writings on the social construction of nature (DEMERITT 2002: 782), his claim that *nature is produced* emerges as considerably different when relating his work to recent debates over nature (LOFTUS 2012: 3) and urban environmental change in particular.

Drawing on HENRI LEFEBVRE (see e.g. LEFEBVRE 1991 [1974], 1996), who already emphasized the centrality of the *production of space* to the survival of capitalism; NEIL SMITH (1984) developed the *production of nature thesis*. His writings inspired political ecologists to theorize about the *urbanization of nature* (cf. LOFTUS 2012, SWYNGEDOUW & KAÏKA 2000, 2013). To understand the notions of *urbanization of nature* as well as the later discussion on hybrids (see section 3.2.4), requires a short review of what SMITH meant by claiming that nature is *socially produced*. The following section briefly sketches the influence of SMITH's thesis on UPE in order to outline the grounding for the research framework of this study.

The central critique in SMITH's argument is that following the enlightenment-based worldview and the Cartesian dualism, ontologically separating body and soul, capitalist ideology views nature as to be universal and external. Against this critique, he insists that nature is an integral element of the production process and is therefore inseparable from society. SMITH's writings closely follow MARX, and therefore it has been his aim to show, from a geographical view point, how capitalism is depending on, and forcing, a specific mode of production, which inevitably leads to uneven development (both spatial and socially understood) as well as the destruction of nature. In doing so, SMITH (1984) argues that the externalization of nature must be considered as an 'ideology', which is pervasive and dangerous (for both human and non-humans) because it is not only ontologically wrong, but actually curtains the dominant capitalist interests which try to deny that nature is fully internalized in the existing socio-environmental production process (cf. CASTREE 2008: 278).

Thus, SMITH (1984) does not argue for abandoning the concept of nature, but rather calls for critical research to unveil how nature is produced and by whom (BRAUN 2009, LOFTUS 2012, SWYNGEDOUW 2004, ZIMMER 2010). When engaging with SMITH's thesis, an absurdity seems to emerge at first sight: is not nature widely conceived as something *natural*, in contrast to the human-made, *artificial environments* of streets, canals, houses, industries, cities and the like? Even SMITH (1984), while feeding into his discussion of the production of nature, acknowledges this paradox when he states that "nature is generally seen as precisely that which cannot be produced; it is an antithesis of human productive activity" (SMITH 1984: 32). Yet, his aim is to overthrow this romantic, bourgeois understanding of nature being untouched by humans, which had led to the understanding of the city as the antithesis to nature (cf. LOFTUS 2007: 43).

To clarify his argument, SMITH (1984: 57) refers to what he calls "down-to-earth examples of supposedly unproduced nature" like the National Parks of Yellowstone and Yosemite, which in his eyes are produced as "neatly packaged cultural experiences of environments". In doing so, SMITH (1984: 57) points out that when virtually nowhere on earth the environment is free from either direct or indirect human intervention to believe in pristine nature makes very little sense (cf. LOFTUS 2007: 43, 2012: xxii).[13]

The *green left*[14] within the Anglophone human geography discussed and partly challenged SMITH's thoughts. The most influential writer in this process has been DAVID HARVEY, who highlighted that urban research should focus on the process of urbanization rather than studying the already existing city (HARVEY 1996a: 50). By focusing on the processes of urban environmental change and urban environmental governance, this study follows the request made by HARVEY. HARVEY's writings have also been influential in overcoming the dichotomy of nature and the city when he refers to the city as a "created ecosystems" (HARVEY 1993, 1996b) and states that "in a fundamental sense, there is [...] nothing **unnatural** about New York City" (HARVEY 1993: 28 [emphasis in original]).

Acknowledging this, the city emerges to be "no more or less 'natural' than any other kind of modern landscape" (GANDY 2006b: 63). However, the nature of the city appears to be of special interest, because the metabolic transformation of nature manifests itself most evidently both in the material landscape as well as its socio-ecological challenges and problems of the city (KEIL 2003: 727, SWYNGEDOUW & HEYNEN 2003: 907). Moreover, the urbanization of nature emerges as a "simultaneous process of social and bio-physical change" through which urban space is produced, destroyed and reproduced (GANDY 2006b: 63). This process includes the creation of technological networks (e.g. transportation, water, sewage, energy) or "new appropriations of nature" through making the city beautiful and green by creating, for example, recreational spaces such as public parks or riverfront walkways (ibid.).

As GANDY (2002, 2006a) further outlines, both networks of urban infrastructure and recreational amenities "create their own distinctive spaces or landscapes within the fabric of the city" (2006a: 140) and reshape nature in specific ways. In adopting this understanding, Delhi's riverscapes should be understood as produced landscapes. Environmental change along Delhi's riverscapes is then to be inter-

---

13 A similar view is also reflected in the debate of renaming the geological epoch from 'Holocene' to 'Anthropocene' to acknowledge the tremendous human impact on the Earth's system (cf. CRUTZEN 2002, CRUTZEN & STOERMER 2000, EHLERS 2008, ZALASIEWICZ et al. 2010)
14 I borrow the term *green left* from NOEL CASTREE (2002) and include especially Anglophone geographers like ERIK SWYNGEDOUW, MATTHEW GANDY, NIK HEYNEN, and others, but also NOEL CASTREE himself. Especially, UPE literature illustrates the myriad connections between the production of nature and urbanization (see among others BRAND & THOMAS 2005, BUNCE & DESFOR 2007, DESFOR & KEIL 2000, 2004, GANDY 2002, 2006a, 2008, HEYNEN 2006, HEYNEN et al. 2006a, KAÏKA 2005, 2006a, 2006b, KEIL 2003, 2005, SWYNGEDOUW 1996, 2004, 2006b, SWYNGEDOUW & HEYNEN 2003).

preted as the expression, and outcome, of the (re)production of the urban environment along the river – or in Gandy's words the "fabric of space" (GANDY 2014). The complex underlying structure of Delhi's riverscapes fabric is perceived, interpreted and described differently by the different actors. For some actors the river emerges as a domesticated resource of water for the city, for others it is a holy goddess. Delhi's riverscapes are therefore always simultaneously socially produced as well as socially constructed.

Acknowledging this, SMITH's work helps to interpret and deconstruct the motivations of the different actors. In this regard CASTREE (2008) highlights that the production of nature thesis is useful to deconstruct the rationalities of the major strands in environmentalism and their understanding of nature. Both, technocentric and ecocentric environmentalism share the hypothesis of an external nature (ibid.). The technocentrists argue from an explicit anthropocentric perspective and want to manage and improve nature for the sake of human interest (e.g. ensuring the flow of a river as a water resource). In contrast, the ecocentrics argue for "a more harmonious human-nature relationship" aiming to "save nature" from the destruction by humans from a rather implicit anthropocentric perspective (CASTREE 2008: 276). Considering that almost all nature is produced, rebuts both strands' arguments and even questions their very existence (ibid.). These thoughts emerge as highly relevant for a study of a degraded but still comparably wild (or natural in the sense of an inherent force) river like the monsoonal Yamuna in Delhi. Acknowledging that Delhi's riverscapes are produced nature, challenging questions arise with regard to the material actions and practices as well as their discursively produced spatial imaginaries. One particularly pertinent question is who produces Delhi's riverscapes, through which practices and actions?

The argument that this study develops is that environmentalists and all other actors, particularly planners and engineers need to recognize that Delhi's riverscapes are hybrids. They are a riverine landscape that has been altered, modified and produced by humans but also determined by the agency of nature in 'gestalt' of a monsoonal, highly unpredictable river. The next sections therefore theoretically conceptualize the understanding of socio-natural hybrids.

3.2.4 The urban environment as a hybrid

"Is it our fault if the networks are simultaneously real, like nature, narrated, like discourse, and collective, like society?" (LATOUR 1993: 6).

To overcome the dichotomy of nature and society, a series of single terminologies have been introduced and debated from DONNA HARRAWAY's *cyborgs* (1991a), to BRUNO LATOUR's (1993, 2004) and SARAH WHATMORE's (1999, 2002) *hybrids*. Drawing briefly on the discussion of these terminologies, and the notion of urban metabolism, the following section outlines the understanding of the urban environment as a *hybrid*.

For LATOUR (1993: 10) modernity is designated by "two sets of entirely different practices", which are "hybridization" and "purification". Hybridization results in "mixtures between entirely new types of beings, hybrids of nature and culture" (ibid.). At the same time, the separation of nature and culture in modernity is based on the purification of these hybrids attributing them either to the pole of nature or culture. Following LATOUR (1993: 11) this results in a first dichotomy of nonhumans/humans or nature/culture, but is also responsible for a second dichotomy separating these dichotomous categories from the hybrids, which are not, or not yet, *purified* either as nature or culture (cf. FORSYTH 2003: 87). Therefore what is defined as nature or as culture is always dependent on the context and discursive (rhetorical) resources of the involved actors (ZIERHOFER 1999: 11). Deconstructing this process of *purification* requires relational thinking which acknowledges the existence of hybrid networks (ibid.). LATOUR's writings but also SARAH WHATMORE's book 'Hybrid Geographies' (2002) have certainly resulted in much debate on the nature/culture dichotomy in geography, which cannot be outlined here. The following paragraphs concentrate solely on outlining why it is appropriate to conceptualize the urban environment as a hybrid, and Delhi's riverscapes in particular.

Hybrid (but also cyborg) urbanization theories have been applied in geographical research, and within the UPE literature, drawing closely on the concept of urban metabolism. The term *metabolism* is to be understood here in the sense of the German word *Stoffwechsel*, referring to the circulation, exchange, and transformation of material elements in living organisms (cf. MONSTADT 2009: 1926). HEYNEN (2014), who considers the conceptualization of urban metabolism to be the most important theoretical contribution of UPE, condenses the understanding of urban metabolism to:

> "a dynamic process by which new sociospatial formations, intertwinings of materials, and collaborative enmeshing of social nature emerge and present themselves and are explicitly created through human labor and non-human processes simultaneously" (HEYNEN 2014: 599).

In theoretical and empirical work of UPE, the concept of urban metabolism emerged as being especially inspiring for the work on urban water (see among others BAKKER 2003a, BUDDS 2009, GANDY 2004, KAÏKA 2005, LOFTUS 2007, SWYNGEDOUW 2004, 2005a, VIEHOFF 2009, ZIMMER 2015a) and other urban infrastructures (MCFARLANE 2008, MONSTADT 2009).

In the field of industrial ecology, the concept of urban metabolisms is used in order to quantify the material and energy flows through industrial systems. Such kind of material flow analyses are also applied for cities (cf. BACCINI 1997, KENNEDY et al. 2011). In contrast, the notion of urban metabolism in UPE points much more to a theoretical framework conceptualizing the interactions between humans and nature in the city originating largely from Marx. Marx used the metaphor of metabolism to capture both the conditions imposed on humans by nature as well as humans ability to affect and transform nature; and further to describe the intimate dependency on, and exploitation of, nature as well as of human labor in capitalism (PINCETL 2012: S34).[15]

Metabolization is therefore not just a metaphor. Rather, it "offers an ontological way through the all too stifling dualisms that have historically plagued the discussions of nature *and* society" (HEYNEN 2014: 599 [emphasis in original]). Political ecologists have focused on the various processes of urban metabolization to study who produces what kind of city for whom (GANDY 2004, HEYNEN et al. 2006a). They outline that different actors, based on an uneven distribution of knowledge and power, are able to differently utilize, and shape, the diverse socio-natural flows in the city (WACHSMUTH 2012). In doing so, unequal urban landscapes are produced, which are often shaped by the interests, ideas and discourses promoted by urban elites compromising on the rights of the poor. Urban environmental change is thus to be understood as a political economic process (PINCETL 2012: S35). As SWYNGEDOUW (1996: 68) termed it, in the metabolizing process of urbanization "nature provides the foundation [while] the dynamics of social relations produce nature's and society's history."

In order to better understand the ontological value of the concept of urban metabolism, it seems appropriate here to draw again on the two processes of hybridization and purification, which are characteristic when dealing with socio-natural hybrids (LATOUR 1993). Hybrids are the result of hybridization through the metabolism of human practices (in Marxist terms: human labor) and 'natural' processes (see Figure 9).

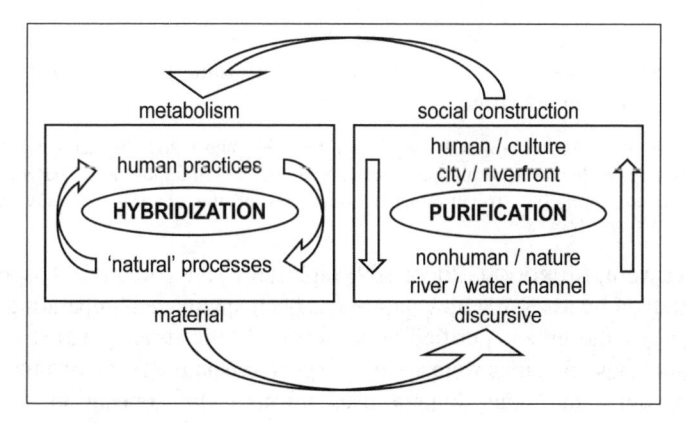

Figure 9: The practices of 'hybridization' and 'purification'
Source: Graphic by author

---

15 For a detailed review of Marx's notion of metabolism see LOFTUS (2012) and SWYNGEDOUW (2004). Due to the strong influence of the production of nature thesis and urban metabolism, UPE literature often draws much closer on Marxist's thoughts than political ecology in general (cf. GABRIEL 2014, HEYNEN 2014, LAWHON et al. 2014, ZIMMER 2010). The notion of urban metabolism is further not new in urban research. ABEL WOLMAN (1965) described the metabolism of cities as a circulation of materials going in and out of the city to sustain the city.

Delhi's riverscapes (as with all riverscapes in the world) therefore need to be understood as hybrids. The materiality of the riverscapes is the result of the urbanization of nature through the process of urban metabolization. However, through the practices of purification, which draw on the modern dichotomies of human/nonhuman, culture/nature, Delhi's riverscapes are purified into the categories of city/river and land/water. This purification and splitting of a hybrid is based on social constructions. Yet, as LATOUR (2004: 24) highlights, hybrids "have no clear boundaries, no-well defined essences, no sharp separation between their own hard kernel and their environment".

As "assemblages of different entities" hybrids therefore "cannot be divided in two poles" (ZIMMER 2010: 345). Since modern thinking is based on dichotomies (e.g. human/nonhuman, culture/nature, spirit/body, us/them, center/periphery), these hybrids create contradictions, tension, and conflicts for modernist, dichotomous understandings of the world. Therefore modernist social construction constantly attempts to separate the different ontological spheres by drawing boundaries. Boundary-making "allows the establishment of an order and framework from which to proceed" (FORSYTH 2003: 89). Since there are always different ways of drawing boundaries, boundary-making emerges as a political exercise, which is shaped by truth claims and power struggles. (Natural) science often acts as a powerful advisor in the process of boundary-making, while other truth claims remain largely unheard in the decision-making process (see among others BLAIKIE & MULDAVIN 2004, BUDDS 2009, FORSYTH 2003, KARPOUZOGLOU 2011, KEELEY & SCOONES 2003).

Drawing on LEFEBVRE (1991 [1974]), WHATMORE & BOUCHER (1993: 168) argue that land-use planning is an "important institutional terrain" in the process of boundary-making and purification:

"Land-use planning formalizes the separation between nature and abstract space through the written codes of legal statute and professional conduct which impose a site-based, rather than system-based, narrative structure on its treatment of the environment" (WHATMORE & BOUCHER 1993: 169).

In a similar vein, MURDOCH (2006: 109) highlights "the paradoxical aspect of environmentalism by assessing how nature has been spatialized" separating the "two spatial zones"; the urban (purified as society) and the rural (purified as nature). These spatial classifications were applied to preserve nature from urbanization (e.g. suburbanization) and other human uses through the concept of zoning (cf. MURDOCH 2006, WHATMORE & BOUCHER 1993). Zoning and land-use planning deals with topographies, territories and land, but generally fails to be sensitive to ecological processes and the fluidity of the multiple relations between different spaces and places (MURDOCH 2006: 127).

The changing morphologies of rivers and their floodplains due to their inherent fluidity have always posed difficulties for spatial classifications whether with regard to land ownership (BLOMLEY 2008) or land-use zoning (HAIDVOGL 2012, HARTMANN 2013, LÜBKEN 2012). Furthermore, as MURDOCH (2006: 131–138) outlines, land-use planning is about "demarcating territories" and the "'taming' of space", and making space 'legible' for the state through simplifications (cf. SCOTT 1998).

> "[...] planning is not only a technology of spatial management it is a political arena also. Planning decisions are made on the basis of political calculations and this too can result in very partial assessments of space being made. The upshot is that planning has considerable difficulty in 'representing' the complex and heterogeneous spaces in which it is inevitably immersed" (MURDOCH 2006: 131).

Land-use planning is thus highly political (FLYBBJERG 1998) and environmental planning in the age of the overarching discourses of sustainable development and climate change has produced its own spatial imaginaries in the city (cf. BENTON-SHORT & SHORT 2013, KRUEGER & GIBBS 2007, MURDOCH 2006, RYDIN 1998). It is this understanding of land-use planning (and environmental policy-making) which has directed DESFOR & KEIL (2004) and others to concentrate on the interactions between state and non-state actors in urban environmental governance both with regard to land-use and environmental policies. This study adopts this critical view of land-use planning and environmental policy-making in order to analyze the ways purification has influenced the materiality of, and discourses associated with, Delhi's riverscapes.

At this point, it is important to highlight that both knowledge and practices of different actors "are always situated in the web of social power relations" (SWYNGEDOUW 2004: 22). Due to the actors' different knowledge, powers, and resources, the means and opportunities of deploying the processes of both hybridization (materially) and purification (discursively and administratively) according to their interests vary among the actors. Knowledge and power are not static, but are constantly modified and altered by the different actors through both material and discursive practices. Discourses and practices are therefore dialectically interwoven with the knowledge and powers of the different actors; the relations among the actors are therefore dynamic rather than static (SWYNGEDOUW 2004: 22, ZIMMER 2010: 346).

3.2.5 From land- and waterscapes to hybrid riverscapes

For human geographers the term *landscape* refers to more than just the physical environment or the surface of the earth (COSGROVE & DANIELS 1988, WINCHESTER et al. 2003). The term clearly denotes something real and material in the sense of a certain pattern of physical features including, for example distinct landform configurations (relief), vegetation and human settlements. Furthermore, the term *landscape* is also to be understood as something socially constructed and discursively produced over time (DUNCAN 1993, OLWIG 1996, WINCHESTER et al. 2003, WYLIE 2007). The term has therefore a historic dimension, too.

Both cultural landscapes (*Kulturlandschaften*) being shaped by human and nature as well as supposedly untouched natural landscapes (*Naturlandschaften*) – if the latter even still exist today – "are invested with cultural meaning through representation of them" (WINCHESTER et al. 2003: 4). Like the broader understandings of landscape, the notion of the waterscape refers to both the physical environment – materially produced through the metabolisms of nature and culture – and the social construction of the physical environment – representationally produced through discourse, expressions, perceptions and ideas of humans (SWYNGEDOUW 2004).

SWYNGEDOUW (1999, 2004, 2006b) used the concept of the waterscape to bring together the material hydrological cycle of water with the (nonmaterial) landscapes of power constituting together the *hydrosocial cycle* of water in the city.[16] Elaborating on the concept of urban metabolism, SWYNGEDOUW outlines how water is metamorphosed from its sourcing to the tap. This metamorphosis is not restricted to the waters' physic-chemical characteristics but also its social characteristics and its symbolic and cultural meanings are transformed. Drawing on the notion of *landscapes of power* (ZUKIN 1991), SWYNGEDOUW uses the concept of the waterscape to outline the embedded political, social and cultural power relations among the different actors and scales involved in producing the visible, material environment of water in the city. Focusing on water, SWYNGEDOUW (1999, 2004, 2006b) coined the term waterscape, while others, being influenced by his writings, refer to *hydroscapes* (BAGHEL 2014, HEYNEN 2014, LORD 2014, NÜSSER 2014, NÜSSER & BAGHEL 2010).

When transferring and applying SWYNGEDOUW's thoughts to river control works in India, other authors prefer to use the term "hydroscapes" to grasp "the complex geographical spaces that are produced through large river control projects" (BAGHEL 2014: 16). While one might argue that this is only another name for the same idea, they highlight that by using the term hydroscapes instead of waterscapes the analytical scope is widened by avoiding "privileging water as the central lens" (NÜSSER & BAGHEL 2010: 231).[17] As BAGHEL (2014: 17) further argues, these hydroscapes "are created through modern practices of river control". Since controlling monsoonal rivers like the Yamuna completely emerges to be a utopian notion, the river always retains agency in itself, and river control works "simply *alter* the expression of this agency, *without removing it*" (ibid. 18 [emphasis in original]).

While *well-behaved* rivers are envisioned to be "linear features clearly defined by banks and levees, their flow regulated and largely predictable" (COSGROVE 1990: 3, cf. LAVAU 2011: 246), monsoonal rivers like the Yamuna remain still largely unpredictable and their flooding patterns regularly challenge their space assigned to them by humans. Furthermore, besides all the positive effects of human interventions with regard to the use of the river as a resource and the reclamation of land for the growth of the city, these interventions have induced that the river

---

16 ERIK SWYNGEDOUW coined the term waterscapes in his analyses of the role of water in Spain's early-twentieth century modernization process (SWYNGEDOUW 1999) and his analysis of urban water in Guayaquil, Ecuador (SWYNGEDOUW 2004). The notion of the waterscape was later adopted by others focusing especially on water extraction and supply in the Global South (BAKKER 2003a, BUDDS 2009, LOFTUS 2006, 2007, LOFTUS & LUMDSEN 2008).
17 Authors have used the waterscape terminology also for large dams, see among others SWYNGEDOUW (1999), HARRIS (2006), LOFTUS (2007), and BUDDS (2008).

"behaves differently" (BAGHEL 2014: 19) in the form of flash floods or increased sedimentation. These underpin that an urban river, like the Yamuna in Delhi, is anything but passive, but is actively responding to the altered conditions with a different behavior. Thus, while agreeing with the above quoted authors that the waterscape notion is to some extent too centralized on water to grasp the multiple dimensions of urbanized rivers, the author prefers the term *riverscapes* since it underlines the agency of the river in *scaping* its territory within and outside the city. The suffix *scape*, when understood as socio-nature, highlights that this creation is a result of a process of both human practices and the actions of the river (nature) itself. As indicated in the Introduction, the term riverscape is chosen to provide further distinction between the common terminology of the *riverfront* and plentiful literature on urban rivers focusing almost solely on the liminal space *somewhere between* the river and the city.

However, the connections between cities and their rivers are complex. An efficient urban water distribution system has proved to be essential in the growth of cities around the world. In order to ensure reliable and sufficient supply of water both for cities and its supporting hinterlands (e.g. agricultural food production) rivers have been tapped, rectified, diverted and dammed and their waters have been impounded, extracted and pumped onto the agricultural fields and into the city (CRONON 1992, WESCOAT JR & WHITE 2003, WORSTER [1985] 1992). When the water-health connections became known through the discovery of bacteriological epidemiology, water provision became a key task of municipalities in order to ensure the "sanitized city" (CASTONGUAY & EVENDEN 2012a, GANDY 2002, 2014, SHARAN 2014).

Complex engineered networks for urban water supply have been developed almost globally. These were generally accompanied by the emergence of large, engineer-led, often highly bureaucratic authorities which govern urban water provision as well as water resource extraction. As a result, the great importance of water provision for the growth and health of urban agglomerations resulted in the emergence of "the water industry" being an important part of the public utility sector resulting in infrastructure investments like dams, canals and technological networks to pump water into and through the cities (SWYNGEDOUW 2004: 39). The city-river relationship was materially manifested by a network of pipes, pumps and treatment plants through which rivers became interwoven with "the geographical fabric of the city" (CASTONGUAY & EVENDEN 2012a: 5). In such a framing, rivers have generally been reduced to a basis of resource extraction in terms of water provision and disposal of wastewater. The governance of rivers has therefore largely been designed for resource extraction, too. In this context, MOLLE et al. (2009) argue that in order to gain "full control of the hydrologic regime" not only large-scale infrastructural developments (e.g. dams and embankments) needed to be achieved, but further modernity's hydraulic mission resulted in centralism in governing rivers with large "state bureaucracies".

In sum, the urbanization of rivers, as "Modernity's Promethean Project" (KAÏKA 2006a: 277), has engaged in the mastering of the rivers 'wild nature' in order to harness their resources especially water but also hydropower (cf. MOLLE

2007, NÜSSER & BAGHEL 2010). River engineering was therefore viewed as modernity's progress. Concrete was its most important materiality. In the city itself, channelization[18] was about concretization of rivers to enable the concrete jungle of the city to expand its territory (cf. GANDY 2014: 173). Channelization thus means establishing clear physical boundaries between water and land, the river and the city, nature and society.

In India, the fundamental principals in this context were established during the British colonial times when resource extraction was the key priority (cf. D'SOUZA 2003, 2006b). The colonial hydrological interventions created canal-based, perennial irrigation systems in the semi-arid plains of north-western India involving the construction of canals, barrages and weirs across rivers and embankments for flood control. In post-colonial India, big dams in the Nehruvian sense were deemed as the temples of modern India and in the course of the green revolution even more water was to be diverted from rivers in order to increase agricultural productivity, foster economic development and eventually contribute to the growth of cities (cf. ALLEY 2012, IYER 2009a, LAHIRI-DUTT 2000). The empirical part of this study will outline this process for Delhi's riverscapes and highlight the multiple relations between the urbanization of nature and mega-urbanization.

3.2.6 Riverscapes in neighboring disciplines

The concept of riverscapes has been deployed in geography's neighboring fields of landscape ecology and environmental anthropology. These two fields are interesting, since they engage with riverscapes from different perspectives and work on different scales.

Landscape ecology is a discipline which can be considered to lie at the intersection of geography, ecology and social anthropology. The term riverscape is maybe most widely accepted and relatively frequently used in this field (see among others ALLAN 2004, HASLAM 2008, MALARD et al. 2000, SULLIVAN et al. 2007, WIENS 2005). In simplified terms, landscape ecologists use the term riverscape to emphasize upon "a landscape perspective of streams and rivers" (SULLIVAN et al. 2007: 1169, cf. MALARD et al. 2000). Traditionally, landscape ecologists have considered rivers largely as "internally homogenous element within a broader terrestrial landscape" demarcated by a seemingly clear boundary of aquatic ecosystem from the terrestrial ecosystem (WIENS 2005: 503). Such a view of rivers correspondents with the representation of a river in a map, generally a single line or flat depiction (sometimes with a line as boundary). More recently, landscape ecologists have started to analyze "the landscape within a river" (WIENS 2005: 504) and their research has underlined that the landscape surrounding a river is an "integrated ecological unit", which encompasses the river itself, its floodplain and the riparian area

---

18 Channelization is generally understood as the widening, deepening, and/or straitening of rivers in order to control flooding, develop river channels for navigation, control bank erosion and/or improve river alignment.

(SULLIVAN et al. 2007, WIENS 2005). While the internal structures and dynamics of rivers historically remained a 'black-box' for landscape ecologists, nowadays they are interested in the internal heterogeneity of rivers and their multiple connections to their surroundings. Landscape ecologists have, thus, *rescaled* their approach towards rivers to become progressively finer. In doing so, clear boundaries of water and land have been considerably blurred.

Of course, the interest of landscape ecologists is very different from the focus of this study. Despite that, the landscape ecology perspective on riverscapes emerges as a helpful starting point in developing a research framework for this study. Translating landscape ecologists' thoughts into a broader spatial framework promises a better understanding of urban rivers and their multiple nexuses with the city. In order to understand and explain these multiple nexuses, geographical research on urban rivers needs to analyze internal structure and dynamics in the riverbed and, thus, is depending on a high definition resolution in describing and analyzing the changing spatial patterns. In analogy to the landscape ecology approach, the aim of conceptualizing riverscapes here is to take the analysis *into* the riverbed (in contrast to remaining concerned with the riverfront) in order to draw a more holistic picture of both the internal dynamics as well as the dynamic relationship with the *rest* of the city. In doing so, geographical research is able to engage and deconstruct the seemingly 'natural' boundaries between water and land, the river and the riverfront, nature and the city.

In contrast to the small scale perspective of landscape ecologists, an environmental anthropology perspective on riverscapes, as applied by ANNE RADEMACHER (2005, 2007, 2009, 2011) explicitly *zooms out* from the physical nature of the riverscape to overarching interrelations between socio-political and environmental change. Analyzing the dynamics of competing understandings of river restoration in Kathmandu, Nepal, RADEMACHER's work sheds light on the larger environmental and political transformation in the city to better understand the multiple relations between socio-political and environmental changes in a volatile period of Nepal's history. She considers "environmental and political change as being mutually produced" and reveals how Kathmandu's riverscape serves as a stage, a "discursive and practical terrain", on which ecological restoration emerges as a strategy "for both a state seeking to retain its legitimacy and for those who sought to challenge or reshape that very state" (RADEMACHER 2011: 15). The riverscape, in her words, is a "facet of urban nature", which is "produced at a nexus of cultural and biophysical processes" (RADEMACHER 2011: 16). Her findings highlight that a multiplicity of actors, "differently positioned and differently powerful" are engaged in this discursive and material production of riverscapes simultaneously (re)mapping urban space, (re)ordering urban social life and thereby implicitly and explicitly (re)making the state (RADEMACHER 2011: 178).

RADEMACHER's work is inspiring for research in urban environmental governance since it draws connections between socio-ecological problems and conflicts (e.g. river pollution, demolition of settlements in the riverbed) and larger changes in the political system which are reflected in changing governing approaches and shifting environmental politics. Overall, her research highlights that changing political settings and socio-cultural conditions have a tremendous influence on the way ecological problems are discursively framed and subsequently how and what action is taken. At the same time, urban environmental change along the river emerges as a powerful trigger to induce larger socio-political turmoil. Following theses insights, geographical research on hybrid riverscapes should also *zoom out* in order to be able to relate environmental changes in the riverbed to the underlying driving forces. In a similar way HEIKKILA (2011) has outlined the importance of cultural tradition with regard to river restoration projects in China.

Geographical research on urban environmental change seems to benefit immensely from incorporating both landscape ecology's and environmental anthropology's understandings of riverscapes. While insights of landscape ecology highlight the material complexity of riverscapes, environmental anthropology's findings suggest a deeper debate of the socio-environmental underlying conditions of urban environmental change. The aim of this section is therefore to develop a *UPE-governance perspective* which allows the conceptualization of the many relationships between the material and underlying patterns of political decision-making while simultaneously taking into account the multiple discursive, socially constructed representations of hybrid riverscapes (cf. BUNCE & DESFOR 2007: 253).

3.3 Riverscapes as hybrids – a framework for research

> "[…] ideas of hybridity can open up the possibilities of wet theory, a new methodology, to explain lands that are soaked in water" (LAHIRI-DUTT 2014: 508).

As the reflections on the use of the concept of riverscapes in the fields of landscape ecology and environmental anthropology have shown, analyzing riverscapes can be done on very different scales.

While landscape ecologist's use a fine grain analysis, environmental anthropologists draw multiple connections on different scales. By introducing an explicitly geographical perspective on riverscapes this study sits within these two insights of neighboring disciplines. The emphasis here lies on a change of focus and change of perspective in geographical research by shifting the interest from the riverfront, which embodies a rather city-centric perspective and a clear boundary between land and water, to the fluid concept of riverscapes. Outlining the framework for research, this section further combines the theoretical insights from UPE with analytical governance research.

3.3.1 Hybridity of riverscapes and the interplay of materiality and discourse

Geographical research on riverscapes needs to pay special attention to the fluid, complex and highly political (re)production of riverscapes as socio-natural hybrids (cf. LAHIRI-DUTT 2014). The constantly changing discursive/representational configurations of the riverscapes as well as material environmental changes must be taken into account. This requires an analysis of the material and discursive practices of the involved actors constituting Delhi's riverscapes, which is sensitive to time, scale and place.

The research framework presented here is grounded on a relational understanding of the city-river relationship (BAGHEL 2014, LAVAU 2011). As mentioned, this study introduces the *riverscapes* as a single terminology in order to overcome binary conceptualizations of river/city, water/land, river channel/riverfront, etc. In dissolving these dualisms, the riverscapes are understood as *hybrids*. Figure 10 provides an overview of the multiple dimensions of hybridity of riverscapes.

RIVERSCAPES	meta level		conceptual level	
	human	non-human	metabolism	social construction
	culture	nature	material space	discursive representation
	body	soul		
	spatial		temporal	
	land	water	present	future
	city	river	existing	envisioned
	riverfront	water channel		

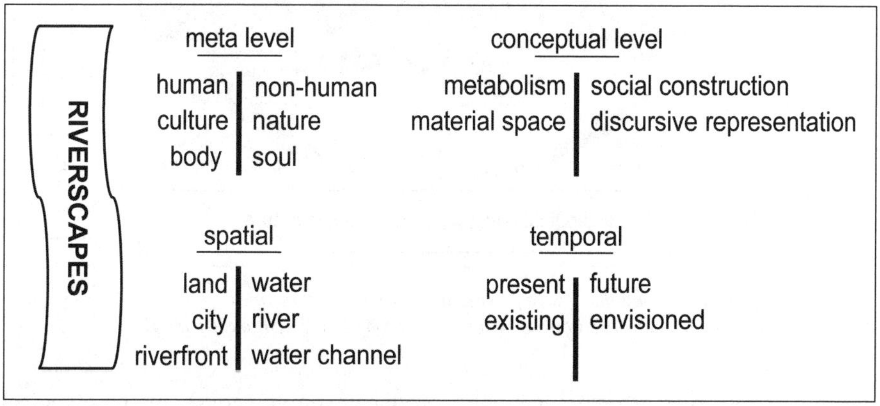

Figure 10: Hybridity of riverscapes
Source: Graphic by author

Thinking beyond these binaries, and dissolving these dualisms, emerges as a challenging task. The deconstruction of the multiple dimensions of hybrid riverscapes is especially associated with methodological and epistemological challenges. This is because "our claim of the socio-natural production process of socio-nature is itself caught in a representational discourse that produces nature/society (socio-nature) in a particular partial fashion" (SWYNGEDOUW 2004: 20).

Keeping these challenges in mind, the concept of *hybrid riverscapes* acknowledges that Delhi's riverscapes, as social constructions carry social meanings, power relations and the different actors' intentions. Delhi's riverscapes are therefore also hybrid in the sense of being both material and immaterial, produced through metabolism as well as social construction. Delhi's riverscapes are thus always hybrids of the material space and the discursive representation. The analysis of riverscapes also needs to pay attention to the simultaneously existing and the envisioned riverscapes – the real and the imaginary riverscapes. Overall, the concept of hybrid riverscapes aims to underline "the impossibility of separating 'representation' from 'being', the signs from the signified, the discursive from the material" (SWYNGEDOUW 2004: 14). In doing so, referring to riverscapes embodies a consciousness that social constructions are inevitable when discussing, analyzing, and writing about urban environmental change along the river (see Figure 11).

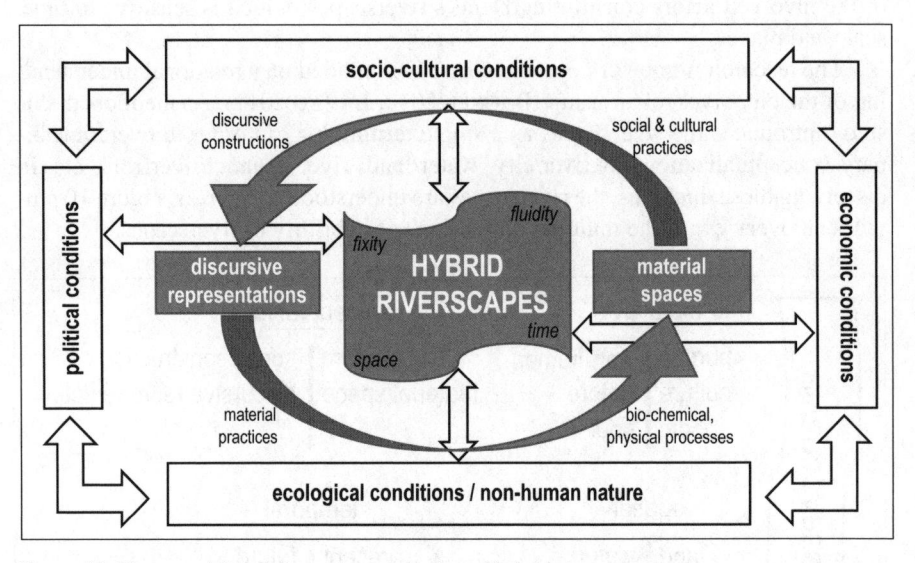

Figure 11: The material and discursive production of hybrid riverscapes
Source: Graphic by author based on SWYNGEDOUW (2004: 22), ZIMMER (2010: 347)

Delhi's hybrid riverscapes are inevitably produced simultaneously by bio-chemical material practices and discursive constructions. Human practices include all material practices generated by humans (e.g. construction of embankments, streets or buildings, washing of clothes, farming, etc.). Natural processes refer here to the 'natural', nonhuman influences in (re)shaping the riverscapes (fluvial processes, erosion, vegetation, etc.). Together these processes (re)make the riverscape as a hybrid, material landscape. Importantly, as outlined above, through modernity's underlying dominant social constructions, these hybrid riverscapes are constantly *purified* in the discourse, and therefore broken up into nature and culture, river and city, water channel and riverfront. Additionally, both the material and discourse practices of different actors are influenced by broader conditions, which include the ecological conditions (non-human nature), socio-cultural conditions, political settings (including the governance set-up) and economic conditions (SWYNGEDOUW 2004, ZIMMER 2010).

These broader interconnected conditions are never constant but they change dynamically over time. They are mutually connected to the discursive and material practices constituting the riverscapes. It becomes clear that the riverscapes' physical conditions are also never fixed, but rather fluid. The altering ecological conditions of the river (including floods, meandering of the river course) emphasize the riverscapes' high level of fluidity. The same is the case for their discursive representations, as perceptions and ideas might also be constantly changing. Therefore what is of major interest here is not the actual state of the riverscapes at a certain moment of time, but rather the perpetual remaking of both their physical characteristics and discursive framings over time.

While riverscapes are largely fluid, certain elements of their material composition from rocks and vegetation to human construction like bridges, embankment or barrages are also bound to a certain degree of fixity. They exist for a long time in a rather static form, shaping the physical character of the riverscapes. Discursive representations of a place can also emerge as fixed when a place has been carrying a certain (bad) reputation for a long time and struggles to improve this reputation (e.g. neighborhoods with high crime rates, industrial brownfield site). Riverscapes might, for example, carry a negative image associated with a dangerous act of nature, such as floods, given the association as a breeding zone for mosquitos or for being highly polluted.

As riverscapes are further connected to the global deterritorialized flows described by (APPADURAI 1990: 297), the dialectic of flow and fixity emerges as insightful to study urban environmental change in Delhi's riverscapes. APPADURAI distinguishes five dimensions of flows (APPADURAI 1990: 296–297, 1996: 33–34):

1) *Ethnoscapes* (global migration and tourism);
2) *Technoscapes* (global circulations of technology);
3) *Financescapes* (different forms of global capital flow);
4) *Mediascapes* (global production and dissemination of information by the media); and
5) *Ideoscapes* (complex landscapes of ideas).

Urban and environmental planning processes are very much part of these globalized technoscapes and ideoscapes which is reflected in travelling ideas like the world-class agenda or planning tools like master planning or land-use zoning (HEALEY 2013, MCCANN & WARD 2011, ROBINSON 2006, 2011). Globalized technoscapes and ideoscapes also shape the socio-ecological imaginations of urban rivers mobilizing certain desires and visions to make use of the special character of the land at the water's edge of a *well-behaved* river (BAGHEL 2014, BAVISKAR 2011c, HOLIFIELD & SCHUELKE 2015, LEIGH BONNELL 2010, SHARAN 2015).

When referring to an urban river like the Yamuna in Delhi, these social meanings, power relations and intentions narrate stories about structure and development of the river and city, while simultaneously provide insights in the river-city nexus, both historically and geographically. The next sections outlines how the theoretical thoughts derived from UPE-literature are merged into a UPE-governance approach to study hybrid riverscapes.

3.3.2 A UPE-governance approach to study riverscapes

As indicated above, UPE research can benefit from a refocusing on actor-oriented and policy-oriented governance research (GABRIEL 2014, MONSTADT 2009, ZIMMER 2010). Likewise, governance research can theoretically be broadened and strengthened by integrating conceptual ideas from UPE. This combination is broadly based on the premise that such a *UPE-governance perspective* offers a more holistic approach towards urban environmental change and governance than normally found in both urban studies and environmental science. This study therefore suggests a UPE-governance approach, which is methodologically rested upon the classical interpretative/hermeneutic paradigm focusing on actors and extended by a discourse analytical methodology focusing on policy-making as outlined by MAARTEN HAJER (1993, 1995, 2006).

In the conceptual framework of this study urban environmental governance is embedded in the previously described broader ecological, economic, socio-cultural, and political conditions (see Figure 12). Furthermore, these conditions influence material practices by humans and bio-chemical, physical practices by non-humans as well as social and cultural practices and discursive constructions by humans. Together these produce the dialectical arrangement of material spaces and discursive representations which frame the making of urban environmental policies. Therefore the interactions among the different actors in urban environmental governance (state, private, civil society) are embedded in this dialectic between the changing material spaces and discursive representations.

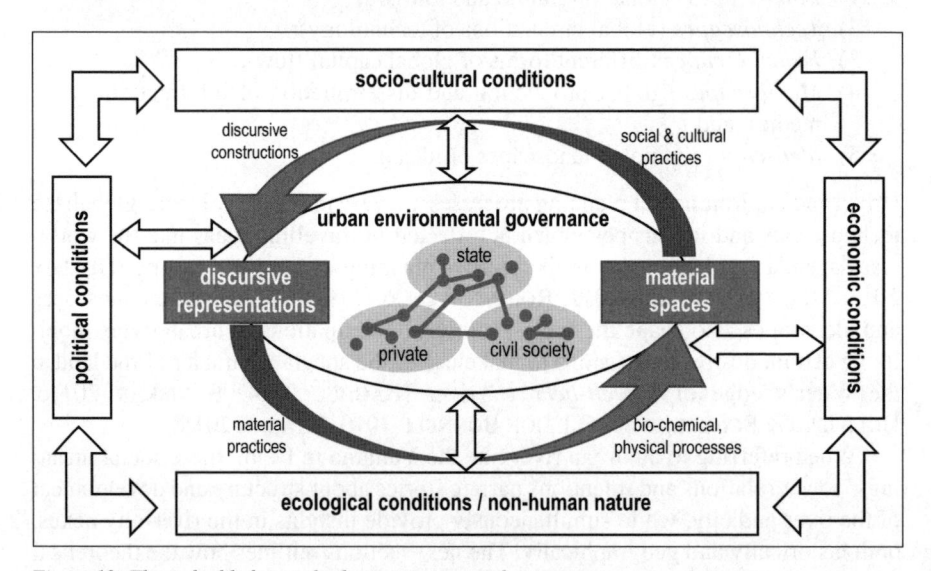

Figure 12: The embeddedness of urban environmental governance
Source: Graphic by author based on SWYNGEDOUW (2004: 22), ZIMMER (2010: 347)

The discursive representations will be analyzed in this study using the concept of story-lines and discourse-coalitions (discussed below in detail based on HAJER 1995, 2006). The material changes in Delhi's riverscapes will be examined through a detailed review of the urbanization of Delhi's riverscapes including a multi-temporal mapping of land-use changes (for details regarding the used methods see Part III).

3.3.3 Redefining the research questions

Based on the theoretical and conceptual reflections presented in the proceeding sections, this section redefines the research questions. In addition to the four key research questions outlined in the introduction (see 1.5), further research questions have been identified in the course of the theoretical chapter. These include aspects with regard to the production of Delhi's material riverscapes, their discursive framing as environmental orthodoxies and hegemonic truths, and the actual practices and actions of the remaking of Delhi's riverscapes and their relation to discursively produced spatial imaginaries. The dialectics of purification and hybridization further request a deeper engagement with these identified questions.

The empirical parts of this book are dedicated to provide the knowledge basis for answering these questions. The Part IV of this book therefore starts with the situating of the case study in the megacity of Delhi (chapter 6) and the introduction of the river Yamuna and Delhi's riverscapes in chapter 7. These chapters outline the broader ecological, economic as well as social and cultural conditions influencing urban environmental change – addressing the first research question:

What are the urban environmental changes in Delhi's riverscapes?

Chapter 8 then explicitly engages with the political conditions and the governance set-up by outlining the existing institutions including the actors and legal provisions involved in governing Delhi's riverscapes. The central question of chapter 8 is:

How are Delhi's riverscapes governed?

Part V then focuses on the historical-geographical production of Delhi's riverscapes outlining the historical urbanization of the monsoonal river (chapter 9) and engages with the discursive framing of channelization and riverfront development (chapter 10). This leads into chapter 11, analyzing the policy-making of the Zonal Development Plan (ZDP) for the River Zone. This part offers a detailed analysis and interpretation of the 'co-production' of urban environmental policies and addresses the following question:

How are environmental knowledges discursively co-produced by different actors, and how are these discourses reflected in the urban environmental governance of Delhi's riverscapes?

This question can be broken down into a series of questions:

- Which discourses and practices have produced Delhi's riverscapes?
- Which ideas and imaginaries have dominated the discourses around Delhi's riverscapes?
- Whose knowledges are involved in urban environmental policy-making?
- What role did scientific knowledge play in the policy-making and how have scientific 'facts' been negotiated between science, planning, politics and judiciary?
- To what extent did policy-making involve hybridization and purification in defining the city-river relationships?
- How are non-state actors (especially environmental NGOs) involved in the (re)making of urban environmental polices?

Part VI shifts the focus from policy-making to projects. The basis for this is laid in the multi-temporal analysis of land-use change in chapter 12. Chapter 13 engages with the reclaiming of Delhi's riverscapes from slums and other 'undesired' land uses. In contrast, chapter 14 highlights the governance practices around the 'desired' land uses by analyzing the political ecology of the construction of new embankments and urban mega-projects.

The closing empirical Part VII then engages with the most recent remaking of the policy-framework for the River Zone:

How has urban environmental policy-making for Delhi's riverscapes responded to recent development pressures of dynamic urban transformation and restructuring?

Part VII (synthesis and conclusion) brings the different insights from Delhi's riverscapes together and outlines the overarching discursive formations highlighting the main challenges for urban environmental governance. The concluding chapter then reflects on the transferability of the findings from Delhi's riverscapes with regard to other ecological sensitive spaces in (mega)cities of the Global South and evaluates the conceptual approach. Before engaging with the empirical knowledge, the next sections outline the methodology and methods used in this study to gain the empirical insights.

III METHODOLOGY AND METHODS

This Part outlines the methodological approach of the study including the author's strategy of approaching the complex research area and the evolving relationships with different actors (especially environmental activists and planners) during the different phases of field work in Delhi. Further, by presenting the details of the applied methods, specifying the different sources of data and explaining the processing of these different sets of data, the author aims to unveil the research process.

4 METHODOLOGY

4.1 Research philosophy

This study is rooted in the interpretivist tradition of emphasizing the subjective views of different actors as well as on discourses interpreted as "inter-subjective shared views of communities of actors" (KHAGRAM et al. 2010: 391). Urban environmental change is highly complex. The study therefore follows a relativists understanding of social science, and the author strongly believes that understanding reality is always dependent on particular sets of perceptual and cognitive templates. Thus, there is not the *one* truth or dominant (not to say: *correct*) interpretation of reality, but rather multiple ways of comprehending reality (cf. CASTREE 2005: XX, WINCHESTER & ROFE 2010: 8).

Having said this, the aim of this study is not to search for the *one* final truth about the riverscapes or to distinguish between right or wrong, good or bad governance. In fact, this study rather is a narrative and analysis of how different actors view, interpret and treat the riverscapes. These different viewpoints are influenced as well as manifested in what is understood as 'discourses' about Delhi's riverscapes. The research is thus interested in how different actors interpret urban environmental change and environmental problems. Consequently, this study analyzes "the knowledges of the non-human world that circulate within and between real-world groups and organisations" and finally produces "knowledge about other people's knowledges of the environment" (CASTREE 2005: 24). On these grounds, this is as much a study of Delhi's riverscapes as it is an analysis of the knowledges of the different actors being involved in (re)producing Delhi's riverscapes.

According to FLICK (2014a: 12) empirical research needs "sensitizing concepts for exploring and understanding" in order to be able to develop "locally, temporally, and situationally limited narratives". Social science research is therefore forced to make use of inductive approaches, since in many cases the knowledge of the specific topics is too limited to start with theories and hypothesis to be tested. This study has been highly explorative using inductive reasoning throughout to develop an understanding of Delhi's riverscapes. In addition to initial explorative interviews, the author spent several weeks walking along the river, crossing its bridges, observing people's activities, mapping land uses, smelling the blackish waters and feeling the spray of the water. In discussing developments along the river with planners, bureaucrats, environmental activists, or farmers, the author actively contributed to the circulation of knowledge about Delhi's riverscapes. Accordingly, this study as well as the author himself became an element of the discourses. This should not be viewed as a conflict, but needs to be made explicit.

The applied research methodology of this study needs to cover the main interests of this study. These include:

– The (historical) material practices leading to environmental change along the river;
– The discursive practices giving insights into the production of knowledge(s) and the making of urban environmental policies; and
– The co-evolution of both spheres.

Here emerges a major methodological challenge in terms of how to bring together socio-spatial practices and discourses. RICHARDSON & JENSEN (2003: 21) discuss a "dialectical tension between material practices and symbolic meanings" which needs to be overcome by a detailed engagement with both. Keeping this major challenge in mind, the following sections outline the methodological groundings of this study.

4.2 The chosen discourse analytical approach

This study deconstructs different knowledges and conceptions of hybrid riverscapes in order to better understand socio-environmental change. A discourse analytical approach emerges as suitable for urban environmental governance research because it helps avoiding normative categories during both fieldwork and interpretation.

Although the term *discourse analytic research* is not clearly defined and draws on a diverse body of theory (RICHARDSON & JENSEN 2003: 16), an interdisciplinary comparison shows a common focus on the analysis of the use of language either in oral or written (textual) form. Further, besides the heterogeneity in the different approaches, KELLER (2013: 3) highlights several common features of discourse analysis. These include the overall understanding that "meanings of phenomena are socially constructed" and "that individual instances of interpretation [of these phenomena by different actors] may be understood as parts of a more comprehensive discourse structure that is temporarily produced and stabilized by specific institutional-organizational contexts" (ibid.). As the "use of symbolic orders is subject to rules of interpretation and action" discourse, analytical research aims to reconstruct these relationships by interpretative patterns and actions (ibid.). Thus, the foci and methodological tools of discourse analytical research in social sciences differ distinctly from its counterpart in linguistic sciences. In this vein, HAJER (2002: 62) even questioned if it has been really a 'linguistic turn' in social science and suggests to rather speak of an "argumentative turn" because social science is generally more interested in the larger stories people tell, and the arguments they make, than in the specific wording or phrasing of what they say.

Nevertheless, there is no single method for discourse analysis in social sciences, but rather many different approaches to analyze discourses. Significantly, the work of the French philosopher MICHEL FOUCAULT (and other post-structuralists) produced important contributions for discourse analytical research (KELLER 2013: 8). Despite their significance, neither FOUCAULT nor anyone else writing in the basis of his work has developed a coherent methodology of discourse analysis (MATTISSEK & REUBER 2004: 230). The field of discourse analytical research is rather characterized by a large heterogeneity, which results in a variety of theoretical concepts and research approaches being used, including in geography and urban studies (BAURIEDL 2007, GLASZE & MATTISSEK 2009, JACOBS 2006, LEES 2004, MATTISSEK 2007, MATTISSEK & REUBER 2004).

Three major lines of discourse analytical approaches can be distinguished in the social science which have also found their way into geographical research: 1) structuralist approaches (e.g. the Critical Discourse Analysis); 2) poststructuralist approaches; and 3) sociology of knowledge approaches which include among other the Sociology of Knowledge Approach to Discourse (SKAD) as well as different approaches to policy (process) analysis rooted in political science. While these strands show different theoretical foundations they are closely intertwined and have often been mixed (LEES 2004: 103).

The chosen discourse analytical approach applied in this study draws on *policy process analysis* as developed by KEELEY & SCOONES (1999, 2003) and the *argumentative discourse analysis* developed by (HAJER 1993, 1995, 2006). Drawing on different bodies of theoretical literature from political science, sociology of knowledge and science studies, KEELEY & SCOONES (1999, 2003) have used policy process analysis to better understand the making of environmental policies. They understand polices to be "co-constructed" by different actors embedded in certain networks reaching across scales from the local to the global (KEELEY & SCOONES 2003: 3).

In the Latourian style, they aim to "trace the actors and networks, which are creating the 'facts' that make up policy" (ibid. 5). HAJER (1995) emphasizes the role of so-called "story-lines" and "discourse-coalitions". HAJER as well as KEELY & SCOONES draw on FOUCAULT, but they hold on to a more structuralist understanding of actors. While sometimes criticized for mixing structuralist and post-structuralist approaches (GLASZE & MATTISSEK 2009: 32), both approaches offer a clear terminology, which has been adopted for this study. Policy process analysis, as deployed by KEELY & SCOONES, draws on HAJER's work and emphasizes the *making* of the policy as a *process*. The main features of the *argumentative approach to discourse* as developed by HAJER are therefore outlined in the following.

4.3 The argumentative approach to discourse

HAJER's argumentative approach to discourse acknowledges that discourses are "related to the social practices" in which discourses are produced but also emphasize a "clear institutional dimension" of discourse analysis (HAJER 1995: 44). Following his understanding, any "discussion of a typical environmental problem involves many different discourses" (HAJER 1995: 45) and a wide range of different forms of knowledge resting with a large number of different actors shapes policy discourses. HAJER is interested in how different actors – who operate in different spheres of discourses, e.g. public discourse, scientific discourse, etc. – communicate with each other on an "inter-discursive level" (which brings the different discourses and actors together) and how different strands of knowledge come together forming "authoritative narratives" and eventually find their way into policies (HAJER 1995: 46).

Drawing on FOUCAULT, HAJER acknowledges that the actors are not 'free' in manipulating the discourse based on their interests, but rather the actions of the different actors are restrained by the overall structures and rules of the discourse (HAJER 1995: 49–50). Whether actors are able to participate in a discourse, or manage to shape a discourse in certain ways, is conditional to the "way in which institutions and actors are implicated in discourses" (HAJER 1995: 49). Therefore, the 'power' of an actor is defined relationally depending on the overall situatedness of the actor within the discourse and the existing institutions (ibid.).

Overall, HAJER develops his discourse analytical approach on Foucaultian discourse theory in respect to discourse formations, but sees a need to develop so-called "middle-range concepts", which "can be related to the role of individual strategic action in a non-reductionist way" (HAJER 1995: 52). By introducing "story-lines" and "discourse-coalitions" HAJER outlines two middle-range concepts to analyze inter-discursive communication focusing on how discourses are maintained, shaped and eventually transformed by certain (groups of) actors. HAJER (1995: 53–55) argues that his approach is a "correction to Foucaultian discourse theory" by laying emphasis on the "argumentative interaction" in the discourse formation. Influenced by the work of BERGER & LUCKMANN and GIDDENS, he emphasizes that the "(institutional) structures are both constraining and enabling" (HAJER 1995: 48).

4.3.1 Story-lines and discourse-coalitions

In clear contrast to FOUCAULT, HAJER (1995: 55) stresses that it is important to study the subject (or actor) in shaping the discourse. HAJER views environmental politics as an "argumentative struggle in which actors not only try to make others see the problems according to their views but also seek to position other actors in a specific way" (HAJER 1995: 53). Accordingly, he studies the argumentative interaction between the different actors as the constituting element of discourse formation (HAJER 1995: 54). In doing so, he is interested in specific story-lines, which he defines as "a generative sort of narrative that allows actors to draw upon various discursive categories to give meaning to specific physical or social phenomena" (HAJER 1995: 56). Story-lines are important for the overall understanding of a discourse as well as for the coalescence of different discourses on the way to formulating environmental policies.

> "The key function of story-lines is that they suggest unity in the bewildering variety of separate discursive component parts of a problem [...]. The underlying assumption is that people do not draw on comprehensive discursive systems for their cognition, rather these are evoked through story-lines. As such story-lines play a key role in the positioning of subjects and structures. Political change may therefore well take place through the emergence of new story-lines that re-order understandings. Finding the appropriate story-line becomes and important form of agency" (HAJER 1995: 56).

Thus, story-lines refer to a common understanding and allow actors to overcome existing fragmentation within one discourse as well as constitute important points of reference for merging different discourses (HAJER 1995: 62).

As discourses around environmental problems are complex, story-lines are able to reduce complexity and make things understandable for a certain group of actors. Story-lines function as a 'short hand' note summarizing and simplifying complex narratives (HAJER 2006: 69). Based on the reduced complexity, the actors are able to better comprehend the problem, realize that the other actors actually also understand the problem, relate their own and the other actors' actions to address the problem, and eventually collectively develop problem-solving strategies. In order to do so, story-lines need to sound right for a certain group of actors, who then use them to cluster their knowledge and position themselves in the discourse, resulting in coalitions amongst actors (HAJER 1995: 63, 2006: 69). Story-lines therefore make the "communicative miracle" possible by which different actors eventually reach mutual understanding and urban environmental policies are framed in a certain way (HAJER 1995: 46). Accordingly, the second middle-range concept introduced by HAJER is defined as discourse-coalition. Discourse-coalitions are characterized by

> "unconventional political coalitions, each made up of such actors as scientists, politicians, activists, or organizations representing such actors, but also having links with specific television channels, journals and newspapers, or even celebrities" (HAJER 1995: 12–13).

By drawing on particular story-lines over a specific period of time, groups of actors emerge as discourse-coalitions, which "develop and sustain a particular discourse" (ibid.). The interest of discourse analysis is therefore to identify these discourse-coalitions and their interlinking story-lines.

4.3.2 Discourse structuration and discourse institutionalization

In addition to identifying these discursive structures and ways discourses are reproduced, HAJER emphasizes that in order to be able to link discourse to power and dominance, and the ways in which discourses shape policy-making, research needs to reveal how certain discourses start to dominate (HAJER 2006: 70). Following HAJER, the influence of discourses on policy-making can be assessed by taking into account "a simple two-step procedure" of *discourse structuration* and *discourse institutionalization* (ibid.). Discourse structuration occurs when a group of actors draws on certain ideas, concepts, and categories to gain credibility and acceptability. A certain discourse starts to dominate if these story-lines and rhetorical powers of a certain discourse are accepted and adapted by central actors (ibid. 71). Such a dominating discourse then gets translated into concrete policies (rules and laws) and is thereby institutionalized. A hegemonic discourse thereby also becomes dominant in the framing of policies and the practices of organizations implementing these policies.

4.4 The application of the argumentative approach in this study

The author suggest that a study like this, which understands governance mainly as a process of negotiation, is only practical if holding on to actors as (more or less strategically) acting subjects. Ecological, socio-cultural, economic and political conditions, as well as their discursive framing (e.g. through certain story-lines), produce institutions and structures which constrain and enable the different actors to participate, act and shape the discourses around Delhi's riverscapes.

Since the argumentative approach to discourse draws on both structural and post-structural elements, and acknowledges strategic actions of different actors within discursive structures, the approach emerges as suitable for this study. HAJER's mid-range concepts (story-line and discourse-coalition) are particularly useful. The concept of story-lines highlights that policies are not a straightforward solutions identified by (natural) science, but rather policies are made based on "discursively negotiated 'facts' or convenient points of agreement between different actors" (FORSYTH 2001: 4). Since story-lines often simplify complex issues and reduce them to an understandable level they also remove uncertainties and blind out alternative readings, which in turn make them acceptable to certain actors to push through their interests.

Working out story-lines and discourse-coalitions is often a difficult task, since discourses on urban nature are characterized by "all sort of slippages, ambiguities, and incoherencies" (HARVEY 1996b: 89). Urban environmental discourses thus emerge as "porous" (DESFOR & KEIL 2004: 10) and their formation is dependent upon the mobilization of different practices (e.g. surveys, mappings, etc.) by different actors (EVANS 2007: 146). Analyzing urban environmental discourses therefore needs to be done carefully when drawing interdiscursive connections; especially when looking at expert versus lay, and technical versus political discourses (DESFOR & KEIL 2004: 10). Further, locating a certain actor within a specific discourse is not an easy task, since an actor who, for example, usually opposes the commodification of urban nature (e.g. an environmental NGO) might be in favor of such an approach when it emerges helpful in changing the state authorities' general approach to urban nature (DESFOR & KEIL 2004: 10–11).

In the case of the Yamuna in Delhi, environmental NGOs opposing the commercialization of the land along the river might be in favor of framing water as a commodity and pricing domestic water uses higher in order to achieve water savings and larger flows in the river. This example further highlights the "constitutive role of discourse in political processes" (HAJER 1995: 58) and shows that "neither identities nor interests are fixed in discursive processes" (DESFOR & KEIL 2004: 11). Identifying story-lines and discourse-coalitions therefore helps to understand how intra-discursive differences and inconsistencies are levelled and how inter-discursive communication works on the way to formulate urban environmental policies (DESFOR & KEIL 2004: 11, HAJER 1995: 46). Table 1 provides an overview of the basic terminology used in this study.

actor(s)	individual or collective producers of statements; those who use specific rules and resources to (re)produce and transform a *discourse* by means of their practices
discourse	overall focus of discourse analytical studies, which is investigated with regard to institutionally stabilized common structural patterns, practices, rules and resources for meaning-creation
discourse-coalition	group of actors whose statements can be attributed to the same discourse (e.g. through the use of the same *story-line*), formation does not have to take place consciously or strategically
story-line	common thread in a discourse, generally simplification of complex reality
discourse structuration	a certain idea or concept to gains credibility and acceptability by employing e.g. the same *story-lines*
discourse institutionalization	a *hegemonic discourse* gets translated into concrete policies (rules and laws) and is thereby institutionalized
hegemonic discourse	a certain discourse emerges to dominate if *story-lines* and rhetorical powers of a certain discourse are accepted and adapted by central actors, e.g. a *discourse-coalition*

Table 1: Terminology used for discourse analysis
Source: Based on HAJER (1995, 2006), KELLER (2013: 72–74)

5 METHODS AND FIELDWORK

5.1 Data collection and fieldwork

Qualitative empirical methods have been used throughout this study to collect and interpret the data. The applied qualitative methods for primary data generation include a series of in-depth interviews, a focus group discussion (in form of a *future workshop*), observation, document analysis and multi-temporal mappings. Combining a classical geographical suite of methods for data collection with a discourse analytical approach for the interpretation (and to some extent validation) of the data, the study links and compares the physical realities (based on mappings and observations) with representations in different discourses. Data collected by the author himself (e.g. non-verbal data from observation and mapping) merges with knowledges from interviewees (verbal data) and information gathered from different text corpora (documents as data).

The author conducted fieldwork in Delhi during six different fieldwork phases between 2009 and 2014:

- October 2009 – March 2010 (6 months): explorative fieldwork;
- September 2010 (3 weeks): explorative fieldwork, in-depth interviews, mapping flood situation;
- March – April 2011 (7 weeks): in-depth interviews, mapping;
- November 2011 – December 2011 (6 weeks): in-depth interviews, mapping of land-use changes, observation during court proceedings;
- February 2013 – March 2013 (4 weeks): in-depth interviews, mapping of changes, future workshop; and
- September 2014 (2 weeks): informal meetings with interview partners, observation during court proceedings, mapping of land-use changes.

The findings of observational research and multi-temporal mapping were discussed with the interviewees while information by the interviewees concurrently influenced observational research (particularly suggestions to visit several sites) as well as the categories for the mapping exercise. Sometimes ambiguous narratives and differences between ground realities evaluated through mapping and narratives by interview partners emerged during fieldwork. These are interpreted as representing *multiple realities* depending on diverse knowledges, subjective perception and agenda setting resulting in different conceptualizations of the same space (CASTREE 2005: XX, WINCHESTER & ROFE 2010: 8).

5.1.1 Triangulation of data and analysis

The multiple realities appearing in both the physical riverscapes and the discursive representation of the riverscapes, asked for a mixed-method approach. Triangulation enables social science research to engage with its research subjects in a more holistic way and explore reality from different perspectives (FLICK 2004: 178).

Triangulation of data has been applied in this study by including primary data collected through mapping and interviewing as well as secondary data like official reports, media reports, minutes of meetings and court documents. Additionally, different interview methods have been used throughout the study including semi-structured, problem-centered interviews and more flexible responsive interviewing techniques as well as time series of interviews. In the final fieldwork phase, the author facilitated a group discussion in form of a *future workshop* inviting all interview partners.

Analyzing triangulated data "may lead to three types of results: converging results, complementary results, and contradictions" (FLICK 2014a: 189). All three types of results are in principle interesting and helpful for analysis and interpretation. While converging results mutually validate the information of the other method(s), complementary and contradictory results in particular reveal the additional value of triangulation, since diverging results may indicate shortcomings and weaknesses of particular methods or their application by the researcher (ibid.).

In a sensitive and politicized research environment, contradictions might further illustrate the complexity of the research problem as well as the subjectivity of the involved actors. Contradictory information shows the diversity of how Delhi's riverscapes are perceived.

In short, triangulation of data and triangulation of methods has been used throughout this study "as a strategy for a more comprehensive understanding and a challenge to look for more and better explanations" (FLICK 2014a: 190). In connection with a discourse analytical approach methodological triangulation allows for a deeper engagement with the *multiple realities*. By this means, triangulation emerges as an important feature of critical, reflexive empirical research (cf. ETZOLD 2013: 101–103).

5.1.2 Primary data: multi-temporal mapping and observation

Understanding land-use changes generally requires an in-depth spatial knowledge of the research area. The author mapped Delhi's riverscapes during the different field work periods in order to document the different land uses and occurring changes. Linking the ground realities retrieved from detailed mapping exercises to multi-temporal satellite data further allowed for establishing a spatial overview of the development of Delhi's riverscapes. In order to portray the dynamics and recent changes in the floodplain, the study draws on a set of satellite images produced within the main study period from 2001 to 2014.

This timeframe has been chosen in order to illustrate two major processes; 1) the large-scale demolition of slum settlements from the river banks between 2004 and 2006 as well as 2) the development of urban mega-projects along the river in the early 2000s. A spatial overview is provided for four specific *moments*[19], which represent different stages of the remaking of the riverscape in the most recent period.

Based on satellite images from sources free of charge (LandSat, Google Earth, Digital Globe), own mappings, and detailed discussions with interview partners, the following four *moments* have been documented:

- 2000/2001: pre-demolition, first mega-projects under construction;
- 2006: post-demolition, mega-projects under construction;
- 2010: year of the CWGs, mega-projects largely completed; and
- 2014: post-Commonwealth Games developments.

The following land-use classifications (and sub-classifications) were used:

- Built-up space (residential, institutional, industry, power stations, under construction);
- Transportation (roads, railways, metro, under construction);
- River;
- Open riverbed with natural vegetation (sand/grassland, wooded area/forests);
- Agriculture (cultivation, nursery);
- Parks and recreational areas; and
- Other uses (embankments, major sites of solid waste and debris dumping, fly-ash ponds).

The ArcGIS-based data set was used for change detection and the calculation of land-use changes for the 2001–2014 period (see chapter 12 for results). In addition, historical maps were digitalized and integrated in the GIS database in order to analyze long-term changes. Similarly, recent planning documents (e.g. Zonal Development Plans, information on land ownership, flood zonation) were integrated in the data base.

5.1.3 Primary data: interviews

Semi-structured, in-depths interviews remained the core fieldwork tool used in this study. In the course of the study a total of 72 interviews with government officials, politicians, scientists, lawyers, farmers and environmental activists were conducted by the author (see list of interviews in Appendix I). In conjunction with discourse analysis, interviews provided opportunities to critically discuss discourses and

---

19 By using the term *moment* the author likes to emphasize that the different maps do not show the research area at a specific date but rather combine information of various sources (different satellite images free of charge, own mappings, interview data) for giving a spatial overview of the situation of Delhi's riverscapes in the different years.

thereby, allowed for discovery of counter discourses and differing claims to reality (DUNN 2010: 102).

Interview techniques chosen for this study adopted features from problem-centered interviews (WITZEL 2000, WITZEL & REITER 2012) and expert interviews (BOGNER et al. 2009a). Both interview techniques are specific forms of semi-structured interviews and both are more open than focused interviews but more closed than narrative interviews (cf. FLICK 2014a, HOPF 2004). As a result, these techniques ensure enough openness and flexibility to be responsive to new and interesting points raised by the interviewees, which could also eventually reveal new theory-generating insights. Such an attitude towards doing in-depth interviews is also described by RUBIN & RUBIN (2012) as "responsive interviewing".

This study aimed to give a voice to as many actors from the research field as possible. The selection of the respondents for arranging interviews was therefore closely connected to an actor mapping done during the first fieldwork phase. This first map of actors grew constantly based on informal talks, interviews, media analysis and other secondary data like court documents, government websites, and materials shared by environmental NGOs.

As a guest researcher at the School of Planning and Architecture (SPA) in Delhi the author benefited from personal contacts between academia and government officials. Crucially, SPA's high reputation and excellent network permitted access to contacts inside various government agencies (especially DDA, which is located across the street from SPA). In the later fieldwork phases the author aimed at interviewing different individuals within the different government agencies.

Multiple problems of access and willingness to talk complicated the desired diversification of voices from within different government agencies. Therefore the author concentrated on conducting at least one interview with high-ranked officials from each of the major involved government agencies. Further, the author decided to focus on the DDA, as the key agency, and interviewed 10 officials (including three retired planners who were involved in preparing earlier schemes for the river Yamuna). Contacting officials in the central ministries proved to be difficult. Unfortunately, all contacted officials from the National River Conservation Directorate (NRCD) in the Ministry of Environment and Forests (MoEF) denied any interview on the river Yamuna.

In addition to these practical problems of access (LITTIG 2009), the questions of whom to interview and whom to consider as an expert were challenging. The term 'expert' is commonly used with varying meanings in different contexts. Even within the field of human geography, different perceptions of how to define experts occur (as experienced by the author in many discussions with colleagues).

> "The answer to the question, who or what are experts, can be very different depending on the issue of the study and the theoretical and analytical approach used in it. ... We can label those persons as experts who are particularly competent as authorities on a certain matter of facts." (DEEKE 1995: 7–8, translated by FLICK 2014a: 227)

Many empirical studies use expert interviews in the early and final stage of fieldwork to gain explorative insights into the research field and to validate other empirical data. The distinction between expert interviews and other semi-structured interviews with actors from the research field is often unclear. While a horizontal division between expert and lay knowledge, including referring to the "expert as agent of truth and authority", had been dominant for a long time, the acknowledgement of the social construction of knowledge and new forms of knowledge production demystified the expert and their specific role in decision-making (BOGNER et al. 2009b: 3).

Due to the proliferation of expert knowledge and multiplication of expert systems based on societal change (including new forms of governance, information technology, rising role of the civil society, etc.), scientific expert knowledge is confronted with different forms of counter-expertise, counter-knowledges and counter-discourses and therefore inherently faces a crisis of legitimacy (MEUSER & NAGEL 2009: 21). 'Expert' knowledge is produced outside the traditional professional contexts and different forms of knowledge merge inevitably (ibid.). The work and expertise of environmental NGOs being only one example. One might even add that the shift from government to governance would not have been possible without new and pluralistic forms of knowledge production by new actors involved in policy-making and implementation. Overall, the identification of the experts (who is an expert and who not) thus "depends on the researcher's judgement" (MEUSER & NAGEL 2009: 19).

In the course of this study all interviewed individuals representing certain actors within urban environmental governance have been viewed as experts. All of them have different kinds of expertise depending on their engagement with the river whether as officials in a government department, environmental activists or farmers from along the river.

A narrow understanding of 'the experts' would exclude the farmers and also to some extent the environmental activist; both important voices in this study. However, from a methodological point of view it is important to state that conducting an interview with a farmer is fundamentally different to interviewing decision-makers (e.g. high-ranking government officials).

Individual semi-structured interview guidelines focusing on several issues of Delhi's riverscapes were prepared, depending on the specific interview partner. Due to the sensitive research environment, the short questionnaire to collect additional information proposed as one of the instruments of problem-centered interviews (WITZEL 2000, WITZEL & REITER 2012) was substituted in meetings with government officials by a rather unstructured introduction at the beginning of the interview, often accompanied by a cup of *chai*. Most interviewees agreed to tape recording and the author drafted short postscripts including main points, atmosphere and open questions in the aftermath of the interviews. After ensuring anonymity and explaining the background of the study the interviews commonly started with open questions like these:

– How would you describe the state of the river Yamuna in Delhi?

- Do you see changes within the last years?
- What are the major concerns/problems of the river Yamuna in Delhi?

These opening questions were designed as incentives for the respondents to unfold the personal view of the state of the river (the symptoms of the socially relevant problem) before shifting to more specific questions. Most interviewees replied to these questions extensively. These narratives were the starting points for joined problem centering (cf. WITZEL & REITER 2012: 24) between the author and the respondents. In some cases, the interviewer used short narratives, maps and pictures from the field to set further incentives. After discussing the overall situation in Delhi, often some agreement was reached on several points regarding Delhi's riverscapes. The interviewer then tried to shift the focus of the conversation more to the involved actors in governing Delhi's riverscapes exploring the respondent's viewpoints on the different agencies' duties and responsibilities. Triggering this shift towards governance aspects and keeping the focus was often a major challenge during the interviews.

Despite some difficulties, all interviewees revealed that they perceive the current state of the river Yamuna as problematic, but the narratives indicated that the problems are differently perceived, articulated and managed depending on subjective views. In many cases, extensive narratives became only meaningful to the author in retrospective when transcribing, coding or analyzing. Ex post, narrative parts of the interviews emerged consequently to be especially insightful to analyze the different perceptions and framings of the problem(s) of Delhi's riverscapes as well as policy options.

Semi-structured interviews are characterized by complex power asymmetries between the interviewer and the interviewee (DOWLING 2010, KVALE 2007). Related to the expert status of the interviewees and particularly evident when talking to senior government officials, some interviewees reacted markedly distressed when the author raised critical questions regarding the work of their organization. These occasions illustrated that building up trust in such a sensitive and politicized research environment coupled with complex interviewer-interviewee relations (e.g. age differences, highly hierarchical society, foreign background of the researcher) proved to be extremely difficult. Confrontations generally provoked either ad-hoc justifications, skipping of the question or even a complete refusing to answer any more questions. Based on the experiences from the first two fieldwork phases the author reconsidered the design of the interviews to ensure even more flexibility and a more sensitive questioning (see Table 2).

The problem-centered interview nevertheless provided a helpful tool in the beginning of the research in structuring rather explorative interviews. It was later perceived as over-emphasizing on keeping the interviews focused. Especially the shortcutting of narratives as well as confrontational questions proved to be counterproductive in the Indian context in which story telling emerged to be very informative. The style of responsive interviewing offered a more flexible and appropriate tool, since in contrast to the problem-centered interview approach, which explicitly includes confrontational questions towards the end of the interview, responsive interviewing aims to completely avoid confrontation (RUBIN & RUBIN 2012: 37).

	Problem-centered expert interviews	**Responsive expert interviews**	**Follow-up interviews (time series)**
Fieldwork phase	Fieldwork phase I + II	Fieldwork phase III-IV	Fieldwork phase II-IV
Opening question	state of the river	state of the river	recent developments
Incentives by the interviewer	combination of narratives and questions by the interviewer	maps and pictures main questions probes follow-up questions	maps and pictures main questions probes follow-up questions
Main challenges	shifting of focus from the symptoms of the socially relevant problem to the research problem (governance aspects)	time constraints keeping the focus	time constraints
Main strengths	helpful in structuring interviews in the explorative phase	detailed narratives keeping the focus	building up of trust

Table 2: Overview and types of conducted in-depth interviews

Based on first very positive follow-up meetings with interviewees in the second fieldwork phase, during which the interview partners seemed to acknowledge the effort and interest put in by the author to come to follow-up the discussion, the author decided to conduct a time series with key individuals including planners and environmental activists, which have enabled the author to trace the developments over a longer period of time.

Benefitting from the gaining of trust, the follow-up interviews appeared more intense in terms of critical debate and information shared. Through a series of interviews the author was able to add a temporal component. This has been instructive for a better understanding of the changing physical environment but also of the further developing urban environmental governance. The follow-up interviews generally focused either on recent events or newly emerging topics (media reports, court orders, floods, etc.). The outlined approach enabled the author to discuss not only recent changes, but also certain expectations, which in following visits were correlated to the recent developments.

In sum, the follow-up interviews proved to be most informative and emerged as a highly recommendable tool for governance research in sensitive research environments. The approach is certainly dependent on the time and interest of the interviewees in follow-up meetings.

5.1.4 Primary data: Yamuna Future Workshop

To bring together planners, scientist and environmental activists for a group discussion, the author organized a *Yamuna Future Workshop* on March 8th, 2013, 4pm–9pm, in cooperation with the Max Mueller Bhawan, New Delhi. In total, 22 people participated in the workshop including among others planners from the Landscape and Environmental Planning Unit Delhi Development Authority (DDA), environmental activists from different NGOs (INTACH, Yamuna Jiye Abhiyaan, PEACE Institute, Meri Delli Meri Yamuna, Centre for Sciene and Environment (CSE), Natural Heritage First, Toxic Links), and colleagues from GIZ Delhi, Department of Geology, University of Delhi, University of Lausanne and Jawaharlal Nehru University (JNU).

The future workshop is a method developed to find future ideas and solutions for social problems in a highly participatory process (JUNGK & MÜLLERT 1987). The method has especially been used in urban planning to involve citizens at an early stage of the planning process. The method shares similarities with other participatory workshop settings like Open Space, Bar Camp or Unconferences. Normally a more or less clearly defined problem, which needs in-depth planning to find a solution in the future, is the focus of the future workshop method.

In contrast to using the future workshop as a participatory planning tool to involve normal citizens, the participants of the future workshop on the Yamuna were all 'experts' in the field with a wide knowledge about the developments along the river as well as its problems and concerns. The author suggested six major questions for the workshop:

- What are the major problems of the Yamuna in Delhi?
- Who is responsible for the state of the river today?
- How should a positive scheme for the Yamuna look like?
- What are the common goals for the future of the Yamuna in Delhi?
- What will be the major challenges and difficulties?
- How can environmental NGOs and local experts 'promote' a positive scheme and influence future decision-making?

In the group discussion the different interrelationships and dependencies of the different problems were discussed. As expected, the examination of the different problems and ideas quickly moved into an in-depth discussion about the major challenges and difficulties regarding their practicability. The future workshop (indicated with the coding 'YFW-2013-03-08' in the text) was in particular insightful with regard to the direct interactions of the different actors. It also revealed certain conflicts, which had not become evident in the interviews. The workshop further confirmed many findings from the interviews and the author received positive feedback from participants regarding the presented findings.

5.1.5 Secondary data: the text corpora

In addition to primary data, a series of secondary data (text documents) were incorporated in the analysis of this study. The following section gives a brief overview of the different sources and documents.

Planning documents and minutes of meetings

The planning documents made available were:

- The three Master Plans for Delhi approved in 1962, 1990, and 2007: 'Delhi Master Plan' (MPD) (DDA 1962); 'Delhi Master Plan 2001', short: MPD-2001 (DDA 1990); 'Delhi Master Plan 2021', short MPD-2021 (DDA 2007);
- Three draft versions (DDA 1998, 2006d, 2008) and the final version (DDA 2010b) of the Zonal Development Plan (ZDP) for the River Zone (Zone O);
- Internal planning documents of the DDA on the Yamuna (DDA - SPECIAL PROJECT CELL 1998), annual reports (DDA 2011, 2013a), minutes of meetings and public notices.

Additionally, planning and policy documents of the National Capital Region Planning Board (NCRPB) gave insights into the regional and interstate level (see e.g. NCRPB 2013b). Official documents included further policy documents produced by various ministries (e.g. Ministry of Urban Development, Ministry of Environment and Forests) and information obtained from various government websites.

Minutes of meetings from different state authorities revealed insights into the discourse on the Yamuna within the state authorities. These included, among others, the minutes of the Yamuna Standing Committee (YSC) for a period of almost 40 years from the 33rd meeting of the YSC in 1975 to the 83rd meeting in 2013, minutes of the Upper Yamuna River Board (UYRB), minutes and reports of the Yamuna River Development Authority (YRDA). For a detailed list of the minutes and the used codes for reference see Appendix II.

Court proceedings: petitions, affidavits and orders

An important source of information for this study were the different court cases related to the Yamuna. Court proceedings give detailed information on processes of decision making as well as offer an interesting opportunity to study discourses by analyzing petitions and affidavits by the different involved actors.

A starting point for gaining an understanding of the legal proceedings with regard to the Yamuna in Delhi has been RITTWICK DUTTA's (2009) report 'The Unquiet River. An Overview of Select decisions of the Courts on the River Yamuna' published by the PEACE Institute Charitable Trust, which is involved campaigning for the Yamuna (Yamuna Jiye Abhiyaan). Original petitions, several affidavits by the applicants and respondents, as well as orders and judgements were reviewed.

Additional materials were accessed through the involved environmental NGOs or their lawyers. These materials are cited in footnotes.

Newspaper articles

A media analysis covering the main English-speaking daily newspapers (Times of India, The Hindu, and Hindustan Times) was done in the explorative phase of the study for the years 2000 to 2009 to gain an insight in the public discourse on the Yamuna. The analysis for this period focused on dominant discourses like water shortage and pollution levels, and particularly on important events such as the slum demolitions and related court cases (2004–2006), flood events, the mega-projects (Akshardham, Delhi Metro, Commonwealth Games Village), as well as the activities of certain actors like the environmental NGOs (e.g. protest camp 'Satyagraha' 2007–2009) or politicians like Jagmohan (prominent public figure behind the evictions).

About 500 relevant articles on the Yamuna were included in the analysis for the pre-study period. For the main study period from January 2010 to December 2014 about 1,200 relevant newspaper articles were found by searching for the keyword 'Yamuna' in the online archives of the three main English-speaking daily newspapers. All newspaper articles from the online archives were converted into pdf-documents and integrated in the qualitative data analysis software MaxQDA. Newspaper articles are used at different points of this study to illustrate the public discourse. Their analysis was further important for a deeper understanding of the role of the media as an actor filtering and releasing information and thereby shaping discourses over a long time-span.

Academic and other literature

The academic literature on Delhi's urban development is burgeoning. Different aspects of urban governance in India, using Delhi as a case study in particular, are thermalized in a series of edited books (see among others BAUD & DEWIT 2008a, DUPONT et al. 2000, HUST & MANN 2005, RUET & TAWA LAMA-REWAL 2009, ZÉRAH et al. 2011). Recent journal articles and monographs discussing governance issues in Delhi relevant for this study include the work on slum relocation (BHAN 2009, DATTA 2012, DUPONT 2008, DUPONT & RAMANATHAN 2008, GHERTNER 2008, 2010, 2011b, 2011c, 2011d, MENON-SEN & BHAN 2008) and questions of neoliberal urban change connected to the idea of the 'world-class city' (AHMED 2011, BAVISKAR 2003, 2007a, 2011a, 2011b, DUPONT 2011). The work of VERONIQUE DUPONT on Delhi's slums and her research on the development of former slums sites (including sites along the riverfront) and GAUTAM BHAN's (2009) article on the evictions of the poor from the banks of the Yamuna mark starting points for this study.

An increasing number of research is published focusing on specific features of the Yamuna including, among others, the water quality, pollution levels and sewage treatment plant capacities (BHARGAVA 1985, FARAGO et al. 1989, GAUTAM et al. 2013, JAIN 2004, KAUSHIK et al. 2008, MANDAL et al. 2010, MEHRA et al. 1998, MISRA 2010a, PARMAR & KESHARI 2012, SEHGAL et al. 2012, SHARMA & KANSAL 2011a, 2011b, 2011c, SHARMA & SINGH 2009, SINGH et al. 2006, SINGH & KUMAR 2006, WALIA & MEHRA 1998a), as well as drinking water augmentation (LORENZEN et al. 2010, NARULA et al. 2001, SANDHU et al. 2010, SHARMA & KANSAL 2011c, SHEKHAR & PRASAD 2009, SONI et al. 2009, SONI et al. 2014, SONI & SINGH 2013). The book 'The Yamuna River Basin' by RAI et al. (2012) gives the most detailed overview

Despite this broader literature on Delhi and the Yamuna, more specific literature on the development of the Yamuna, is virtually only produced by former planners associated with the DDA including A.K. JAIN (1990, 2009a) and R.G. GUPTA (1995) who outline some aspects of the ideas for channelization and riverfront development. Both authors only give a limited insight into the planning discourse from the perspective of the DDA. Further, these books lack in reviewable references. For additional perspectives the author had to rely on 'other', grey literature including work published by civil society organizations (e.g. MEENAKSHI DHAWALE's 'Narratives of the Environment of Delhi' published by Indian National Trust For Art and Cultural Heritage (INTACH) in 2010 (DHAWALE 2010) or reports by the Hazards Centre on the eviction of slums along the river (HAZARDS CENTRE 2004a, 2004b)).

Unheard voices – learning from ethnographic work on the river

A detailed ethnographic study or quantitative survey of the people living along the river would have gone beyond the focus and scope of this study. The conducted interviews with environmentalists working together with 'riverfront people' for many years and selected key persons from the farmers' association and villages from the riverbed are taken to represent these voices for the purpose of this study. The author focused here on the farmers, since they play an important role in shaping Delhi's riverscapes. Other groups like (former) fishermen and boatmen, washer men (*dhobis*), nursery operator, cattle herders, wrestlers from the *akhadas* (wrestling grounds and exercise centers on the riverside), coin divers or temple priests and worshippers from the *ghats* remain largely unheard in this study.

SARANDHA JAIN (2011) offers a detailed narrative of these voices in her book 'In search for Yamuna - Reflections on a River Lost'.[20] Another excellent source referring to these hidden stories and religious practices of the pilgrims, priests, and worshippers of the Yamuna is offered by DAVID L. HABERMAN's book 'River of Love in an Age of Pollution – The Yamuna River of Northern India' (2006).

---

20 SARANDHA JAIN accompanied the author on 6th of March 2013 on field work talking to the farmers in the riverbed.

HABERMAN's interest lies on the "overlap between religion and ecology" and the contradiction between Yamuna's importance as a river goddess and its high levels of pollution. The focus of his study lies on the Braj region south of Delhi where the river is most extensively worshiped in places like Vrindavan and Mathura. Exploring the reactions of religious communities on river pollution, HABERMAN provides deep insights into the nature of "Indian river environmentalism", as he has termed it, and outlines that it is "quite different from its Western counterpart" (HABERMAN 2006: 2).

5.2 From data processing to analysis

Handling, sorting, coding and retrieving of qualitative data on the way to generating analytical ideas are inherently time-consuming activities in qualitative analysis, because of the sheer amount of empirical data gathered in the research process (cf. GIBBS 2007). Further, in qualitative research using in-depth interviews data collection and analysis are inseparably connected and proceed for large parts of the research simultaneously (FLICK 2014b). As outlined above, the design of the discourse analytical approach chosen in this study deliberately evoked the amalgamation of data collection and interpretation in actively getting involved in the discourse through in-depth interviews and the organization of the future workshop.

All data, which either was available in digital textual form (newspaper articles, letters, etc.) or has been transformed into textual form through transcription (recordings of interviews and future workshop) has been analyzed using the qualitative data analysis software MaxQDA. Several documents (e.g. minutes of meetings, material from court proceedings) had to be analyzed manually using highlighter and notepads. The author prepared digital memos and summaries of these materials, which then were incorporated in MaxQDA.

In order to analyze the materials as objectively as possible, the initial coding was done without any pre-defined theoretical categories. The codes merely included the different actors, mentioned planning documents or court cases, identified problems as well as different place names or projects (e.g. Commonwealth Games Village). Therefore, the first round of coding was primarily focusing on an understanding of the governance set-up and the identification of the main problems identified by the different actors. The emerging code system developed from document to document. A first set of identified *categories of problems* related to the river emerged by combing the identified problems from the analysis of the newspaper analysis and the problem-centered interviews. These included in alphabetical order the following main codes (examples for sub codes in brackets):

- Agriculture (use of pesticides and fertilizers, buffalo bathing);
- De facto channelization of the river (pseudo bridges, embankments);
- Disposal of religious offerings (from poojas and festivals);
- Encroachment (including slums, unauthorized colonies, sand mining);
- Floods (reduced carrying capacity of the river channel);

- Flow (including water extraction upstream, upstream dams and barrages, diversion of water);
- Pollution / sewage (capacity of STPs, non-functioning of STPs, open defecation);
- Solid waste dumping (construction debris); and
- Water augmentation / water supply (groundwater level, water treatment capacity).

The organizing of the identified problems into these main categories already *produced* certain overlaps, which indicated that the multiple connection between the different problems started to challenge any further categorization. Therefore, the process of coding the problematized issues was a good starting point, but for a more detailed analysis of the complex interactions more analytical categories were needed to be established for a second round of coding. This second round of coding was best overwritten by the *extraction of emerging discourses*. The codes included, but were not limited to:

- Alternative schemes / visions for Delhi's riverscapes;
- Concepts of nature (e.g. biodiversity);
- Expert knowledge / role of science and experts;
- Formal uses versus informal uses;
- Multiplicity of agencies (e.g. undefined responsibilities); and
- Perceptions of the riverscapes (e.g. wasted land).

Based on these second-tier codes combined with the coding of actors and places dominant story-lines and discourse-coalitions were extracted to structure, and understand, the discursive and material production of Delhi's riverscapes.

5.3 Reflections on the research process

ISRAEL & HAY (2006: 2) convincingly argue that since social scientists generally belief that their work contributes to 'making the world a better place', protecting informants and minimizing possible harm to them should have top priority, because it would otherwise jeopardize social research's own legitimation to serve socially desirable ends. They outline *informed consent*, *confidentiality* and *beneficence and non-maleficence* as the three basic principles, which are key to ethical practice in social research besides dealing with the multiple problems relating to research relationships (ISRAEL & HAY 2006).

The author outlined the research focus before any interview through e-mail or telephone contact as well as at the beginning of the interview, to inform the interviewee about the purpose of the interview as well as the intended outcomes of the research project. Recording was only used if the interviewee agreed to be recorded and confidentiality was granted by assuring that interview data (recorded or non-recorded) will be anonymized. Confidentiality and anonymity is however problematic when doing research in a specific setting (FLICK 2014a: 59). Notably, many of the people interviewed during this study are also quoted in media reports and, therefore participate and to some extent influence the public discourse. This implies that most of the interviewed experts are public figures to some extent and in some cases insiders might be able to identify certain informants based on their specific knowledges.

Based on their role as experts, the author assumed that the interviewees were aware of which information they were able to share and which information they were not. Nevertheless, the author has tried to minimize the possibilities of identification as much as possible. The data collected through interviews is therefore anonymized by codifying the interviews including the interviewees' affiliation or background, consecutive numbers (interviews with the same interviewee carry the same number) and the date of the interview (year, month, day). For example, the interview code 'DDA-01/2009-12-02' stands for an interview conducted with a high-ranked planner from the Delhi Development Authority (DDA) on 2nd December 2009.

Interviews conducted with environmental activists can be identified by the acronym 'NGO'. This broader code system ensures anonymity also for interviewees from smaller NGOs as well as environmental activists, who work in the NGO-context but are not explicitly associated with a specific NGO.

When dealing with significant public figures, such as the former Lieutenant Governor of Delhi, Tejendra Khanna, ensuring anonymity is both not possible and not desirable, since otherwise the importance of the words would be lost. The interview with Tejendra Khanna was not recorded, but the interviewer noted certain important phrases and prepared a detailed postscript.

The interviewees of this study typically knew each other. Referring to specific public meetings or information from media reports helped the author to discuss apparently problematic topics during the interviews without revealing other interviewees as source of information. The author therefore followed the standards of ethical research by ensuring informed consent and confidentiality.

The principles of beneficence and non-maleficence are challenging and need further explanation. In general, these principles should be followed in social science research to "minimize risks of harm or discomfort to participants" and if possible "promote the well-being of participants" (ISRAEL & HAY 2006: 95). Referring to the principles outlined by ISRAEL & HAY (2006), HACKENBROCH (2013: 93–94) highlights the difficulties arising when following such a principle in ethnographic research dealing with planning issues in slum settlements. She explicitly acknowledges the danger that findings derived from social research (in her case in slums of Dhaka, Bangladesh) being co-opted by policy-makers "might entail a reaction and intervention into people's lives" based upon these findings (HACKENBROCH 2013: 94). Certainly, revealing informal practices of slum dwellers might put their very livelihoods (deeply depending on these informal practices) at risk.

While the research focus and environment of this study is completely different, and also the interviewer-interviewee relationships were very different, sensitive information was discussed (e.g. informal practices by farmers or informal channels of information tapped by environmentalist activists). Interestingly, when discussing delicate issues, the interviewees indicated that they were very much aware of the possibility of negative consequences after this information was disseminated. When the interviewees asked the researcher not to write about certain information discussed, the researcher's moral integrity as being independent was considerably challenged. Based on the principle of beneficence and non-maleficence the author respected the request made by interviewees not to release information when requested. By critically reflecting upon those issues faced, the author would like to highlight that following the principle of beneficence and non-maleficence is always a balancing act between protecting participants by not disclosing information on the one hand, and risking harm to certain groups (or informants) by disclosing certain information on the other hand.

IV SITUATING THE CASE STUDY

6 THE MEGACITY OF DELHI

The following sections introduce the physical features, the growth, and the current socio-spatial mosaic of the megacity of Delhi. The chapter ends with a review of the ambitious vision to make Delhi a 'world-class' city.

6.1 The physical features of Delhi

India's capital city stretches from 28° 24' to 28° 53' North and 76° 50' to 77° 20' East. Delhi borders on the state of Haryana to the north, west and south as well as the state Uttar Pradesh (U.P.) to the east. The topography of Delhi is dominated by its distinct natural features; the river Yamuna running from north to south and the Delhi Ridge encompassing the central parts of the city in the west (see Figure 13).

The quartzite rocks of the Delhi Ridge make up the north-western part of the Aravalli Range entering Delhi from south-west and expanding towards the Yamuna in a north-easterly direction up to Wazirabad. Three main physiographical features characterize the city: the older alluvial plains (*bāngar*) and the younger alluvial plains (*khadar*) of the floodplain of the river Yamuna as well as the Delhi Ridge.

Referring to the soil characteristics and flooding patterns the Indo-Gangetic plains are differentiated into *khadar* (also spelt *khādir*) and *bāngar* making up the landscape between two rivers (also referred to as *doab*, meaning between two rivers, interfluve). The *khadar* areas comprise of low-lying areas prone to flooding. The annual flooding deposits new, fine-grained, fertile sediments on the *khadar*. The *khadar* is naturally traversed by oxbow lakes, swamps and marshes (GEOLOGICAL SURVEY OF INDIA no date). The *bāngar* plains are of higher elevation, less prone to flooding and made up of older sandier, darker and less fertile alluvial materials. Agricultural patterns vary between *khadar* and *bāngar* due to different fertility and water availability (STANG 2002: 27).

In the Delhi region the *khadar* is further characterized by a number of cut-off meanders which suggest an oscillatory shifting of the river. East Delhi sits entirely within *khadar* land with overall low lands. In the western part of the city the transition from *khadar* to *bāngar* is determined by a three meters to four meters bluff roughly along a line from Okhla village, Humayun's Tomb, Red Fort around the northern spur of the Ridge towards Model Town and Narela in the north (DHAWALE 2010, GEOLOGICAL SURVEY OF INDIA no date). The younger alluvium of the *khadar* areas have a thickness of about 70 meters in the northern part to about 30–40 meters in the southern part of the floodplain (SHEKHAR & PRASAD 2009: 1557).

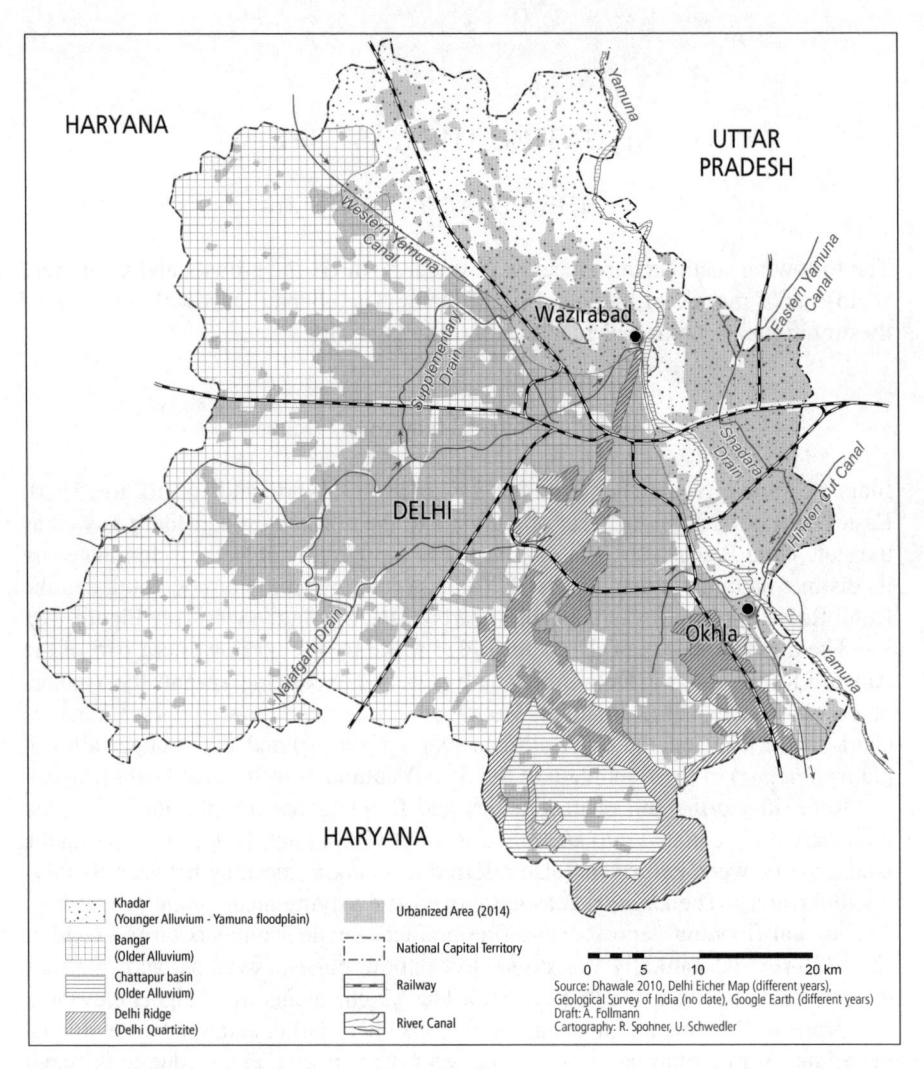

Figure 13: Overview of the physical features of Delhi

The Delhi Ridge, which reaches the height of 318 m above sea level (ASL), and the floodplain of the Yamuna determine the drainage pattern of Delhi. A number of perennial and non-perennial streams follow the natural gradient as local tributaries of the Yamuna. Aside from the Delhi Ridge the topography is gentle in slope oriented in the flowing direction of the Yamuna (ca. 210 to 195 meters ASL from north to south).

The city's subtropical climate is dominated by the Indian south-west monsoon. Delhi experiences an average annual rainfall of about 716 mm with great seasonal variability. The monsoonal months of July to September contribute to about 75%

of the annual rainfall whereas the rest of the year remains fairly dry (monthly rainfall of less than 21 mm). The mean daily maximum temperature vary between 39.8 °C in May and 20.8 °C in January (WORLD METEOROLOGICAL ORGANIZATION 2015).

6.2 The growth of the city

Delhi has historically been a center of power in South Asia. Due to the protection from the Ridge and the Yamuna, the location of the city has always been of strategic importance in terms of defense, logistics and economic opportunities in Northern India (MANN & SEHRAWAT 2009: 544). The city is referred to as the gateway to the Indo-Gangetic Plains. Delhi is an ancient city with a long history of continuous human settlement from the mythological *Indraprastha*[21] built on the banks of the Yamuna as the capital city of the kingdom led by the Pandavas in the Mahabharata epic, to the Mughal's Shahjahanabad, the Imperial Delhi of the British colonial times, to today's capital city of India (SINGH 1989). The ruins of the ancient cities are intrinsically a part of the cultural heritage of the city of Delhi (TRUMPP & KRAAS 2015).

The Mughal dynasty ended with the suppression of the Indian rebellion in 1857 and the city was put under direct control of the British crown. Before the proclamation of New Delhi as the new colonial capital, the territory of Delhi was part of the Punjab Province and known as the "Imperial Delhi Estate" (SRIRANGAN 1997: 95). The colonial period left its mark on the city's structure, including the older civil lines and cantonment areas along the Ridge north of Shahjahanabad (Old Delhi) and Lutyen's and Baker's new capital, 'New Delhi' south of the walled city. The colonial period has also reshaped Delhi's waterscapes and the city-river relationship (for details see 9.1).

After independence, the city of Delhi[22] experienced rapid urban growth and transformation. Following partition, Delhi had to accommodate waves of migration, resulting in a tremendous growth of the city's population from 0.95 million in 1947 to about 1.74 million in 1951 (JAIN 1990: 74). Further, partition resulted in a change of population structure with more than 300,000 Muslims leaving for Pakistan and half a million non-Muslims arriving in the city (LEGG 2006b: 194, PANDEY 2001: 122). The tents of the migrants developed into colonies, but basic civic services were inadequate. The state agencies in charge of urban development at that time, consisting of the existing municipalities as well as the Delhi Improvement Trust (DIT), failed to manage the growth of the city, predominantly due to lacking resources and public control over land (MANN 2006: 134, SRIRANGAN 1997: 95).

---

21   The authenticity of Indraprastha is still contested (SINGH 2006: xvii).
22   When talking about Delhi one tends to confuse the different administrative and political entities. In this study, if not explicitly mentioned otherwise (e.g. with regard to statistical data), Delhi always interchangeably refers to the city of Delhi (urban area) and the political entity of the National Capital Territory (NCT) of Delhi. For details regarding political and administrative set-up see chapter 8.2.

As a result the Delhi Development Act, the DIT was replaced by the DDA in 1957. The DDA prepared the first Master Plan for Delhi, which was sanctioned as the legal development plan for the city in September 1962. It was envisioned to guide the urban development of the city to the year 1981 with an estimated population by the end of the period of 5 million people. Its major goal was to "check the haphazard and unplanned growth of Delhi", which already materialized itself in "an estimated 50,000 dwelling units in bustis [slums] scattered all over the city" (DDA 1962) and about 110 so-called unauthorized colonies (JAIN 1990: 77).

Migration played a major role in Delhi's demographic growth and continuous to do so (DUPONT 2000, 2004, 2011). The population of the National Capital Territory (NCT) of Delhi grew by more than 50% each decade between 1951 and 1991 (CENSUS OF INDIA 2011a). While the percentage figures in decadal growth declined significantly, the NCT of Delhi is still growing by about 300,000 people each year in absolute numbers (see Table 3). Population estimation in the first Master Plan (5 million in 1981) was short by 1.2 million and the city's growth largely exceeded the proposed development schemes leading to an overburden of basic infrastructures and large-scale informal urbanization.

Year	Population (in million) NCT	West	East	Population Growth (in million) NCT	West	East	Decadal Growth (in %) NCT	West	East
1951	1.74	-	-	-	-	-	-	-	-
1961	2.66	2.49	0.17	0.9	-	-	52.4	-	-
1971	4.07	3.60	0.46	1.4	1.12	0.29	52.9	45.0	167.2
1981	6.22	5.11	1.11	2.1	1.51	0.64	53.0	41.8	139.8
1991	9.42	7.31	2.11	3.2	2.20	1.00	51.4	43.1	89.7
2001	13.85	10.61	3.23	4.4	3.31	1.12	47.0	45.2	53.3
2011	16.75	12.80	3.95	2.9	2.19	0.72	21.0	20.6	22.2

Table 3: Population growth NCT of Delhi 1951–2011
Source: own calculation based on CENSUS OF INDIA (2011a)
Note: East Delhi includes the census districts North East and East; West Delhi includes the districts North West, North, New Delhi, Central, West, South West and South.

Before 1950, the urbanized zones were concentrated around British New Delhi and Cantonment areas as well as the historic core of Shahjahanabad (Old Delhi) and its surroundings, whilst the spatial extension of the city increased enormously in the following decades (see Figure 14). Urban growth spread over the Yamuna, an area which was only sparsely populated and urbanized prior to the early 1960s. In 1961, the entire East Delhi had a population of only about 173,000. Large-scale relocations of slums from central parts of the city including the right banks of the river contributed to the fast growth of East Delhi (DUPONT 2004, TARLO 2000, 2003). The first Master Plan for Delhi envisioned development of East Delhi (still referred to as Shahdara at the time) for a population of about 700,000 to 750,000 in 1981, which underestimated the actual growth of East Delhi by about 50% (DDA 1962).

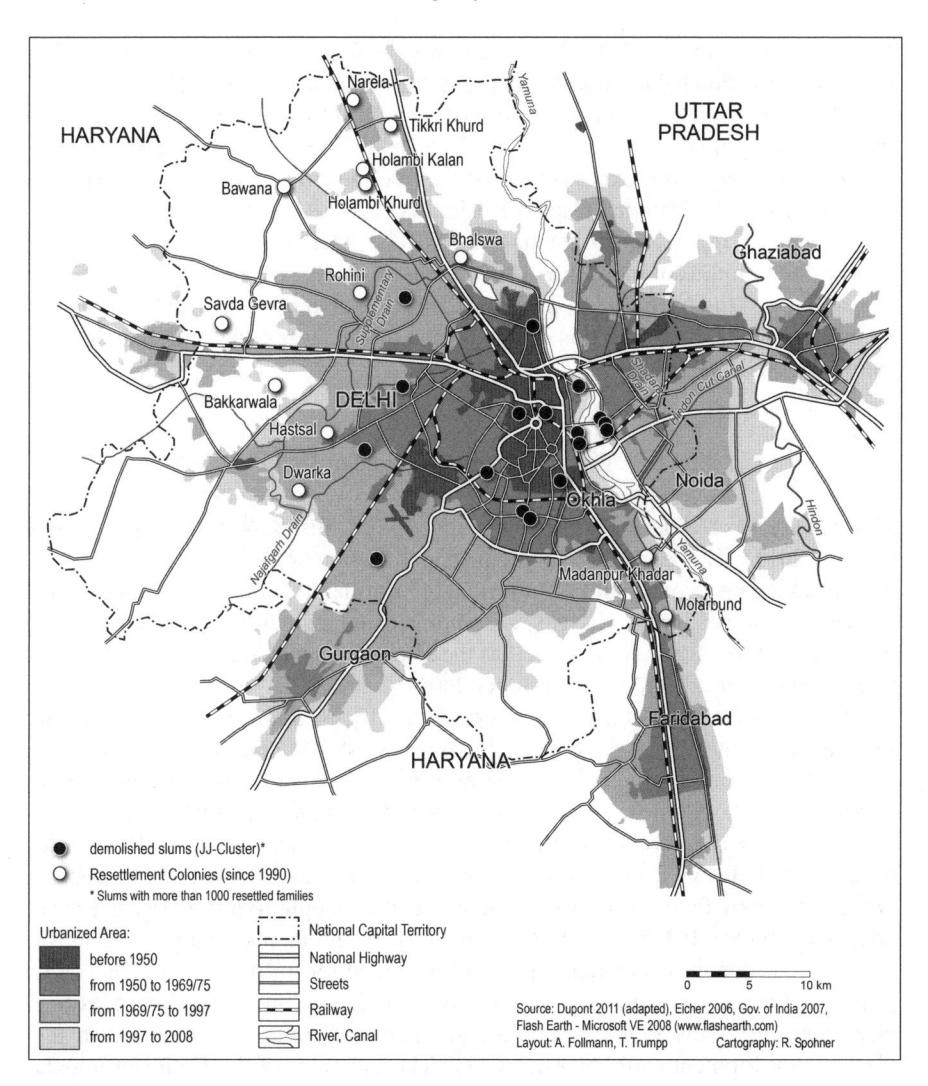

Figure 14: The growth of Delhi

In the last decade, the population growth in Delhi has been concentrated in the South West District (including the area around the international airport towards Gurgaon and the planned sub-city of Dwarka), the North West District (due to a large number of resettlement colonies) and the North East and South District (CENSUS OF INDIA 2011a). As Figure 14 shows, the multi-directional urban growth extended beyond the border of the NCT of Delhi "leading to the development of a multi-modal urban area" (DUPONT 2011: 6). In addition to the capital core, the conurbation comprises of a number of satellite cities in the adjoining states including

Noida and Ghaziabad in Uttar Pradesh as well as Gurgaon and Faridabad in Haryana with an estimated population of about 24 million people (DUPONT 2011: 7).

This ring of satellite cities accommodates much of the regional growth and even shows higher growth rates than the NCT of Delhi (MISTELBACHER 2005: 6). Regional growth, however, exceeds beyond the satellite cities (JAIN et al. 2013). As a result, by 1985, the National Capital Region (NCR) was constituted as an administrative unit on the regional/interstate level. The NCR encompasses an area of 34,144 km^2 (a radius of more than 100 km around Delhi). In 2011 about 46 million people lived in the NCR and population projection predicts a growth to more than 60 million by 2021 making this region to one of the largest metropolitan areas in the world (NCRPB 2013a: 32).

6.3 The socio-spatial mosaic of Delhi

Major parts of Delhi developed 'unplanned' (cf. BHAN & SHIVANAND 2013, DUPONT 2011). This resulted in a socio-spatial mosaic of different types of settlements making up a continuum from formal to informal including the DDA-planned residential colonies, urban villages, unauthorized and regularized unauthorized colonies, slum-designated areas, as well as the JJ clusters and JJ resettlement colonies (see among others BHAN 2013, DUPONT 2004).

Jhuggi jhopdi or JJ cluster[23] is a term generally used for the slums or squatter settlements of the urban poor in Delhi. In contrast, the commonly known term 'slum' refers from a legal perspective only to settlements, which are notified as 'slums' under the Slum Areas (Improvement and Clearance) Act from 1956. However, the last notification of a settlement as a 'slum' under this act in Delhi dates back to 1973 (BHAN 2009: 131). As a result of the lack of low-income housing (which the DDA failed to provide), the number of people living in JJ clusters increased continuously for decades. Consequently, the urban poor were forced to live in precarious conditions on land that was unclaimed for formal developments, due to its unattractive location, exposure to noise, pollution, and natural hazards (e.g. along highways, railway tracks, drains and the river Yamuna) or simply before any planned development could be established. The estimates of the JJ cluster population vary from about 2.1 million or 14.8% of the total population in 2000 (BHAN 2009: 132) to about 3 million or 27% of the total population in 1998 (DUPONT 2008: 81).[24]

The relocation of these JJ clusters from central areas has a long history. Large resettlement colonies were developed on the outskirts both in West and East Delhi in the 1960s and 1970s (DUPONT 2004). After the Emergency Period, slum demolition and resettlement was suspended until the 1990s. In the younger history of the city, more than 60,000 families were evicted and displaced (BHAN & SHIVANAND

---

23 The term JJ cluster is also referred to as 'jhuggi jhompri', 'jhuggi jhonpri', or 'jhuggi jhopri'.
24 JJ cluster surveys generally counted or estimated the number of jhuggis as units of households. Population estimates derive from an approximation of five members per household.

2013, DUPONT 2008, DUPONT & RAMANATHAN 2008). As shown in Figure 14, the most recent resettlement colonies are located at the urban fringe of Delhi in a distance of 30–40 kilometers away from their original location and the city center.

Even though the population figures are only a rough estimation, due to the evictions, the number of people living in JJ clusters is significantly declining since the last survey of JJ clusters in 1994 when 1,080 JJ cluster with more than 480,000 *jhuggis* (huts) were recorded and a population of 2.4 million was estimated (DUPONT 2008). 2014 data registered 672 JJ clusters with about 304,000 jhuggis and an estimated population of about 1.5 million (DUSIB 2014).

Not only the poor, but also a large proportion of Delhi's middle class and even rich population groups built their own colonies, non-conforming to the Master Plan, or simply settled on land which was outside the area of the Master Plan and in the time of development not yet notified as land for urban development (BHAN & SHIVANAND 2013). All these areas are known as unauthorized colonies, of which some have been regularized over time (BHAN 2013, DUPONT 2005b, ZIMMER 2012a). When aggregating these figures, less than one quarter of Delhi's population is living in planned colonies making Delhi largely an informal or even illegal city (cf. DATTA 2012).

Overall, Delhi's socio-spatial mosaic is complex. Besides the heterogenic mix of settlements, Delhi's socio-spatial characteristics reveal a major east-west contrast.[25] The Yamuna clearly defines a geographical, social and political boundary. East Delhi is also referred to as *Jamuna par* or *Trans-Yamuna*. The term *Trans-Yamuna* further reveals "the 'West-is-Centre' logic of the city" (TARLO 2000: 53).

Today, *Trans-Yamuna* is characterized by the highest population densities in Delhi (North East District: 37,346 people per km^2, East District: 26,683 people per km^2) (CENSUS OF INDIA 2011a) and is considerably poorer compared to the western parts of the city (MISTELBACHER 2005: 25). All larger green areas (excluding Delhi's riverscapes) are located in the western part of the city, which appears to be a rather green city. In sharp contrast, East Delhi is lacking adequate open green spaces. Additionally, East Delhi's urban areas directly merge with the surrounding neighborhoods of Noida and Ghaziabad, while a green belt still exists in the western part and separates it from its neighboring districts, even though the green space is encroached upon in many parts (JAIN & SIEDENTOP 2014: 35). East Delhi is further lagging in terms of centrality (e.g. all major institutional areas are located west of the Yamuna). This dichotomy of West and East Delhi needs to be taken into account because it influences the planning process and discourses also with regard to the Yamuna. Even if an increasing number of bridges and new metro lines connected both parts of the city, the river still acts as a major physical barrier between these two sides of the city.

---

25 Delhi also shows a north-south contrast (e.g. income disparities and infrastructure provision), which is less important for this study.

6.4 The 'world-class' agenda – Delhi as a global city

The "Dream of Delhi as a Global City" (DUPONT 2011) is emblematic for the "speculating on the next World City" in the Indian context and beyond (GOLDMAN 2011a). In the new millennium, the city has witnessed the development of new 'world-class' amenities like a high speed metro network, an expansion of the road transportation system with new fly-overs and elevated road corridors, the creation of shopping and urban entertainment complexes, and new integrated residential mega-projects. The drive for global recognition led to the "cleansing" of central parts of city from slums and other "undesirable elements" (BATRA & MEHRA 2008, DUPONT 2011).

A major boost for the world-class agenda in Delhi emanated from the award of the Commonwealth Games (CWGs) in 2003. The mega-sports event enhanced a dynamic process of urban restructuring and face-lifting across the city (BAVISKAR 2011b, UPPAL 2009, UPPAL & GHOSH 2006). The ambition to transform Delhi into a global city directly found its way into the capital's urban development policy, namely the Master Plan 2021 (MPD-2021). The Master Plan's vision is "to make Delhi a global metropolis and a world-class city, where all the people would be engaged in productive work with a better quality of life, living in a sustainable environment" (DDA 2007). The DDA declared that the MPD-2021 and "the planning process itself" are "the cornerstone for making Delhi a world-class city" by providing the major guidelines for reshaping the megacity in the beginning of the 21st century (DDA 2007). In doing so, the DDA aims to transform Delhi's physical urban landscape to a world-class standard – howsoever this standard is defined – and to customize the city in such a fashion that it attracts (foreign) business investments and economic elites (cf. DUPONT 2011, TRUELOVE & MAWDSLEY 2011).

Delhi's rising population and its vital economic growth had already resulted in increased development pressure on remaining 'vacant' land. The world-class agenda has certainly added to this pressure. Furthermore, in contrast to declaring land owned by the central government as 'development area' and from time to time developing it (as it had been done for decades by the DDA), the "centrally planned land management was replaced by the neoliberal mantra of public-private partnerships" (BAVISKAR 2007a). As in many other Asian cities (SHATKIN 2007), the increasing role of the private sector is manifested in the urban landscape of Delhi in the form of urban (infrastructure) mega-projects (BON 2015, FOLLMANN 2015, KENNEDY 2015). Thus, raising Delhi to global standards as quickly as possible resulted in an urban restructuring, where apparently underutilized or barren land became promising for future major urban developments. These former "unclaimed spaces" – urban niches often settled by the poor – became the "new enclosures" for a redevelopment of the city (BAVISKAR 2007a). Furthermore, Delhi's world-class aspiration is closely linked to city beautification movements, which are without doubt interconnected with the demolition of slums. The 'world-class' discourse, only briefly explained here, played a vital role in the transformation of the city in the new millennium and its narratives and logics traverse the analysis of Delhi's riverscapes in this study.

7 INTRODUCING DELHI'S RIVERSCAPES

The following sections introduce the river Yamuna focusing on the stretch of the river in Delhi. The chapter ends with outlining the major challenges for governing Delhi's riverscapes.

7.1 The river Yamuna

The Yamuna River originates from the glaciers of the Mussourie range of the lower Himalaya near Yamunotri in the district of Uttarkashi (Uttarakhand) at an elevation of about 6,320 meters ASL (RAI et al. 2012: 14). It flows for more than 1,300 km almost parallel to the Ganga before merging with the Ganga at the city of Allahabad in Uttar Pradesh at an elevation of around 100 meters ASL. The Yamuna is the largest tributary of the Ganga and of major mythological importance as one of the sacred rivers of India. As the daughter of Sury (the god of the sun) and the sister of Yama (the god of death) Yamuna is considered to be a holy river in Hindu mythology. Furthermore, the Yamuna is connected to Lord Krishna and his incarnation. The river has been worshipped "as an aquatic form of divinity for thousands of years" (HABERMAN 2006: 1).

The long history of continuous human settlement in the Delhi region has always been closely connected to the river Yamuna. Like the river Ganga and other sacred rivers in India, the Yamuna has been a "perennial source of inspiration for creativity" and a site of a "plethora of cultural activities and spiritual discourses" in the city of Delhi (JAIN 2011: 15–16). For Hindus, the river is a goddess and spiritually linked to Lord Krishna. In prayer rituals (*pooja*) they honor and worship the river. Many temples and ghats – the series of steps leading down to the river and the traditional bathing places – along its bank give utterance to the religious importance of the river. They are an expression of the deep-rooted and multi-faced connections between nature and society, the river and the city. The burning ghats (cremation grounds) further highlight the connection of the river to human life and death.

After winding its way from the Garhwal mountain range, which is the dominant mountain range of the central Himalayas separating the watershed of the Yamuna from the Ganga, the Yamuna makes its way through the foothills, called Shivaliks, before entering the Indo-Gangetic plains at the town of Dakpathar (Uttarakhand). The character of the Yamuna changes with its emergence in the Indo-Gangetic plains from a rather wild and comparably free-flowing river, which is eroding its bed into the mountains, to a highly regulated and managed river of the plains depositing its suspended load due to descending slope and velocity. Like other South Asian rivers originating from the Himalayas, the Yamuna carries a high amount of sediment, which accumulates along its banks creating alluvial lands.

At the town of Dakpathar, a first barrage across the river diverts water into a parallel canal for hydroelectric power production (run-of-the-river scheme). The foundation stone for this project was laid in 1948 by India's Prime Minister Jawaharlal Nehru. The project was delayed due to funding problems and disputes between the states of Punjab and Uttar Pradesh. It started operation in 1965 (HABERMAN 2006: 5). From Dakpathar, the Yamuna flows in south-west direction through the plains for about 200 km before entering the NCT of Delhi.

After meandering for about 48 km through Delhi from Palla in the north to Okhla in the south, the Yamuna forms the border between Haryana and Uttar Pradesh. About 270 km south of Delhi the river reaches the city of Agra flowing along the Taj Mahal. From there the river turns south-east until its confluence with the Ganga at Allahabad. On its way, the river flows through five states (Uttarakhand, Himachal Pradesh, Haryana, Uttar Pradesh, and the NCT of Delhi) and its entire basin with all its tributaries consists of a catchment area of 366,233 km² and covers parts of Rajasthan and Madhya Pradesh (see Table 4). Yamuna's catchment area makes up about 40% of the entire Ganga basin and covers about 10.7% of India's total landmass (CPCB 2006: 10). In 2001, a population of more than 127 million lived in the river's catchment area consisting of more than one eighths of India's total population (RAI et al. 2012: 176).

	Catchment Area*		Population of the Catchment Area (2001) **	
	km²	%	population	%
Uttarakhand	3,771	1.1	1,033,572	0.8
Uttar Pradesh	70,437	20.4	42,227,667	33.1
Himachal Pradesh	5,799	1.7	1,107,815	0.9
Haryana	21,265	6.1	14,733,152	11.6
Rajasthan	102,883	29.7	24,213,470	19.0
Madhya Pradesh	140,208	40.5	30,363,417	23.8
NCT of Delhi	1,485	0.4	13,844,958	10.9
Total*	**345,848**	**100**	**127,524,051**	**100**

Table 4: Area and population of the river basin
Source: *CPCB (2006), **RAI et al. (2012: 176) based on Census of India (2001), Ministry of Home Affairs, Government of India. Notes: ***The total catchment area is 366,233 km² including 20,375 km² water body of the river Yamuna.

In the Himalayan stretch, the Tons River is the major tributary carrying even more water than the Yamuna itself. In the plains, the most important tributary to the Yamuna is the Chambal River (CPCB 2006: 4). The Hindon River, originating also from the Shivalik hills, joins the Yamuna south of Delhi. In Delhi, the Sahibi River (commonly known as the Najafgarh Drain) originating from the district of Jaipur (Rajasthan) joins the Yamuna after flowing through major parts of West Delhi.

The Yamuna is a perennial river naturally fed by glaciers and precipitation. Influenced by the Indian south-west monsoon, the river's natural flow varies extremely. The annual rainfall in the basin varies spatially between 200 mm in the north-western parts of the catchment (Haryana) and 2,350 mm along the Shivalik range. The central parts of the catchment receive between 300 and 600 mm rainfall per year. Monsoon rainfall accounts for about 50% in the Himalayan parts of the basin and more than 80% in the plains. Therefore in the non-monsoon period the natural flow of the river is fed by glacier melt runoff, rainfalls in the Himalayan stretch and base flow recharge (RAI et al. 2012). The base flow recharge results from groundwater seeping into the riverbed. Accordingly, the river flow is highly variable during the course of the year and about 80% of the river's flow occurs during the months from July to September (CPCB 2010: 84). Notably, Yamuna's flow regime has been altered greatly through human intervention in the form of dams, barrages, embankments, canals and inter-basin water transfer (GOPAL & CHAUHAN 2007, RAI et al. 2012).

7.2 The river Yamuna in Delhi

As indicated in the introduction, the ecological importance of the Yamuna of Delhi is acknowledged by several environmental policies and legislations, but the river and its 'remaining' floodplain are not decisively protected by law. The defined River Zone (Zone O) comprises of about 97 km^2 and is the largest remaining natural feature of the megacity making up about 6.5% of the total area of the NCT of Delhi. The River Zone is an important groundwater recharge zone (CGWB 2012, CHATTERJEE et al. 2009, SHEKHAR & PRASAD 2009, SONI 2007, SONI & SINGH 2013). Despite its significance, the discharge of untreated sewage (CPCB 2006, RAI et al. 2012, SHARMA & KANSAL 2011a, 2011c) and the dumping of fly ash from coal-based power plants along the river (DUBEY et al. 2012, MEHRA et al. 1998, WALIA & MEHRA 1998b) as well as the dumping of debris and solid waste, has resulted in increasing levels of pollution of the river and contamination of its groundwater (SEHGAL et al. 2012, SEHGAL et al. 2014).

Taking a closer look at the morphology of Delhi's riverscapes, the river in Delhi can be divided in three major segments: a northern stretch up to the Wazirabad Barrage, a central stretch, and a southern stretch south of the Okhla Barrage. In the northern stretch the river flows through predominantly rural areas for about 21 km before entering into the dense urban area of Delhi at Wazirabad, where the waters of the river are impounded by the Wazirabad Barrage for the drinking water supply of the city. Embankments along the river in this stretch narrow the river to a width of two to 3.5 kilometers. The land use in this area is dominated by agriculture.

Towards the Wazirabad Barrage, the river is further restricted by forward embankments to secure urban areas. The width of the floodplain reduces to about one kilometer at the Wazirabad Barrage. The river water in this stretch is comparably clean, since it lies north of all major settlements and the drains from the city enter the river only south of the Wazirabad Barrage. The growth of unauthorized colonies and existing villages behind the embankments has resulted in major construction activities and the dumping of solid waste into existing wetlands in this stretch (documented by the author during field work, see also EXPERT COMMITTEE MoEF

(2014)). Furthermore, groundwater extraction through a series of wells is practiced around the village of Palla (information based on own fieldwork, for details see CHATTERJEE et al. 2009, SHEKHAR & PRASAD 2009).

The central stretch is about 22 km long and defined by the Wazirabad Barrage in the north and the Okhla Barrage in the south. The floodplain in this stretch is about 800m (at Old Railway Bridge) to 3 km wide, defined by a consistent embankment in the east and a series of embankments and the natural slope towards the west. The river in this stretch is surrounded by dense urban areas reaching up to the embankments. The land uses on the floodplain further include different human settlements (unauthorized colonies and urban villages), power stations, fly ash ponds, cremation grounds (burning ghats), bathing ghats, water works, and sewage treatment plants. In recent decades the floodplain has further been reduced by major development projects. Furthermore, due to an increasing number of bridges and their guide bunds the river's meandering process has been restricted. Today in the central stretch two railway bridges (including one rail-cum-road bridge), three Metro Bridges and five road bridges cross the river. The minimum river width bounded by these bridges varies between 455 m and 800 m. Due to the dynamic land-use changes in the central stretch in the last decades this area is chosen as the focus area of this study. The developments in this stretch will be discussed in detail in chapter 12.

The southern stretch of the river in the NCT of Delhi is about 5km long stretching from the Okhla Barrage to the Haryana Boarder. The width of the floodplain in this stretch is reduced to about one kilometer and defined by embankments. The river in this stretch constitutes the border between the NCT of Delhi and Uttar Pradesh. In Uttar Pradesh, the satellite city of Noida has grown up to the embankment. Extensive sand mining activities are found in this stretch. On Delhi's bank unauthorized colonies, urban villages and a resettlement colony (Madanpur Khadar) are located between the Agra Canal and the river. Furthermore, a large fly-ash dumping site connected to a coal-based power plant (Badapur Thermal Power Station) takes up about 300 ha in this area.

7.3 Water abstraction and flow

The abstraction of water for irrigation purposes has long been an important factor in the urbanization of the river Yamuna. Today more than 10.5 million ha of irrigated agricultural land are found in the Yamuna basin (RAI et al. 2012: 158) and about 94% of the water of the river is directed onto the agricultural fields; only 4% is designated for domestic use and 2% for industrial users (CPCB 2006: 14). Since the Green Revolution, after which cropping patterns and irrigation have significantly intensified in the region, the agricultural system of Haryana and Uttar Pradesh has been highly dependent on year-round water from the river. The Yamuna is Delhi's most important drinking water resource.

The first irrigation canal was constructed in the mid of the 14th century. This canal was later improved by Shah Jahan (early 17th century) and rebuilt in the colonial period (early 19th century). Most important for the urbanization of the river on the regional scale was, however, the construction of the Tajewala Barrage under British rule about 220 km upstream of Delhi. Figure 15 (page 128) gives an overview of the major river works (dams, barrages, canals) along the Yamuna up to Delhi. The Tajewala Barrage has been constructed to divert Yamuna's waters into the Western Yamuna Canal and Eastern Yamuna Canal. The Western Yamuna Canal still supplies water for irrigation purposes to Haryana and the Eastern Yamuna Canal to Uttar Pradesh. Delhi's share of Yamuna water is also transferred via the Western Yamuna Canal, which flows through the industrial towns of Yamuna Nagar, Karnal, Panipat and Sonepat. Delhi's water is discharged back into the river via Drain No. 2 about 80km upstream of Delhi. In 1999, the historic Tajewala Barrage was replaced by the Hathnikund Barrage, located about 3 km upstream. The barrage was constructed between 1996 and 1999 with financial assistance of the World Bank. It enables Haryana to withdraw 50% additional water from the river (WAPCOS 2003). Decadal average flow in the river downstream of Tajewala/Hathnikund has decreased accordingly (PANWAR 2009: 11). In the non-monsoon season hardly any water is released at Hathnikund resulting in major dry stretches of the river downstream of the barrage (CPCB 2006: 4, PEACE INSTITUTE CHARITABLE TRUST 2010: 38).

Interstate water disputes are omnipresent in the Yamuna basin today. In 1933 the former Government of the United Provinces (today Uttar Pradesh) complained about an increasing withdrawal of water from the Yamuna by the city of Delhi, arguing that it was having adverse impacts on the already insufficient water availability for irrigation of lands along the Agra Canal (SHARAN 2014: 44). Since 1954 a water sharing agreement allocated the flow of the Yamuna to the states of Punjab (later Haryana) and Uttar Pradesh. This agreement was based on "the contributing area of each state to the basin" (RAI et al. 2012: 18) and, thus, did not take into account the existing population or even population growth. Based on this agreement Delhi had no formal share and was therefore dependent on Haryana and the Bhakra-Beas Management Board (BBMB) for raw water supply.

In the early 1990s, Delhi approached the BBMB to allocate more water to Delhi. This move resulted in a Memorandum of Understanding (MoU), which was signed 1994 between Himachal Pradesh, Haryana, Uttar Pradesh, Rajasthan and Delhi regulating the allocation of flows of the river up to and including the Okhla Barrage in South Delhi. The Upper Yamuna River Board (UYRB) was constituted as a subordinate office under the Ministry of Water Resources (MoWR) to oversee the allocation of available flows, to monitor the return flows, to monitor the quality of surface and groundwater, and to review the overall watershed management in the basin. Delhi's annual share of Yamuna water is about 6% (0.724 BCM), but corresponding with the natural run-off regime, about 80% is allocated in the monsoon season between July to October.

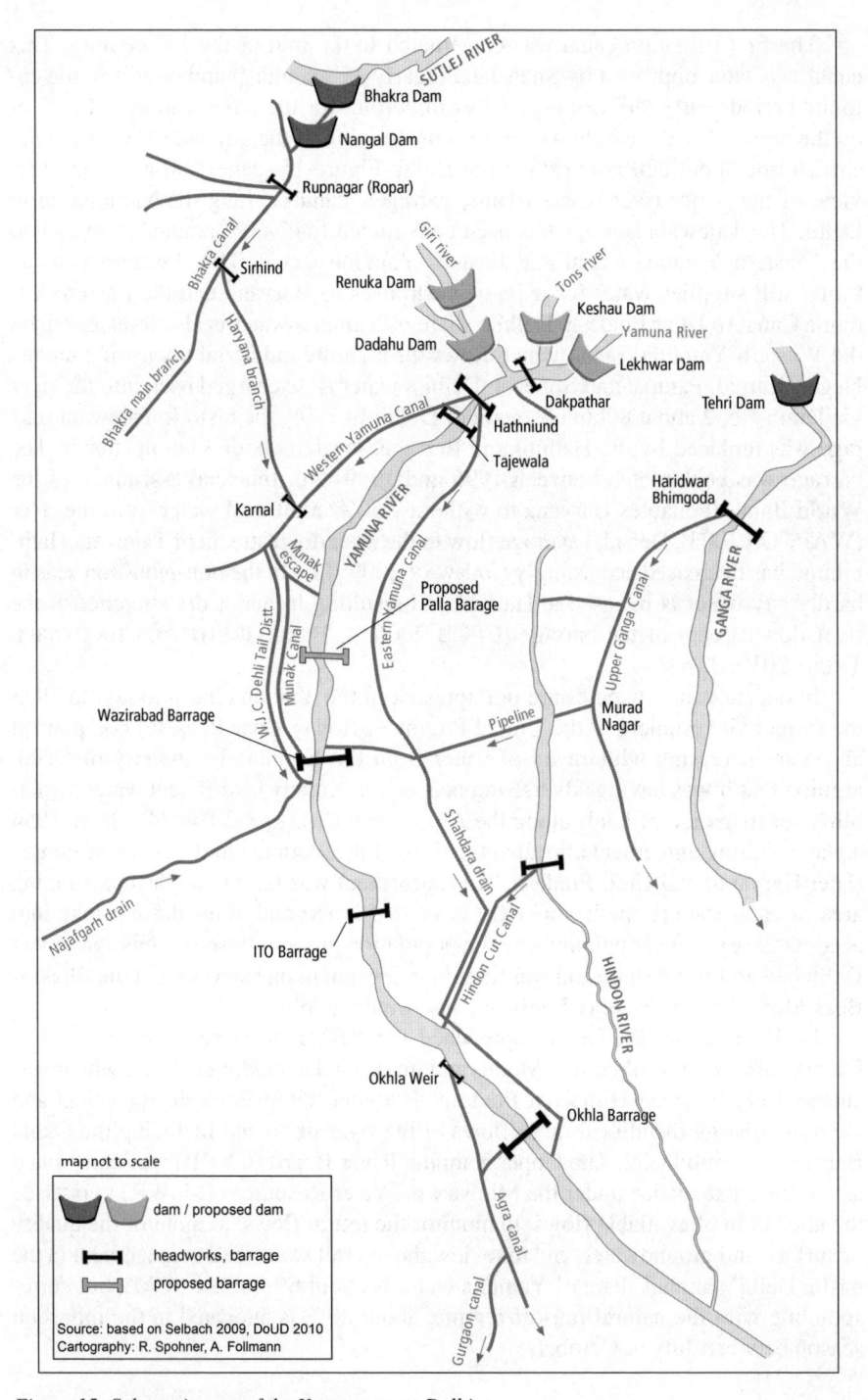

Figure 15: Schematic map of the Yamuna up to Delhi

About 350 million cubic meters water are extracted per day for the water supply of the city at Wazirabad (PLANNING COMMISSION 2009: 216) and hardly any water is released during the dry season at Wazirabad Barrage (CPCB 2006: 9). Further, the city today sources water from the Ganga. Therefore the water supply system of the megacity of Delhi is derived today from three sources outside the city including the Yamuna River (via WYC), Sutlej River (from Bhakra Dam via WYC) and the Ganga River (from the Tehri Dam via the Upper Ganga Canal and pipeline).

The surface waters of the Yamuna today make up about 40% of Delhi's water supply. Additionally, about 10% of Delhi's water is sourced from groundwater (DJB 2014a). In sum, Delhi's waterscape is supported by a large network of dams, reservoirs and canals, which reflects the urbanization of nature on the regional scale.

7.4 Pollution and environmental degradation

"Myth has it that the river Yamuna has been named after Yama, the god of death in the Indian pantheon. With alarming levels of pesticides, heavy metals and definite accumulations of carcinogenic chemicals [...], mythology could well have a touch of prophecy here" (BANERJI & MARTIN 1997: 48).

Due to high levels of pollution, particularly in the Delhi stretch and downstream of India's capital city, the Yamuna River has been referred to as a dying or even dead river (cf. HABERMAN 2006). Both daily newspaper articles as well as reports by environmental NGOs (BANERJI & MARTIN 1997, CSE 1999, 2007, 2012, JAIN 2009b) have highlighted that the river is not only dead but is also causing death due to high levels of pollution. Interestingly, as outlined by HABERMAN (2006), the Yamuna traditionally has not been viewed like this but rather in the opposite. The Yamuna was seen "as a nurturing and life-enhancing goddess" and the "radical change in Yamuna's theological portrait", is a rather recent reinterpretation by environmentalists and the media in the last decades (HABERMAN 2006: 76).

The question of river pollution is complex due to the multiplicity of sources of wastewater (point source/non-point sources) and a range of factors influencing the self-cleaning capacity of rivers (e.g. flow, siltation, oxygen levels, etc.). Further, pollution levels in monsoonal rivers like the Yamuna vary especially between monsoon and non-monsoon period due to completely different runoff characteristics and levels of dilution (CPCB 2006, GOPAL & CHAUHAN 2007, MANDAL et al. 2010, PARMAR & KESHARI 2012, SHARMA & KANSAL 2011c).

Water quality management in India for surface waters follows the concept of "designated best use" (DBU) developed by the Central Pollution Control Board (CPCB). CPCB's "designated best use" for surface waters in India includes the following five classes (CPCB 2010):

A) Drinking water (without conventional treatment but after disinfection);
B) Outdoor bathing;
C) Drinking water (after conventional treatment and disinfection);

D) Propagation of wildlife and fisheries; and
E) Irrigation, industrial cooling and controlled waste disposal.

The classification A to C are suitable for drinking water (with or without conventional treatment) and the lowest water quality (Class E) is only suitable for irrigation, industrial cooling and waste disposal uses (CPCB 2010). The observed water quality at Wazirabad is D (desired C) and after passing through Delhi the river's water at Okhla is classified as E (desired C). Based on the CPCB's best-use classification, the status of the river water emerges to be "much below the desired water quality" (GOVT. OF U.P. 2013: 166).

About 85% of the total pollution load of the Yamuna River comprises of domestic sources, containing mostly organic materials (CPCB 2006: 16). Delhi is the major polluter of the river Yamuna; almost 80% of the pollution load of the river Yamuna is generated by the city of Delhi (CPCB 2006: 19). Since almost no water is released from Wazirabad Barrage during the dry season, whatever water flows downstream of the barrage is untreated or partially treated sewage. Similarly, whatever water makes up the river downstream of Okhla Barrage is wastewater generated from East Delhi and Noida.

Besides domestic pollution, industrial pollution is responsible for high levels of toxic materials of various forms. Industrial clusters upstream of Delhi are found in Yamuna Nagar, Panipat, Sonepat (all Haryana), Baghpat (Uttar Pradesh) and in the downstream areas of Ghaziabad, Gautam Budha Nagar, Faridabad including a variety of polluting industries from pulp and paper, sugar, distilleries, textiles, leather, chemical, pharmaceuticals, oil refineries, thermal power plants, and food processing etc. (CPCB 2006: 19). Compared to Delhi the upstream cities contribute only marginal to the total pollution load of the river Yamuna: Panipat 3%, Sonepat 2% and Baghpat 2% (CPCB 2006: 19).

Upstream of Delhi the river Yamuna is comparatively clean, even when the findings of the CPCB show a deterioration of water quality (1999–2005) due to an increasing number of coliform bacteria and ammonia concentration (CPCB 2006). The entire domestic and industrial sewage of the city of Yamuna Nagar (200,000 inhabitants, about 150km north of Delhi) is though discharged into the Western Yamuna Canal and, thus, the canal and not the river gets directly affected (CSE 2007: 38). This is even more problematic since Delhi's freshwater comes from the canal. With population growth and industrial development in the cities upstream of Delhi (e.g. Yamuna Nagar, Karnal, Sonipat, and Panipat) pollution levels might further rise if treatment capacities for domestic and industrial wastewater continue to be inadequate in this cities (CSE 2007: 47).

Due to the release of high quantities of wastewater from industries at certain times (e.g. distilleries), incidents of higher pollution levels have led to massive problems and the temporary shutdown of water treatment plants in Delhi. Industries located in Panipat and Yamuna Nagar are not equipped with sufficient treatment capacity (CPCB/2011-11-16, SCIENTIST-05/2011-11-25). Additionally, diffused sources (non-point sources) of pollution affecting the river include residues from agricultural use of fertilizers and pesticides, the dumping of solid waste, effects of religious and cultural practices (e.g. immersion of idols during *pooja*, dumping of dead bodies), and other direct water uses like bathing and cloth washing (*dhobis*).

Delhi has been largely unable to overcome the discrepancy between sewage generation (about 680 MGD in 2011) and existing sewage treatment capacity (about 595 MGD in 2011) (DJB 2014b: 14–16). The 34 existing sewage treatment plans (STPs) in Delhi are up to 45 years old and run at a utilization of only around 57% (DJB 2014b: 15). This underutilization of STPs results from an insufficient and malfunctioning sewage network characterized by blogged and leaking pipes and silted trunk sewer lines as well as technical problems within the STPs. Approximately 50% of NCT of Delhi is not connected to any sewage system (DJB 2014b: 6).

In the case of Delhi, the upstream-downstream relationship in the regional water governance set-up is ambivalent, since both neighboring states (Haryana and Uttar Pradesh) represent the interests of the upstream and downstream. Being unable to resolve its own sewage problem – which results in high levels of water pollution in the river and the linked irrigation canals downstream of Delhi (see Figure 16 and Figure 17) – Delhi faces major difficulties in demanding unpolluted fresh water from these two states.

Figure 16: Discharge of polluted water at Okhla Barrage, South Delhi
Source: Follmann, March 2011

Figure 17: Polluted water for irrigation, Agra Canal headworks, South Delhi
Source: Follmann, November 2011

The floodplain of the Yamuna in Delhi is further contaminated by arsenic residues from the discharge of coal-based power plants at Rajghat, Indraprastha (closed in 2010) and Badapur (DUBEY et al. 2012). The fly ash from the power stations is disposed by mixing it with water before pumping it into open fly-ash disposal ponds along the river. The overflow of these ponds is discharged into the river since more than 40 years. These fly-ash ponds have been often leaking and a mix of water and toxic fly-ash is discharged directly into the river (CPCB/2011-11-16, SCIENTIST-05/2011-11-25). DUBEY et al. (2012) highlight that „more than 55% of the samples have concentrations above the WHO limit". The contamination of groundwater in the floodplain was also highlighted in a study of the National Environmental Engineering Research Institute (NEERI) in 2005 (NEERI 2005: 3.26). An earlier analysis of metal concentrations in river bank soils and plants was completed in 1989 (FARAGO et al. 1989). In this study, FARAGO et al. (1989) estimated that about 5.5 tons of arsenic materials was discharged into river from the Rajghat power plant per year (ibid.). With regard to human health the study found that stomach, gastric and gastrointestinal problems were evident in the population dependent on hand pump water along the river (ibid.). Therefore, these environmental problems and their negative effects on human health have been known for a long time.

Major investments have been made to rejuvenate the river under Yamuna Action Plan (YAP), a bilateral river restoration program between the Governments of India and Japan. After the start of the Ganga Action Plan (GAP) in 1986, YAP followed in 1993. In Delhi, YAP has focused on the setting up of low-cost toilets, the construction of STPs and sewer lines, as well as the promotion of electric crematoria and bathing ghats (DDA 2010b: 5). To date, YAP has largely failed to clean up the Yamuna (CPCB/2011-11-16, SCIENTIST-05/2011-11-25, cf. CSE 2007, 2012; SHARMA & KANSAL 2011b).

YAP (and GAP) have been criticized for having disregarded the complex ecological factors and solely concentrated on large-scale infrastructure investments especially in STPs. Further, major operation and maintenance problems of these new infrastructures have been documented (GOPAL et al. 2002: 8). The current third phase of YAP (YAP III) is scheduled for completion by December 2018 and proposes interceptor drains, the rehabilitation and modernization of existing STPs, the construction of a large new STP at Okhla, the rehabilitation of sewer lines, and public awareness programs. Three interceptor sewers along Delhi's major drains (Najafgarh Drain, Supplementary Drain and Shahdara Drain) are under construction to intercept the domestic wastewater (especially of the unsewered areas) and divert it to STPs (for details see Engineers of India & CH2M Hill (India) PVT LTD 2008). The Central Government has approved a total cost of 16,560 million Indian Rupees (ca. 240 million Euro) based on a loan assistance from the Japan International Cooperation Agency (JICA) for YAP III in Delhi (MoEF 2014d). About 85% of the funds come from the Central Government (ibid.). The interceptor project has been sanctioned in 2010, but it remains questionable whether the interceptor project will be successful in cleaning the Yamuna (for a detailed critique see CSE 2009).

By concentrating exclusively on the cleaning of the Yamuna, as rightly argued by ZIMMER (2012b: 105), the interceptor scheme "abandons the idea of access to sanitation for inhabitants of Delhi's informal settlements completely". In doing so, the conducted schemes under YAP have almost entirely focused on the water quality in the river, disregarding the wastewater problems in large parts of the city. Thus, cleaning the river dominates the discourse on wastewater in Delhi and, thus, action focused largely on the water quality at the end of the 22 major drains (originally storm water drains or natural tributaries of the river). Delhi's storm water drains carry domestic wastewater and storm water, resulting in contamination of the groundwater and numerous health problems across the city.

7.5 The risk of monsoonal flooding

Delhi has experienced a series of floods in history. Despite these reoccurring events, a detailed assessment of the flood risks remains difficult due to the non-availability of long term data and a non-existent detailed flood risk zoning. Nevertheless, the IPCC (2014: 1346) has noted a "high risk of floods" for Delhi. The following section gives an overview of the risk of flooding associated with the river Yamuna in Delhi.

In the dry season the general water level at the Old Railway Bridge[26] is about 202m ASL. In the monsoon season the Yamuna recurrently crosses the danger level of 204.83m ASL (672 feet ASL). This mark was set in the 1950s and is still used as a reference point for flood warnings and documentation today. The Irrigation and Flood Control Department (IFCD) further distinguishes between low (< 204.22m), medium (between 204.22m and 205.44m) and high (> 205.44m) floods (IFCD 2010). Since 1963, the Yamuna has exceeded the danger level at least once a year in more than three quarters of the time (see Figure 18).

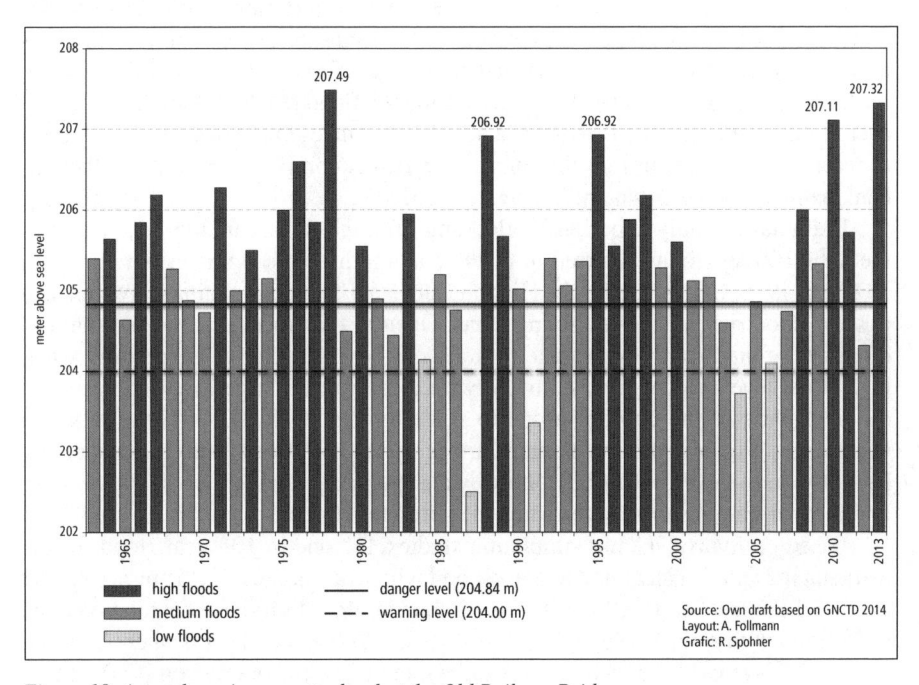

Figure 18: Annual maximum water level at the Old Railway Bridge

With the onset of the monsoon season, the water level in the Yamuna begins to rise in most years in the beginning of July and three to four flood events can be expected

---

26 Flood data in Delhi is collected at several control points. Major reference is given to the water level at the Old Railway Bridge.

in the following months until the end of the monsoon (CWPRS 2007: 3). The highest recorded level in available data was in 1978 when the Yamuna reached 207.49 m, with a discharge of 253,350 cusecs (7,174 m^3/s) at the Old Railway Bridge. The 1978 flood was caused by a major monsoonal storm with heavy rainfall in the catchment area of the Yamuna upstream of Delhi and in Delhi itself. This flood resulted in inundation of urban areas in Delhi, mainly in the north-western part of the city, where a then new embankment breached.

Reliable data is not available prior to 1963, but serious floods were also recorded in 1924, 1933, 1947, 1955 and 1956 (IFCD 2010, SHARAN 2015). The floods of 1924 were caused by heavy monsoonal rainfalls in the catchment area, which considerably exceeded the rainfall of 1978, Subsequently, the discharge in 1924 was estimated to have been higher than in 1978 (YSC/37/1979-04-26).

A major factor when studying the flood intensity in Delhi is the water released at Hathnikund Barrage. It is, however, important to note that because of the absence of any storage capacity at Hathnikund Barrage (like the older Tajewala headworks), the barrage is not meant for flood protection, but only for the diversion of flows to the adjoining canals.[27] The travel time of the water from Hathnikund to Delhi varies between 48 to 72 hours (IFCD/2010-09-27). The timespan is depending on the condition of the river whether it is the first flood or a subsequent flood of the year. In the latter scenario, the riverbed is saturated and the flood reaches Delhi faster. However, if the river is already flowing at a high level, it takes longer for peak discharges to reach Delhi and, therefore the forecast system is complex. Breaches in the embankments can slow down and moderate the floods.

Delhi has not only experienced flooding from the river Yamuna, but also from the Sahibi River (today referred to as the Najafgarh Drain) with serious floods in 1977 and 1978. In East Delhi, the Hindon Cut Canal/Shahdara drainage system also poses a flood risk for the low-lying areas. Overall, the "local drainage system has often been found to be inadequate to drain out heavy discharge" especially when high flood levels in the river coincide with local rainfall large low-lying areas in different parts of the city are prone to waterlogging (GNCTD 2014). The storm water drains flowing into the Yamuna are equipped with flood gates to prevent any backflow in the event of floods. Storm water pumping stations are set up at these drains to pump out storm water when the gates are closed.

Based on hydrodynamic simulation studies, the one in 100 year flood for the Yamuna in Delhi is calculated to correspond with a discharge of 9,111m^3/s (NEERI 2005: 3.10, VIJAY et al. 2007: 383; see Table 5). Based on these calculations, the 1978 flood with a discharge of 7,767m^3/s has to be considered somewhere between a once in 50-years flood (8,077m^3/s) and a one in 25 year flood event (7,034m^3/s) (NEERI 2005: 3.10). Whilst only limited data is available, the existing data suggests that Delhi has not seen a one in 100 year flood or even a on in 50 year flood in its more recent history.

---

27 The storage capacity is the major difference between a dam and a barrage.

Return period / flood event	Flood discharge at Old Railway Bridge (in cumecs, m³/s)	Maximum water level Old Railway Bridge (in m ASL)	Corresponding discharge at Tajewala/Hathnikund (in cumecs, m³/s)
1 in 10 years	5,629	-	-
1 in 25 years	7,034	-	-
1 in 50 years	8,077	-	-
1 in 100 years	9,111	-	-
6th September 1978	7,767	207.49	19,822*
27th September 1988	6,000	206.92	16,353
8th September 1995	7,028	206.93	15,183
22nd September 2010	-	207.11	21,082
19th June 2013	-	207.32	22,800

Table 5: Flood discharge and maximum water level during major flood events
Source: VIJAY et al. (2007: 383), NEERI (2005)
Note: * official discharge, estimated discharge ranged from 18,000 to 26,400 m³/s.

Floods in Delhi, as in the India in general, have been mainly understood as a technical problem, which can be 'controlled' through the adequate construction and good maintenance of embankments and a properly functioning urban drainage system. Several flood protection measures have been implemented to safeguard the low-lying areas of Delhi in the event of monsoonal flooding (see detailed discussion in the following chapters). Available data as well as the narratives of people working on the river indicate that any estimation of a 100 year flood event – and even larger flood events are not unrealistic – is vague, but might have devastating consequences for the city of Delhi.

In conclusion, the megacity of Delhi faces considerable flood risk from the river Yamuna. The detailed analysis provided in the following chapters will highlight that the risk of monsoonal flooding and the complex topic of flood protection has played a major role in the planning and development of the river. Scientific studies and their interpretation by policy makers and the courts will highlight the multiple connections between ecology, science and politics.

7.6 Interim conclusion: major challenges

As introduced above, besides questions of land-use change three major aspects (water abstraction and flow, river pollution, and the risk of flooding) need to be taken into account when discussing urban environmental change of Delhi's riverscapes. These three topics have dominated the discourses around the Yamuna in Delhi as well as on the interstate level for a long time. The huge abstraction of surface flow from the river further affects the groundwater situation. Groundwater levels in the basin area and particularly in Delhi are declining. Nowadays, they fail to support the base flow of the river. On the contrary, much of the released water from the

Hathnikund Barrage percolates into the groundwater aquifers and does not materialize as flow in the river (YFW-2013-03-08). Therefore the abstraction of water further reduces the groundwater recharge capacity. Vice versa, the exploitation of groundwater influences the (minimum) flow of the river, when the soil is literally absorbing the surface water. The water shortage discourse is therefore very prominent in Delhi and supply-side solutions are favored over water conservation measures (MARIA 2006, 2008).

Flow and pollution levels are also mutually connected. Absent flow reduces the self-cleaning capacity of the river ecosystem; an aspect which is largely discussed when referring to missing dilution (a common story-line and excuse of the agencies responsible for sewage treatment in Delhi).

Pollution issues remain unsolved on the city and regional level. YAP has failed so far to address the pollution issue and ongoing infrastructure developments remain focused on large-scale sewage treatment facilities (STPs and interceptor drains). The contamination of the water and the land (soil) in turn also result in the contamination of agricultural products from the floodplain, putting further pressure on the state agencies to act with regard to human health (SEHGAL et al. 2012, SEHGAL et al. 2014). Moreover, high-levels of environmental degradation resulted in a very bad reputation of Delhi's riverscapes among large sections of the population. Due to increasing levels of pollution, the use of riverfront areas for leisure activities has considerably decreased. This resulted in a further marginalization of Delhi's riverscapes. High levels of pollution on one hand result in changing land uses, but on the other also request changes in land-use policies. The mutual connection between land-use policies and flood protection is obvious. The risk of monsoonal flooding in Delhi is enhanced by the construction of embankments in the upstream and the development activities on the floodplain in Delhi. Questions of flood protection (e.g. through floodplain zoning) are discussed in detail in section 8.6.

These major challenges for governing Delhi's riverscapes call for a linking of scales, since solutions for all three aspects are matters to be solved on the interstate level. Yet, water abstraction, (minimum) flow and flood risk management remain major interstate conflicts. The growth of Delhi has further increased the pressure on land and water. The protection of the floodplain in Delhi is therefore a major urban environmental challenge for the future development of the city and the region. The following chapter introduces the involved actors.

8 URBAN ENVIRONMENTAL GOVERNANCE: INTRODUCING ACTORS AND POLICY DEBATES

Due to the status of Delhi as the National Capital Territory with its legal status as a Union Territory, the government framework for Delhi is complex. Therefore it is suitable to shortly review the Indian political system based on the constitutional provision in order to then outline the different levels involved in urban environmental governance.

8.1 Political and administrative organization in India

India is a federal parliamentary democratic republic consisting of 29 states and seven Union Territories. Governmental powers are shared between the Central Government, the States and the local bodies (municipalities/panchayats). The sharing of governmental powers and responsibilities between the Central and the State governments is outlined in Article 246 of the Indian Constitution.

The *Union List* (List I) gives exclusive powers to the Centre to legislate 100 subjects including defense, foreign affairs, air traffic, railways, national highways, national waterways, and interstate rivers. The legislative powers and responsibilities of the State governments are outlined in the *State List* (List II) including 61 (previously 66) subjects, such as public health, sanitation, agriculture, land and water. Water, based on entry 17 of the State List, includes water supplies, irrigation, canals, and embankments. Water as a state subject is subjected to the provision of entry 56 of the Union List regarding the governance of interstate rivers. Furthermore, 52 subjects with overlapping and shared jurisdiction are covered in the *Concurrent List* (List III), which includes forests and wildlife protection, economic, and social planning, among others.

With regard to urban environmental governance the legislative powers of the Centre are therefore limited since the states have control over land and main tasks of urban and regional planning are generally concerned with land. The 74th Amendment Act has provided the states with powers to devolve urban planning to the urban local bodies. The Centre also has certain powers to "evolve policies, guidelines and model laws for adaption by the states" based on entry 20 of the Concurrent List dealing with economic and social planning (KULSHRESTHA 2012: 27).

For half a century the Planning Commission has been the key institution in India's centralized planning system. The Planning Commission used to be the main advisory body in the Government of India (GoI) formulating India's Five-Year Plans including economic planning and overarching policy formulation.[28]

---

[28] Prime Minister Narendra Modi announced the replacement of the Planning Commission in the so-called National Institution for Transforming India Aayog in 2014.

8.2 Delhi's administrative history and current set-up

Delhi's administrative history is complex. Delhi became part of the British Territory (North-Western Provinces) after the victory of the British East India Company over the Maratha Confederacy in the Battle of Delhi at *Patparganj* (located across the Yamuna opposite of Humayun's Tomb) in 1803. After the annexation of the Punjab in 1849, the British formed the Punjab Province to the west of the Yamuna stretching up to the Indus (MANN 2005: 163). Following the suppression of the rebellion in 1857, Delhi was added to the Punjab Province and ruled by the Lieutenant Governor (LG). In connection to the shift of the capital from Calcutta to Delhi in 1911, Delhi was then transferred into a Chief Commissioner's Province. Further 65 villages of the *United Provinces of Agra and Oudh*[29] located across the Yamuna became part of Imperial Delhi (SIDDIQUI et al. 2004: 199). Despite the Government of India Act 1919, the Punjab Government had dominated in governing Delhi until Independence.

In 1947 Delhi came under direct control of the Government of India (ibid.). For a short period from 1952 to 1958 Delhi was a self-governing state with a legislative assembly (also known as Delhi Vidhan Sabha). Following the States Reorganisation Act 1956, Delhi became a Union Territory and the legislative assembly was abolished. The legislative assembly was only re-established in 1993, after the enactment of the 69th Amendment Act declaring the National Capital Territory (NCT) of Delhi. The NCT of Delhi since then has an elected government (Government of the National Capital Territory, GNCTD[30]) with a legislative assembly of 70 members headed by a Chief Minister. However, the NCT of Delhi is a *quasi-state*, since several governing powers remain under direct administration of the Central Government controlling among others land, urban planning and police. Further, the President of India nominates the Lieutenant Governor (LG) as the constitutional head of the GNCTD (see Figure 19).

The city of Delhi is divided into five local authorities: the North Delhi Municipal Corporation, South Delhi Municipal Corporation, East Delhi Municipal Corporation, New Delhi Municipal Corporation (NDMC), and the Delhi Cantonment Board (DCB). The first three local authorities used to be part of the Municipal Corporation of Delhi (MCD) until 2012 when the MCD was divided into three smaller municipal corporations.[31]

---

29 The United Provinces of Agra and Oudh were 1935 renamed in United Provinces. After independence the United Provinces became a state and were renamed into Uttar Pradesh (Northern Province) in 1950.
30 In this book also referred to as the Delhi Government.
31 Since the trifurcation of the MCD took place during the course of this study, the study refers in several parts to the MCD. The trifurcation debate was centred on notions of efficiency, but was mostly backed by political interests and the tension between the Central Government and Delhi Government. While the Congress Party has been ruling on the state level between 1998 and 2012, the MCD was mostly ruled by the Bharatiya Janata Party (BJP), which was in power from 1993 to 2002 and again from 2007.

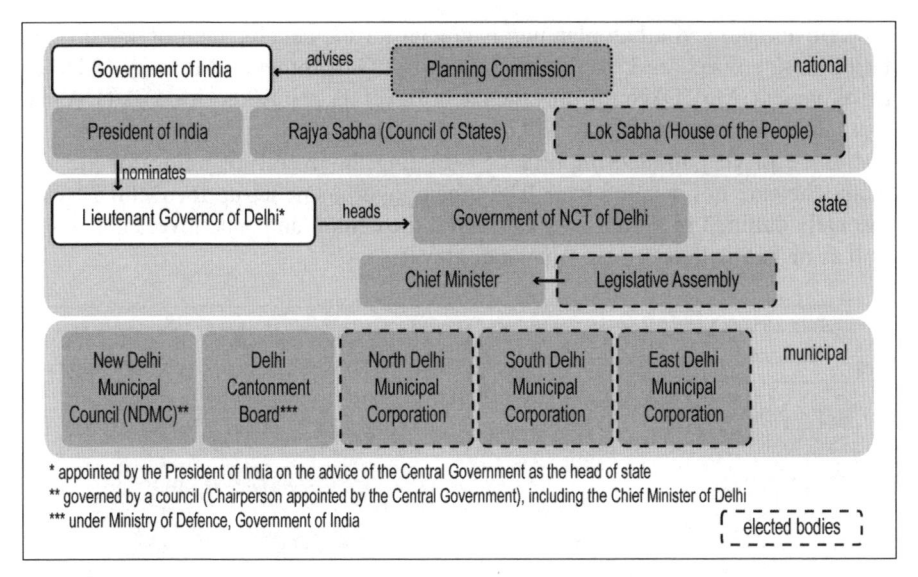

Figure 19: Government structure of the NCT of Delhi
Source: Graphic by author

The former MCD was created as a statutory body under the Delhi Municipal Corporation Act enacted by the Parliament in 1957. Based on the Act, the MCD was in charge of basic service provisions including water supply, electricity and public transport (SIDDIQUI et al. 2004: 199). In the course of the establishment of the NCT of Delhi, basic tasks were removed from the responsibility of the MCD including water supply, sewage, electricity and transport and moved to the state level. After the passing of the 74th Constitutional Amendment Act (1992) the responsibilities have remained largely the same. Indeed, Delhi has emerged as one of the cities where the designated reforms have been reduced to a minimum (cf. DEWIT et al. 2008, TAWA LAMA-REWAL & GHOSH 2005).

In sum, the megacity of Delhi is jointly governed by all three levels: the Central Government, the GNCTD and the five local bodies. However, the sharing of powers between the different levels is highly conflictual. Major conflicts were, for example, evident when the Aam Aadmi Party came to power in 2014 and openly protested against the influence of the Central Government on Delhi Police. The ongoing battles between the Centre and the Delhi Government for the control over land and urban planning further complicate the relations between the two levels.

8.3 Urban governance framework: the state actors

The underlying structure of the urban governance framework in Delhi was shaped in the colonial period, producing post-independence legacies in governing the capital city (JAIN 1989, 1990, LEGG 2006a, 2007, SIDDIQUI 2004). Without doubt, the power over land and urban planning is crucial for the development of any city. In Delhi these powers are vested with the Central Government's Ministry of Urban Development (MoUD) controlling the Delhi Development Authority (DDA). A major focus of this study therefore will lie on the DDA. In contrast, since the municipal bodies are of very limited importance regarding the overall planning for the Yamuna, they are not detailed here. The urban government set-up for Delhi is schematically outlined in Figure 20 focusing on the central and state level agencies as well as including the regional (interstate) level.

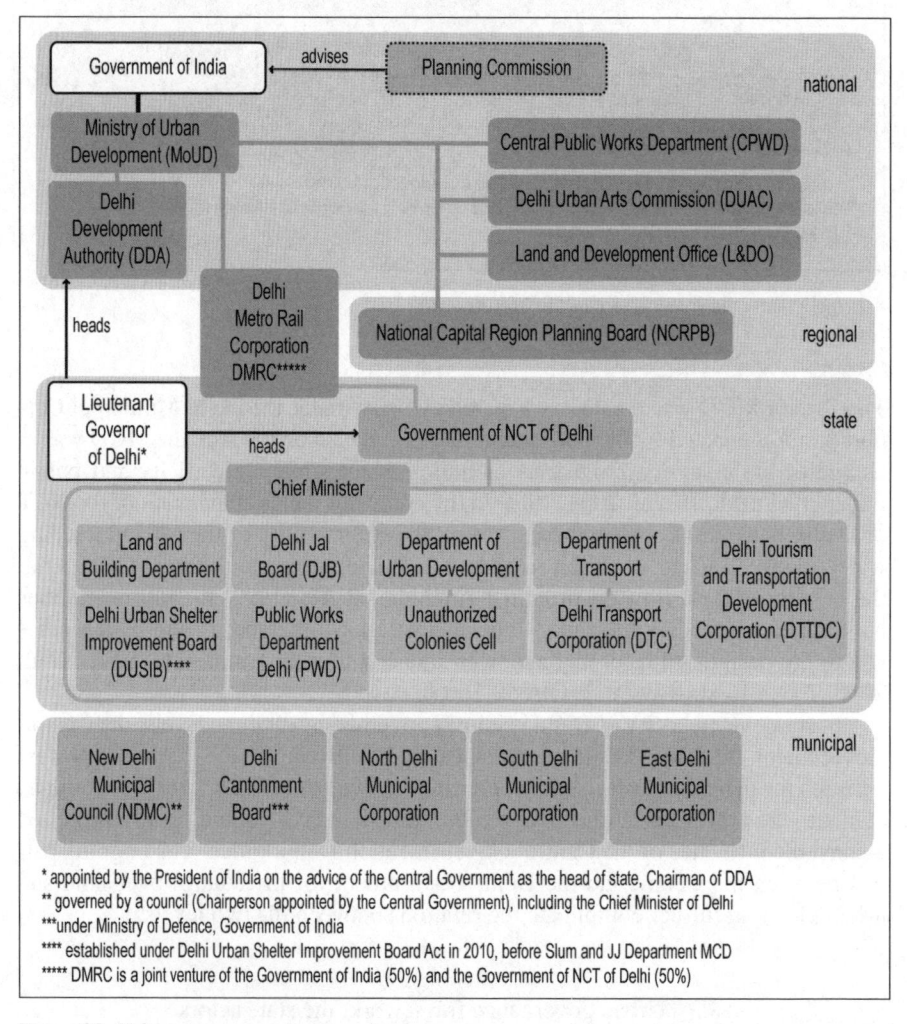

[Figure 20: organizational chart of Delhi urban governance framework across national, regional, state, and municipal levels]

* appointed by the President of India on the advice of the Central Government as the head of state, Chairman of DDA
** governed by a council (Chairperson appointed by the Central Government), including the Chief Minister of Delhi
*** under Ministry of Defence, Government of India
**** established under Delhi Urban Shelter Improvement Board Act in 2010, before Slum and JJ Department MCD
***** DMRC is a joint venture of the Government of India (50%) and the Government of NCT of Delhi (50%)

Figure 20: Urban governance framework
Source: Graphic by author based on various sources and information from interviews

8.3.1 Agencies on the national level

The Ministry of Urban Development (MoUD) is the nodal agency for urban development and planning in India with direct constitutional and legal authority over Delhi. Parallel to the MoUD the Ministry of Housing and Urban Poverty Alleviation (MoHUPA) is entrusted with urban issues on the national level.[32] The Land and Development Office (L&DO) and the Central Public Works Department (CPWD) function as attached offices under the MoUD. The Delhi Development Authority (DDA), the Delhi Urban Arts Commission (DUAC) and the National Capital Region Planning Board (NCRPB) are statutory, autonomous bodies under the MoUD.

Delhi Development Authority (DDA)

The DDA is the principal urban planning authority under central power of the MoUD. The DDA is a large agency with about 16,500 employees working in different departments (DDA 2013a). The departments important for this study are the Planning Department, Engineering Department, the Land Management Department, the Landscape Department and the Horticulture Department. The Lieutenant Governor (LG) of Delhi functions as the Chairman of the DDA. DDA's operations are managed by the Vice-Chairman, who is typically a senior officer of the Indian Administrative Service (IAS) appointed by the Central Government. Therefore, the main functions of spatial planning in Delhi come neither under the GNCTD (and the Chief Minister) nor the municipal bodies.

The DDA Planning Department is responsible for spatial planning including the preparation of the Master Plan for Delhi and the subordinate Zonal Development Plans (ZDPs). Delhi is subdivided in 15 planning zones (named 'A' to 'P', excluding 'I') and ZDPs are to be prepared for these in accordance to the Section 8 of the Delhi Development Act. With the approval of the MoUD the ZDPs become legal planning documents. Prior to the approval of any ZDP or the Master Plan, the DDA needs to prepare drafts for public inspection to obtain objections and suggestions (Section 10 (1), Delhi Development Act). During the time of this study, the DDA initiated a review of the Master Plan for Delhi 2021 inviting suggestion from the public and involving other state agencies. The policy process for the ZDP for the River Zone (Zone O) is reviewed in detail in chapter 11.

In conjunction with the authority's planning functions, the DDA is also in charge of land acquisition and development as well as further allotment of land to other public agencies or private developers for different uses. In doing so, the DDA declares land as co-called 'development areas' (based on Section 12, Delhi Development Act) and develops it according to the above mentioned plans.

Large-scale land acquisition for urban development has generally been done through the Land Acquisition Act 1894. DDA manages different categories of land

---

32 Both ministries have been reformed and renamed a couple of times in the last decades.

including among others *Nazul lands*[33], lands allotted by the Land and Development Office (L&DO) to DDA and land still with the L&DO for maintenance purposes, acquired land (through large-scale land acquisitions) and *Gaon Sabha lands* of urban villages (DDA 2013a: 31).[34] The Land Management Department within DDA is responsible for the supervision of lands with DDA and its function include among others the acquisition of land, protection of land (until its disposal) and execution of demolitions (ibid.). The demolitions and evictions are carried out under the Public Premises (Eviction of Unauthorised Occupants) Act, 1971.

Urban areas developed by the DDA (e.g. new residential colonies developed by the Housing Department) are managed by the DDA unless they are transferred to the municipal bodies. As long as urban areas are with DDA, the Landscape Department (also referred to as Landscape and Environmental Planning Unit) is responsible for the landscape design and planning for the green spaces. The DDA Horticulture Department is in charge of the actual plantation and greenery work including maintenance. These two departments manage the larger green areas (including public parks along the river). This study will emphasize upon the important role of the DDA in planning as well as *not planning* for Delhi's riverscapes.

Delhi Urban Arts Commission (DUAC)

The DUAC was set up as an independent body by an Act of Parliament in 1973 "to advise the Central Government in the matter of preserving, developing and maintaining the aesthetic quality of urban and environmental design within Delhi and to provide advice and guidance to any local body in respect of any project of building operations or engineering operations or any development proposal which effects or is likely to affect the skyline or the aesthetic quality of surroundings or any public amenity provided therein" (Section 11, Delhi Urban Arts Commission Act, 1973). As such, all development projects in Delhi carried out by the DDA or any other local agency are supposed to be reviewed and cleared by the DUAC. The DUAC has the power to approve, reject or modify any (re)development project (KULSHRESTHA 2012: 57). The DUAC is appointed by the MoUD for a period of

---

33 *Nazul land* is freehold land under the control of the Central Government. The term Nazul lands (also Crown Lands or King's Land) goes back to the Mughal period. The British claimed the lands around Old Delhi from the Mughal royalty already at the time of the 1857 mutiny (HOSAGRAHAR 2005: 133). Nazul lands came under DDA either with the transition from the DIT to the DDA (Nazul land I) or they have been acquired after 1957 (Nazul land II & III) (cf. JAIN 1990: 118–110, JOLLY 2010: 144–147). For a detailed discussion of the Nazul lands see also MEHRA (2013: 358).

34 *Gaon Sabha Lands* are the common lands of a village under control of the Gram Panchayat (local self-government of the village) under Panchayati Raj Act, 1954. In the process of urbanization former rural villages have become urbanized and were legally converted into urban villages when declared under Clause (a) of Section 507 of Delhi Municipal Corporation Act, 1957 (DDA 1986). Both rural and urban villages are found along the banks of the Yamuna in Delhi.

three years. Supported by 30 permanent staff and external experts, the DUAC evaluates about 500 projects per year (AGRAWAL 2010: 399–400).

Land and Development Office (L&DO)

The L&DO under the MoUD came into existence with the decision to make Delhi the colonial capital. The L&DO has always, and continues to be, responsible for the administration of land resources of the Government of India in Delhi. The responsibilities of L&DO include, but are not limited to, the maintenance of lease records (about 50,500 leases of government land in Delhi), the allotment of land to government agencies (under direction of MoUD), the selling of vacant land and developed properties, as well as the removal of encroachment on land under its purview. With regard to Delhi's riverscapes, the L&DO plays an important role in allocation of government land for urban development.

Central Public Works Department

The Central Public Works Department under the MoUD is responsible for the construction as well as maintenance of public built environment and infrastructure (excluding railways, defense, communication, atomic energy and airports) all over India.

8.3.2 Agencies on the regional level

On the regional/interstate level, the National Capital Region Planning Board (NCRPB) functions as a statutory, autonomous body under the National Capital Region Planning Board (NCRPB) Act, 1985. The rationale behind the creation of the NCRPB was to promote decentralized growth in the region by diverting growth to the satellite cities (JAIN & SIEDENTOP 2014, JAIN et al. 2013, MOOKHERJEE et al. 2014). The NCRPB is responsible for the preparation of the Regional Plan for the entire NCR as well as to coordinate the enforcement and implementation of the plan with the participating states.

The first Regional Plan-2001 was adopted in 1989, the second Regional Plan-2012 in 2005. In 2013, the NCRPB has revised the Regional Plan for 2021. Notably, sub-regional plans as required under the Act have not been prepared by the involved states to date. Though, the Master Plans for Delhi prepared by DDA need be in alignment with the Regional Plan and the MoUD can request the NCRPB to give advice to the DDA. The NCRPB has however been only partially successful in guiding regional growth (JAIN & SIEDENTOP 2014: 35).

8.3.3 Agencies on the state level

Due to the central role of the DDA, the state level lacks major powers with regard to urban planning and development. The Delhi Government's Department of Urban Development is involved in the planning and implementation of various infrastructure projects and basic service provision, in cooperation with the Delhi Jal Board (DJB), the Delhi Transportation Corporation (DTC), the Delhi Public Works Department (DPWD) and other state agencies. Both DJB and DTC are examples for a centralization of basic service provisions on the state level taking away responsibilities form the municipal bodies. The DJB has been responsible for fresh water supply as well as sewage treatment and disposal for the entire NCT since 1998. In the public transport sector, the DTC is the successor of the Delhi Transport Undertaking (DTU) which was operated by the Municipal Corporation of Delhi prior to 1971.

With regard to land management, the Delhi Government's Land & Building Department is in charge of large-scale acquisition of land under the Land Acquisition Act 1894. The major task of the Land & Building Department is to deal with various proposals for land acquisition by the DDA as well as departments of the Delhi Government. Within the Department of Urban Development, the Unauthorized Colonies Cell deals with the regularization of existing unauthorized colonies (including the unauthorized colonies in the River Zone).

The state level further plays an important role in urban poverty alleviation through the Delhi Urban Shelter Improvement Board (DUSIB). The DUSIB has taken over the responsibility for the slum designated areas and JJ clusters from the municipal bodies who managed these tasks until 2010 (earlier Slum and JJ Wing MCD). The DUSIB is also responsible for basic service provisions (e.g. night shelters) and resettlement schemes. The DUSIB runs a number of night shelters along the river.

With regard to infrastructure one more agency needs to be mentioned: the Delhi Tourism and Transportation Development Corporation (DTTDC). The DTTDC was set-up in 1975 to promote tourism in Delhi. Based on a series of revenue sources (e.g. the selling of liquor), the DTTDC generates funds, which are also spend for other development activities. Since 1989, the DTTDC has been responsible for the construction of flyovers. With regard to the Yamuna, the DTTDC has been involved in different schemes for riverfront development and is currently responsible for the construction of the Signature Bridge over the Yamuna, a flagship development in north Delhi.

8.4 Environmental governance framework: the state actors

In accordance to the urban government structure the following sections introduce the environmental government framework (see Figure 21). The institutional set-up governing the environment in Delhi is complex and reflects the sectoral approach inherent to environmental governance in India. Besides the agencies primarily concerned with environmental issues, agencies primarily involved in urban issues also

Urban environmental governance: introducing actors and policy debates 145

play an important role in governing the urban environment. Here, the limitations of the sectoral and spatial approach in defining certain spheres of governance become obvious. Overlapping responsibilities between urban governance and environmental governance exist and a sharp separation does not appropriately address the multiple connection between the two spheres. For analytical reasons these two spheres (urban and environmental governance) are outlined here separately before merging them in the concept of urban environmental governance.

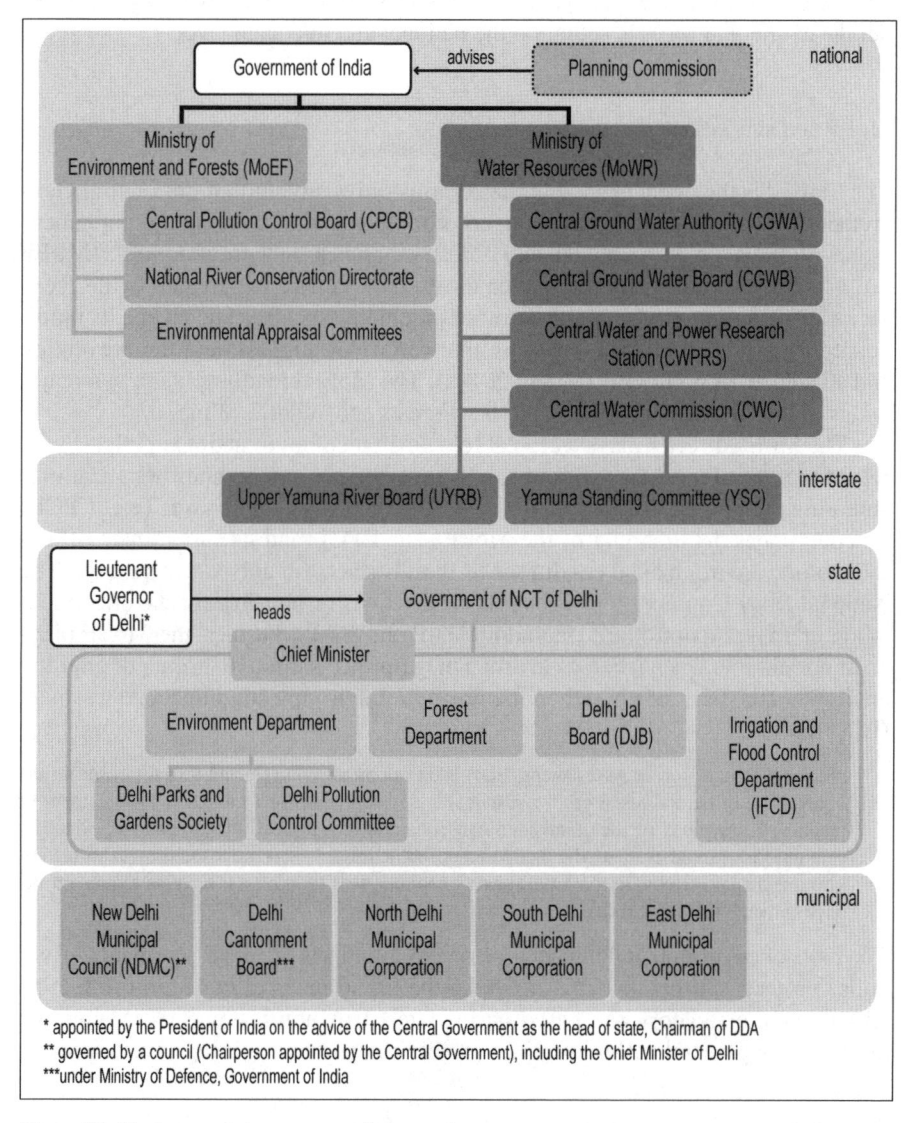

Figure 21: Environmental governance framework
Source: Graphic by author based on various sources and information from interviews

8.4.1 Agencies on the national and interstate level

The Ministry of Environment and Forests (MoEF) and the Ministry of Water Resources (MoWR) are two national ministries and their subordinated agencies are primarily responsible for governing India's rivers.[35] The Ministry of Power also plays an important role with regard to hydropower generation. Taking into account the upstream and downstream, the governing of the Yamuna River always needs to be analyzed from an interstate perspective. This section introduces the agencies under the MoEF and MoWR acting on the national and interstate level.

MoEF and subordinated agencies

The MoEF "is the nodal agency in the administrative structure of the Central Government for the planning, promotion, co-ordination and overseeing the implementation of India's environmental and forestry policies and programmes" (MoEF 2014a). It is responsible for the implementation of policies with regard to the "conservation of the country's natural resources including its lakes and rivers, its biodiversity, forests and wildlife, ensuring the welfare of animals, and the prevention and abatement of pollution" (MoEF 2014a). The MoEF broadly splits in two major wings: the 'Environment Wing' and the 'Forests and Wildlife Wing'.

The MoEF directly plays a crucial role when it comes to national policies with regard to rivers. Further, a series of subordinate offices, autonomous organizations, authorities (e.g. National Ganga River Basin Authority), and boards (e.g. CPCB) function under the umbrella of the MoEF. The CPCB and its subordinate offices monitor the environmental conditions of India's rivers (mainly water quality). The National River Conservation Directorate (NRCD) is responsible for the implementation of River Action Plans (including GAP and YAP). Further, the MoEF plays an important role when it comes to urban mega-projects through its duty of granting environmental clearance based on the projects' Environmental Impact Assessment (EIA) reports.

> "Environmental Impact Assessment (EIA) is an important management tool for ensuring the optimal use of natural resources for sustainable development. Environmental Management or planning is the study of the unintended consequences of a project. Its purpose is to identify, examine, assess and evaluate the likely and probable impacts of a proposed project on the environment and, thereby, to work out remedial action plans to minimize adverse impact on the environment" (MoEF 2014b).

The roots of environmental impact assessment in India lay in the assessment of river valley projects starting in 1978-79. Under the Environmental Protection Act, 1986 EIA became mandatory and based on the latest EIA notification (14th September 2006) developmental activities of 29 project categories involving investments of

---

35 The ministries have been renamed as 'Ministry of Environment, Forest and Climate Change' and 'Ministry of Water Resources, River Development and Ganga Rejuvenation' under the new government in 2014.

500 million Indian rupees and above require environmental clearance from the MoEF (MoEF 2014b).

Applications for environmental clearance are submitted by the project developer to the MoEF, which scrutinizes the documents and requests evaluation by a so-called Environmental Appraisal Committee (EAC). The EAC consists of an independent group of experts, which assesses the impact of the project based on the data provided by the project developer and if necessary conducts site visits. The EAC recommends either the approval or rejection of the project. A decision is made by the MoEF within 90 days – if full information is available (MoEF 2014b).[36] Environmental clearance is generally given based on the implementation of certain conditions, which are then monitored by the MoEF's regional offices.

Central Pollution Control Board (CPCB)

Following the enactment of the Water (Prevention and Control of Pollution) Act in 1974, the Central Pollution Control Board (CPCB) became India's leading statutory organization for pollution control under the MoEF. CPCB's main objectives are monitoring studies for surface waters including rivers, streams and lakes as well as artificial water courses like canals. The functions of the CPCB include further the inspection of polluting industries, the checking of both domestic and industrial sewage disposal as well as the definition of environmental standards for pollution levels (CPCB 2008).

The Water (Prevention and Control of Pollution) Act therefore outlines a series of functions of the CPCB both on an *advisory* and *regulatory* level (for details see KARPOUZOGLOU 2011: 48–49). The mandate of the CPCB is to set binding standards as well as provide advice and coordinate the activities of the state pollution control boards (in the states) and pollution control committees (in the Union Territories). Thus, implementation and enforcement of pollution control is decentralized to the state and regional level (GOLDAR & BANERJEE 2004: 120, IYER 2003b: 21–23, KARPOUZOGLOU 2011: 51). In the case of Delhi, regulatory functions of the CPCB are devolved to the *Delhi Pollution Control Committee* (DPCC). Nevertheless, based on orders of the Supreme Court of India (for details see chapter 8.7) the CPCB remains the main agency for monitoring the water quality of the Yamuna.

National River Conservation Directorate (NRCD)

The National River Conservation Directorate (NRCD) under MoEF is the key agency for the implementation of the National River Conservation Plan (NRCP).

---

36 Special provisions are also given for projects such as mining, river valley, ports and harbours etc. In these cases, site clearance needs to be obtained before applying for environmental clearance (MoEF 2014b). Further, projects located in areas under CRZ or declared forest areas come under special provisions (PANIGRAHI & AMIRAPU 2012).

The NRCD coordinates the Ganga Action Plan (GAP) and Yamuna Action Plan (YAP). The NRCD was formed after completion of the first phase of the GAP in July 1995 as the successor of the earlier Central Ganga Authority. The NRCD is dependent on the data obtained and produced by the CPCB and the technical agencies of the MoWR.

MoWR and subordinated agencies

With regard to the focus of this study, the Central Water Commission (CWC), the Central Groundwater Board (CGWB), Central Groundwater Authority (CGWA), and the Central Water and Power Research Station (CWPRS) are the important government agencies under the MoWR on the national level. On the interstate level the Yamuna Standing Committee (YSC) and the Upper Yamuna River Board (UYRB) also require mention.

The CWC is the main technical organization under the MoWR. In consultation with the state governments, the CWC is responsible for different schemes with regard to the control, conservation and utilization of water resources including flood control, irrigation, navigation, drinking water supply and water power development. In 1961, the YSC was set up as an interstate body under the chairmanship of the CWC to oversee all existing structures and to coordinate flood control works along the Yamuna. Proposals for constructions of flood protection measures (embankments, etc.) and public utilities (bridges, etc.) by the states of Uttar Pradesh, Haryana and Delhi (which are members of the committee) need to be cleared by the YSC.

Technical details of flood control schemes as well as the design of bridges and other structures are examined by the CWC and YSC with respect to their influence on flow pattern and flood management. Detailed physical and mathematical model studies on the hydrology of the river and the impacts of certain projects have been prepared by the CWPRS. These studies have largely constituted the base for the decision-making by the CWC and YSC. The history of CWPRS dates back to the colonial period and was established in 1916 under Bombay Presidency to study irrigation practices. Since Independence, CWPRS has been the principal central technical agency in the fields of water and energy resources development. Reviewing the discourse on channelization, the role of the CWPRS and YSC will be examined in detail in chapters 10 and 11.

While the YSC is concerned with flood protection, the UYRB[37] is responsible for overseeing the maintenance of minimum flow in the river and the compliance of the water sharing MoU between the riparian states. Its mandate includes the allocation of available flows amongst the states, the monitoring of return flows and

---

37 The UYRB consists of a Member of the CWC as Chairman and one nominee from each of the states (Uttar Pradesh, Haryana, Rajasthan, Himachal Pradesh and the NCT of Delhi) and a Chief Engineer of Central Electricity Authority and representatives of CGWB and CPCB (UYRB no date).

conserving/upgrading of the quality of surface and groundwater and the maintaining of hydro-meteorological data for the basin. The UYRB has to review all structural projects affecting water abstraction along the Yamuna up to Okhla barrage.

Established as a technical research body under the MoWR in 1970, the CGWB provides scientific inputs for management, exploration, monitoring, assessment, augmentation and regulation of groundwater resources as well as advising national and state governments. In contrast to the CPCB, which combines both advisory and regulatory functions with regard to pollution, the regulatory functions regarding groundwater are not with the CGWB, but with the CGWA. Constituted under Section 3 (3) of the Environment (Protection) Act, 1986 the CGWA regulates and controls groundwater development and management in India. The CGWA recognized the Yamuna floodplain in Delhi as a critical/overexploited groundwater recharge zone on 2nd September 2000.

8.4.2 Agencies on the state level

The responsibility for the governance of a series of environmental concerns and infrastructure provisions is based on the state level, as outlined in Figure 21. The two most important agencies include the DJB and the IFCD.

The DJB is responsible for water provision, sewage treatment, underground sewer systems, and the operation of the Wazirabad Barrage. Together with the MCD, the DJB is also the local implementation agency for YAP. The IFCD is the nodal agency for flood protection and manages open storm water drains in the city (with a capacity of more than 1000 cusecs). In contrast to other states, the irrigation component of the IFCD is rather marginal, and therefore the IFCD does not have the same important standing as the irrigation departments of the neighboring states (Uttar Pradesh Irrigation Department, Haryana Irrigation Department). Additionally, the Delhi Forest Department is in charge of the forests in the River Zone.

8.5 Overview of the state agencies involved in governing Delhi's riverscapes

With regard to the urban environmental governance of Delhi's riverscapes (including associated environmental concerns and infrastructure provisions) five major aspects need to be considered:

1) Land-use planning and urban development;
2) Water, sanitation and pollution;
3) Flood protection;
4) Solid waste management; and
5) Management of urban green spaces (e.g. parks and forests).

To some extent these five aspects accurately represent the sectoral subdivision and complexity of the urban environmental governance in Delhi (see Table 6).

	Responsibilities		Nodal Agency
land-use planning and urban development	Regional Plan for NCR		NCRPB
	Master Plan for NCT of Delhi		DDA
	Zonal Development Plan for River Zone (Zone O)		DDA
water, sanitation and pollution	groundwater	monitoring	CGWB, CGWA
		extraction	DJB, private actors
	water provision		DJB
	sewage treatment		DJB
	providing drainage/sewage facilities*[1]		DDA (Engineering Department)
	sewer system (underground)		DJB
	open storm water drains*[2]	capacity > 1000 cusecs	IFCD
		capacity < 1000 cusecs	Municipal bodies
		inside JJ clusters	DUSIB (earlier MCD Slum and JJ Department)
		under major roads	DPWD, CPWD
	irrigation (including wastewater irrigation)		IFCD
	pollution control		CPCB, DPCC
	operation of barrages*[3]	Wazirabad Barrage	DJB
		I.T.O. Barrage	Haryana Irrigation Department
		Okhla Barrage	U.P. Irrigation Department
flood protection	marginal embankments (west)		IFCD
	marginal embankments (east)		IFCD, U.P. Irrigation Department
	Akshardham Bund*[4]		DDA (Engineering Department)
	waterlogging in residential colonies		DDA, Municipal bodies
solid waste management	waste collection		Municipal bodies
	allotment of landfill sites		DDA (Land Management Depart.)
	waste utilization		Delhi Energy Development Agency
management of green spaces	parks	landscape design	DDA (Landscape Department)
		maintenance (DDA)*[5]	DDA (Horticulture Department)
		maintenance (ULB)*[6]	Municipal bodies (RWAs, NGOs)
	forests	protected and reserved forests*[7]	Forest Department, DDA, L&DO and other land-owning agencies
		city forests	Forest Department, DDA, other land-owning agencies

Table 6: Agencies involved in governing Delhi's riverscapes
Sources: Based on various sources and information from interviews
*Notes: *[1] only planned colonies, *[2] nearly all open storm water drains carry also sewage, *[3] barrages are operated for water abstraction and have to be opened in the event of floods (no flood control function) *[4] developed and maintained by DDA, *[5] larger green areas (e.g. proposed riverfront parks), which are not handed over to municipal bodies, *[6] e.g. Kudsia Garden (Kudsia Bhag) under Horticulture Department, North Delhi Municipal Corporation, *[7] Reserved Forests under Section 4 of the Indian Forest Act, 1927 (Ridge), Protected Forests under Section 29 of the Indian Forest Act, 1927.*

The urban local bodies are again of rather limited importance for this study. Nonetheless, being responsible for the management of smaller open storm water drains and solid waste collection they deal with important aspects to prevent waterlogging in residential colonies and need to coordinate their actions with other state agencies. They must especially harmonize their actions with the IFCD with regard to flood protection and the DJB with regard to wastewater management.

8.6 India's urban rivers: legal provisions and debates

Water, sanitation, irrigation, embankments and land – all important issues with regard to (urban) rivers – are first and foremost responsibilities of the state governments. In contrast, matters of the environment like forests and wildlife protection are dealt with by both levels. While water is essentially a state subject (entry 17 of State List), based on entry 56 Union List, it can become a matter of the Central Government if the parliament passes a law stating a particular river is an important national/interstate river (IYER 2003b: 21–22). In the view of the overlapping of these constitutional provisions, and the varying governmental agencies on the different levels, the governance of urban rivers in India emerges as inherently complex.

Furthermore, the Indian Constitution, as one of the few written constitutions in the world (DIVAN & ROSENCRANZ 2001: 41), outlines that "[t]he State shall endeavour to protect and improve the environment and to safeguard the forests and wild life of the country" (Article 48-A). Through the 42nd Amendment Act of 1976, environmental protection and improvement was explicitly incorporated in the Indian Constitution (DIVAN & ROSENCRANZ 2001: 44). Further, Article 51-A (g) of the Indian Constitution applies the duty of environmental protection on every citizen: "[i]t shall be duty of every citizen of India to protect and improve the natural environment including forests, lakes, rivers and wild life and to have compassion for living creatures." Additionally, the Indian Constitution provides several fundamental rights including the right to life and personal liberty in Article 21. The Supreme Court has extended these rights and included also "the right to a wholesome environment" (DIVAN & ROSENCRANZ 2001: 49). Therefore if anything jeopardizes the right to a wholesome environment or the quality of life and is in breach of the law, any citizen of India today has the right to make an appeal to the Supreme Court (ibid. 50).

Based on these constitutional provisions the Supreme Court has, through different orders in the last decades, stretched the constitutional mandate of the state to protect the environment (ibid. 41). Furthermore, the Supreme Court has made the various levels of government in India responsible for ensuring inter-generational equity and sustainable development as well as protecting the environment based on the Precautionary Principle, the Polluter Pays Principle, and the Public Trust Doctrine (for more details see among others DIVAN & ROSENCRANZ 2001, GILL 2012, RAJAMANI 2007, RAZZAQUE 2004). On this basis, the Indian jurisprudence has given several orders with regard to Indian rivers. GOPAL et al. (2002: 49) argue that since water, as "the most fundamental right of every individual", needs protection, that rivers "must be protected" also and the regulation of human activities along their banks "is abundantly rooted" in the Indian Constitution.

It is important to note that the Indian legislation encompasses a multiplicity of laws and regulations related to (urban) rivers. On the national level, these include the Easement Act, 1882, the Land Acquisition Act 1894; the River Boards Act 1956; the Interstate Water Disputes Act 1956, the Wildlife (Protection) Act 1972; the Water (Prevention and Control of Pollution) Act 1974 and the Environmental (Protection) Act 1986 (for details of each act see among others CULLET et al. 2012: 5–9). Furthermore, a series of acts has been passed on the state level such as Irrigation and Drainage Acts.

In the case of Delhi, the Delhi Development Act 1957 provides the basis for land-use planning in the NCT of Delhi, including the floodplain. As this study will outline, this Act constitutes the central legal provision for governing land-use changes on the floodplain of the river Yamuna – irrespective of the fact that this Act has solely been set up to govern (or even more narrowly: develop) land without containing any provisions with regard to rivers or their floodplains. Therefore, and significantly, none of the above mentioned Acts and the associated rules and regulations provide comprehensive regulatory measures for governing land uses along rivers whether in the urban or rural context.

In India, the dominating "response to flood damage was to try to 'control' floods through structural means such as dams or embankments" (IYER 2001a: 1118). These structural, engineering solutions in the form of embankments and dams almost unexceptionally dominated the discourse for controlling India's rivers even when they were constantly criticised (AGARWAL & NARAIN 1991, BAGHEL 2014, D'SOUZA 2003, 2006a, IYER 2001a, 2008, LAHIRI-DUTT 2000, MISHRA 2009, THAKUR 2009). After heavy floods in India, a first Flood Control Policy was legislated in 1954. Following this policy formation, a number of committees and working groups were set up to study the problem of flooding in India (MOHAPATRA & SINGH 2003: 138). Based on these committees, in the aftermath of the 1954 flood, 34,000km of embankments and almost 39,000km of drainage channels were constructed in a bid to protect agricultural areas as well as more than 2,400 towns and 4,700 villages across the country (for details see MISHRA 2009: 310 based on data of the MoWR).

Regulatory measures to protect the floodplains of Indian rivers have been a matter of heated debate for decades. In 1975, the CWC, functioning under the MoWR, circulated a model bill for floodplain zoning for adoption by the states. The concept of floodplain zoning aims to regulate land uses and other human activities in floodplains in order to mitigate risks of flood hazard and limit potential damage caused as a result of floods. Floodplain zoning is generally adopted "to prevent undeveloped floodplains becoming unwisely developed, and to constrain the further development of already partly developed floodplains" (PARKER 1995: 343). The basics of these ideas have been discussed in India since at least the 1970s.

The deliberations concerning flood control and management in the National Water Policy of 1987 are an indicator for a slowly changing mind-set already at that time (cf. BAGHEL 2014: 117, SURYANARAYANAN 1997: 174, THAKUR 2009: 214–215).

> "An extensive network for flood forecasting should be established for timely warning to the settlements in the flood plains, along with the regulation of settlements and economic activity in the flood plain zones, to minimize the loss of life and property on account of floods. While physical flood protection works like embankments and dykes will continue to be necessary, the emphasis should be on non-structural measures for the minimization of losses, such as flood forecasting and warning and flood plain zoning, so as to reduce the recurring expenditure on flood relief" (MoWR 1987).

The updated National Water Policy of 2002 is almost identical to its 15 year old predecessor, but adds the word 'strict' in front of 'regulation' (MoWR 2002). The National Water Policy "is not a law; it has only the force of consent" (IYER 2001b: 1242).[38] Nevertheless, the policy indicates an acknowledgement that flood tragedies across the country (especially the breaching of embankments) "are caused by *people* and not just water" (WISNER et al. 2004: 202). As RAMASWAMI R. IYER outlines, floods thus became seen as "man-made" rather than "natural" events and embankments were more and more seen as problematic since they tend to fail in the event of major floods resulting in flash floods (IYER 2001a: 1118). Additionally, it has been realized that embankments are preventing the river from spreading which creates aggregated problems downstream as well as water logging behind the embankment (ibid.).

The so-called 'multipurpose' dams – declared to guarantee perennial water supply for irrigation and power generation, but also mitigate floods – emerged to be ineffective, as they were not initially designed for flood mitigation or "the competing claims of irrigation and power-generation [...] override the flood-moderation function" (ibid., cf. LAHIRI-DUTT 2000). Furthermore, both embankment and dams generate a "false sense of security" (WISNER et al. 2004: 206), which in turn leads to increased human activity along the bank and higher losses in case of breaches (cf. MCCULLY 1996). Therefore a shift from protective, structural flood defense to non-structural preventive/precautionary measures seemed to be predestined. Yet, driven by a set of powerful narratives supporting structural interventions, the construction of embankments and dams continued as the basic approach in flood management in India (for a detailed analysis see BAGHEL 2014).

Furthermore, while "the primary responsibility for flood control lies with the States" (PLANNING COMMISSION 2011: 5), with the exception of Manipur (1978), Rajasthan (1990) and Uttarakhand (2012), none of the States has enacted any floodplain zoning regulation adopting the guidelines of the CWC to date. The reasons for this are quite apparent: floodplain regulation, especially in the flood-prone states of the Gangetic plains, was perceived to (enormously) restrict development options and would thus directly affect a large population (MISHRA 2009: 323–324, cf. PLANNING COMMISSION 2011).

---

38 The MoWR formulated the first National Water Policy in September, 1987. The policy was updated in 2002 and 2012. Experts like Ramaswamy R. Iyer, who was involved in formulating the first version, argue for 'statutory backing' of the National Water Policy.

This has made the concept so far politically unviable and legally not enforceable. DINESH KUMAR MISHRA, author and convener of the environmental NGO Barh Mukti Abhiyan[39] (Freedom from Floods Campaign) in the state of Bihar, commented on this lack of regulation development by the states:

> "If land use is restricted [in the floodplains of the plains] it would lead to a chaotic situation and very little scope would be left to take up any development. This is the reason why most of the flood-prone states are reluctant to adopt Flood Plain Zoning. Unfortunately, this view of theirs [the states] is never communicated to those in authority and the ghost of Flood Plain Zoning haunts the planners always" (MISHRA 2009: 324).

The 'ghost' of floodplain zoning nevertheless remains on the agenda of the MoWR as also the latest update of the National Water Policy in 2012 indicates:

> "Conservation of rivers [...] should be undertaken in a scientifically planned manner through community participation. The storage capacities of water bodies and water courses and/or associated wetlands, the flood plains, ecological buffer and areas required for specific aesthetic recreational and/or social needs may be managed to the extent possible in an integrated manner to balance the flooding, environment and social issues as per prevalent laws through planned development of urban areas, in particular. [...] Encroachments and diversion of water bodies (like rivers, lakes, tanks, ponds, etc.) and drainage channels (irrigated area as well as urban area drainage) must not be allowed, and wherever it has taken place, it should be restored to the extent feasible and maintained properly. [...] key aquifer recharge areas that pose a potential threat of contamination, pollution, reduced recharge and those endanger wild and human life should be strictly regulated" (MoWR 2012).

Since all of these aspects are highly relevant for the analysis of Delhi's riverscapes, the larger debate surrounding Indian rivers must not be left out in the analysis of the local developments – especially since the capital city plays a role model function for whole of India. Therefore this study will pay attention to the wider discussion on demarcating the limits of floodplains as restricted or no development zones.

Another notable example of a policy intent that was not achieved is the so-called River Regulation Zone (RRZ). The RRZ has been discussed in the context of floodplain protection since the early 2000s. It was envisioned that the RRZ would operate in line with the existing regulation for coastal zones[40] (Coastal Regulation Zone, CRZ) and a concept for the RRZ was prepared by the MoEF in 2002 after a workshop in 2001 (GOPAL et al. 2002). The process stalled after that. Only in 2011, the public debate on the topic reassumed, when Jairam Ramesh, environment minister from May 2009 to July 2011, shortly before leaving the MoEF, announced in a press conference that the RRZ would be notified within months. After that an "Expert Group for the formulation of Guidelines for Management of River Fronts through River Regulation Zone" was setup by the MoEF. However, the RRZ remains unnotified to date.

---

39 For a detailed analysis of the work of Barh Mukti Abhiyan see THAKUR (2009).
40 The CRZ, introduced in 1991 under the Environment Protection Act, aimed at protecting India's coastal belt from unregulated development making special CRZ clearance necessary for certain activities. Lacking monitoring of the CRZ and dilution of the notification has however been criticised (GANGAI & RAMACHANDRAN 2010, MENON et al. 2007).

In contrast to these developments, the debates surrounding the 'interlinking of rivers' have shown how questions of the right to life and right to drinking water have been completely differently framed in order to legitimize massive structural interventions to Indian rivers (for details see among others ALLEY 2012, IYER 2002, IYER 2003b, MIRZA et al. 2008). Cutting a long story short, due to the high costs, myriad ecological risks, divergence between scientists and strong opposition from environmental movements, only a few states support the interlinking of rivers (JAIN & ROSENCRANZ 2014). Nevertheless, the Supreme Court in 2002, acting upon a writ petition in favor of the project in a case originally concerned with cleaning of the Yamuna[41], ordered that the Central Government should complete the interlinking of rivers until 2016. In February 2012, the Supreme Court again requested the government to act upon the matter. This case depicts how judicial activism appears to be an uncertain process rather than a straightforward 'amicus environment' with regard to India's rivers.

In sum, a comprehensive legislation for governing river floodplains does not exist in India to date. Rather multiple instruments under different Acts, non-binding policy principles, and often ambiguous judicial decision-making pose several challenges for an integrated governance of India's riverscapes (see CULLET & MADHAV 2009: 511 for a similar conclusion on water legislation). The existing laws nevertheless provide a broad scope in which to protect urban floodplains. However, due to the shortage of land in the urban settings and development interests, the urban floodplains are a "contested space" (LÜBKEN 2012) especially in India's large metropolises. The detailed analysis of the developments along the Yamuna in Delhi will underline this. The specific existing, and debated, policies for governing Delhi's riverscapes are reviewed in the course of this book (chiefly in chapters 10, 11, 15 and 16).

8.7 The courts and the river: overview of selected orders

As indicated in the introduction to this study, Delhi has always been a hot spot of environmental activism within the courts. Landmark cases have been dealing with the cleaning and rejuvenation of urban water bodies, including the Yamuna (MEHTA 2009a, 2009b), air pollution causes and effects (GHOSH 2007, KUMAR 2012, VÉRON 2006), solid waste disposal (GIDWANI 2013, TALYAN et al. 2008), and the protection of the Delhi Ridge forest area. Almost all efforts to clean the Yamuna (together with the Ganga River) and safeguard the floodplain have gone through the courts triggered by PILs filed by environmental activists and NGOs (cf.

---

41   The Interlinking of Rivers has been a brainchild of the NDA government in 2002. In September 2002, the Supreme Court in the so-called "And quiet flows the maili Yamuna"-case on affidavit directed that the interlinking of rivers be treated as an independent writ petition (Writ Petition (Civil) No. 668 of 2002. The Supreme Court's order referred further to a speech of the President of India on the Independence Day Eve.

FOLLMANN 2014, 2015, GILL 2014, MEHTA 2009a, 2009b, SHARAN 2002, 2013, 2014, VÉRON 2006).

The Yamuna has been dealt with by the courts since the mid-1980s. Court proceedings have historically focused either on the pollution or the minimum flow in the river. More recently, cases dealing with land use issues and the protection of the floodplain of the Yamuna in Delhi have emerged. These were also initiated by PILs and are the focus of this study.

The earliest PIL on river pollution was filed by the environmental lawyer MC Mehta[42] in the Supreme Court of India in 1984. Additionally, Sureshwar D Sinha, a retired navy commander and active member of the sailing club at Okhla, filed a case against pollution and for an increased water flow in the Yamuna in 1992.[43] His petition developed into a landmark case and continues to be mandamus in the Supreme Court. After the CPCB outlined that even when all planned STPs under YAP would be made fully operational, the water quality of the river would not meet the defined bathing quality, the Supreme Court ordered that "a minimum flow of 10 cumecs (353 cusecs) must be allowed to flow throughout the river Yamuna" (order of the Supreme Court quoted in DUTTA 2009: 15).

Another important case in the Supreme Court with regard to the Yamuna in Delhi was taken up by the Supreme Court *suo motu* on the base of an article 'And Quiet Flows The Maily Yamuna' published in the Hindustan Times on 18th July, 1994.[44] Since then, the Supreme Court has been dealing with the case issuing directions to various government authorities to ensure flow and the cleaning of the river (cf. DUTTA 2009). Furthermore, in a case filed by BL Wadhera[45] in 1996, the insufficient provision for solid waste disposal in the city of Delhi found to be a major reason for the pollution of the river (cf. DUTTA 2009: 21).

The river pollution has also been a matter seen by High Court of Delhi.[46] The High Court of Delhi dealt with cases regarding to land ownership and land acquisition in the River Zone (see for details section 10.5). Most importantly the High Court intervened in the process of 'reclaiming' Delhi's riverscapes from the slums (for details see chapter 13) and dealing with the mega-projects along the river (see

---

42 M. C. Mehta Vs Union of India and Others, Writ Petition (C) No. 13381 of 1984; also known as 'Taj Trapezium Zone (TTZ) Case'. The case was filed against pollution in Agra and its impact on the Taj Mahal. The case let to the monitoring of the pollution levels in the river by the Pollution Control Boards.
43 Cdr Sureshwar D Sinha Vs Union of India, Writ Petition (C) 537 of 1992.
44 'Maily' (Hindi) means 'polluted'.
45 B.L Wadhera Vs Union of India, (1996) 2 SCC 594. Wadhera originally approached the Supreme Court seeking direction with regard to the collection and disposal of solid waste by the municipal corporations in Delhi. In the case Almitra Patel vs. Union of India (Writ Petition (C) No. 888 of 1996, D/- 15-2-2000) the Supreme Court also dealt with the issue of solid waste dumping (see details in chapter 13).
46 The Himachal Pradesh High Court and the Allahabad High Court (Uttar Pradesh) also dealt with pollution issues (DUTTA 2009: 37–44).

chapter 14). The analysis in this study focuses on the Commonwealth Games Village in the High Court of Delhi and later in the Supreme Court of India.[47]

Furthermore, due to high numbers of PILs resulting in delayed court orders as well as the complexity of environmental issues, "a distinctively green jurisprudence" was setup by the Parliament of India with the National Green Tribunal (NGT) in October 2010 (National Green Tribunal Act 2010) (GILL 2013: 25). Therefore the judiciary framework has transformed during the timeframe of this study. The NGT was set up "for effective and expeditious disposal of cases relating to environmental protection and conservation of forests and other natural resources including enforcement of any legal right relating to environment" (MoEF 2014c). With former Supreme Court justices heading the tribunal and technical experts supporting the judges in decision-making the NGT is dedicated to "provide speedy environmental justice and help reduce the burden of litigation in the higher courts" (ibid.).

The NGT has already dealt with several cases related to rivers in India and the Yamuna in particular. Two cases dealt with by the NGT will be reviewed in detail in chapter 16. These include the case against solid waste dumping along the river (Manoj Kumar Misra & Anr. vs. Union of India & Ors., No. 6 of 2012), which is referred to as the 'Debris Case', and the second case based also on a PIL (Manoj Misra & Madhu Bhaduri vs. Union of India and others. No. 300 of 2013), which is referred to as the 'Drains Case'. Both cases are jointly heard, ongoing cases in the NGT. As is evident, the courts have been playing an important role in the governance of Delhi's riverscapes and their judgements are required to be integrated in the analysis of the urban environmental policy-making process.

8.8 Environmental NGOs working on the Yamuna

This section introduces the environmental NGOs working on the river in Delhi. The environmental NGOs are considered here to be the most important civil society actor in the remaking of Delhi's riverscapes. Environmental NGOs can be setup as different legal entities (e.g. registered societies or public trusts) under a series of laws in India. Table 7 gives an overview of the environmental NGOs working explicitly on the river Yamuna or those who are involved in the wider environmental governance of rivers in India with a more indirect link to the Yamuna.

---

[47] The case in the High Court of Delhi is referred to as Rajendra Singh & Ors. vs Government Of NCT of Delhi & Ors. (WP (C) No. 7506 of 2007 and WP (C) No. 6729 of 2007) and thereafter in the Supreme Court Delhi Development Authority (DDA) vs Rajendra Singh & Ors. (Civil Appeal No. 4866-67 of 2009). The High Court further dealt with the Millennium Bus Depot Case. While also this case is important in the overall development of Delhi's riverscapes, a detailed analysis would have gone beyond the scope of this study. The Millennium Bus Depot Case in the High Court of Delhi is referred to as Vinod Kumar Jain vs Government Of NCT of Delhi & Anr. (W.P.(C) 3479 of 2010) and Anand Arya & Anr. vs. Government of NCT of Delhi & Anr.

environmental NGO (active since)	focus	key activities
Centre for Science and Environment (CSE) (1980)	several different foci, including water management: wastewater treatment, pollution, rain-water harvesting	research, publication, awareness campaigns, knowledge portal, media outreach
Indian National Trust for Art and Cultural Heritage (INTACH), Natural Heritage Division (1984)	several different foci: eco-restoration, conservation and creation of environmental assets (initial focus: water)	research, publication (e.g. "Blueprint for Water Augmentation of Delhi"), PIL on CWGs-Village
Natural Heritage First (NHF) (1995)	nature conservation (Delhi Ridge, Yamuna), water augmentation, groundwater assessment	research, publication, petitioning, talks, presentations, Yamuna Satyagraha
Pani Morcha (1993)	river pollution, environmental flow	research, petitioning (PIL on Yamuna pollution since 1992)
PEACE Institute Charitable Trust (2002/2003)	nature conservation (Yamuna)	environment education, campaigning, research, publication
South Asia Network for Dams, Rivers and People (SANDRP) (1998)	dams, hydro-power projects	monitoring, research, documentation, publications networking, awareness building
Swechha (2000)	environmental education, "We for Yamuna" campaign	awareness campaigns, media outreach Yamuna walks,, documentary film-making, talks, street-plays, photo exhibitions, Yamuna Cyclothon, Yamuna Yatra
Tapas NGO (1996)	augmentation of water resources in Delhi, revival of water bodies in Delhi, pollution of water bodies	petitioning, several PILs on the Yamuna, PIL against CWGs-Village; PIL against Millennium Bus Depot
Tarun Bharat Sangh** (1975)	water, empowerment of communities (Gram Swarajya), temporarily involved Yamuna campaign	Yamuna Satyagraha Rashtriya Jal Biradari (National Water Community)
Toxics Link (1994)	toxic materials (municipal, hazardous and medical waste management and food safety), nature conservation (Delhi Ridge, Yamuna)	research, publication, awareness campaigns, Yamuna-Elbe Seminar (2010), Yamuna-Elbe Public Art Festival (2011), Yamuna Manifesto
Yamuna Jiye Abhiyaan* (YJA) "Yamuna Forever Campaign" (2007)	Yamuna in Delhi and beyond: floodplain protection, pollution, river flow, drains, solid waste dumping	awareness campaigns (e.g. Yamuna Vigil 2009, Yamuna Satyagraha), media outreach, publications, petitioning (different PILs on Time Global Village, Games Village, Millennium Bus Depot, solid-waste dumping, covering of drains)
WWF India (1969)	several different foci, environmental flow assessment, Ganga River Dolphin Conservation Programme (since 1997)	environmental education, capacity-building, policy studies, advocacy

Table 7: Environmental NGOs working on the Yamuna in Delhi
Source: Based on information from interviews and NGO websites
Note: *consortium of originally seven environmental NGOs (including Pani Morcha, Toxics Link, Ridge Bachao Andolan, LIFE, Centre for Media Studies and PEACE Institute) and concerned citizens. Manoj Misra (PEACE Institute) has been the convenor of the campaign since 2007; ** based in Rajasthan.

Just by taking size, two 'subgroups' of environmental NGOs can be distinguished working on the Yamuna in Delhi: 1) 'larger' environmental NGOs who are engaged in wide range of environmental fields operating on multiple scales from the local to the global (e.g. CSE, WWF India) and have a certain number of professional staff; 2) 'smaller' environmental NGOs (sometimes 'one-man shows') working on specific topics more on the local to regional level (e.g. YJA, NHF, Tapas NGO) on a voluntary basis. Yet, all these classification remain vague; there is no clear boundary between the two groups.

Without evaluating their influence and impact in policy-making the larger NGOs could be termed as *environmental think tanks* acknowledging their wide (often scientific) knowledge and professional competency, as well as their engagement on the policy level. Referring to the latter as *local environmental action groups* in contrast, acknowledges their often 'activist nature' operating more from the grass-roots level. As this study is not interested in the internal organization of the NGOs, but rather in their actions and interaction with the state, as well as their impact on urban environmental policy-making, the subgroups are outlined here more as an explanatory insight than based on any conceptual approach. It should be clear that the resources, knowledges and practices of the involved environmental NGOs differ, and the following insights will outline also different strategies and approaches used by them in lobbying for the river.

Judicial activism on the river Yamuna in Delhi has generally focused on three major issues: 1) pollution abatement, 2) minimum flow, and most recently 3) the protection of the floodplain from urban development. Well-known Indian environmental NGOs like the Centre for Science and Environment (CSE) and WWF India have long been working on the issue of floodplain protection (see e.g. AGARWAL & NARAIN 1991). However, a *new wave* of environmental activism on the river in Delhi started over the course of the construction of the Commonwealth Games Village. In contrast to the earlier PILs which focused largely on pollution issues (see 8.7) more recent PILs challenged urban development projects trying to protect the floodplain.

Environmental activism through the courts is only one activity among many different ways environmental NGOs lobby for the river. Environmental (and social) activists in Delhi have convened different events (e.g. Yamuna Cyclothon, Yamuna-Elbe Public Art Festival), but are also involved in the cleaning of the ghats and the strengthening cultural-religious traditions connected to the river. The environmental NGOs draw on a wide spectrum of activities, including several methods of awareness-raising including seminars, vigil campaigns, 'Gandhian-style' protest camps (Yamuna Satyagraha) and journeys along the Yamuna (Yamuna Yatra).

In contrast to other environmental movements in India, Delhi-based environmental NGOs campaigning for the river have not organized larger protest marches or demonstrations. These more 'traditional' forms of civil society protest are, however, used by organizations from the downstream areas of Uttar Pradesh (Mathura, Vrindavan and Agra). They organize frequent protest marches to Delhi demanding actions against river pollution and lobby for an increased flow in the river (e.g.

'Save Yamuna Campaign' from Vrindavan). Some Delhi-based environmental NGOs are closely connected with these movements.

In addition to the listed environmental NGOs, there are also a series of civil society organizations which are not explicitly dedicated to environmental protection, but which work on the river. For example, the International Art of Living Foundation under Sri Sri Ravi Shankar, organized a river (*ghat*) cleaning event in 2010 and convened so-called 'Think Tank Meetings' in 2011 under their 'Meri Delli Meri Yamuna' campaign. With one exception, members of the above listed environmental NGOs were not involved in these 'Think Tank Meetings', which included other civil society organization and aimed to connect civil society action with government agencies.

8.9 Interim conclusion: a complex urban environmental governance set-up

The overall urban environmental governance in Delhi is characterized by predominantly sectoral approaches for environmental upgradation, overlapping responsibilities and ill-defined responsibilities of state agencies leading to a generally reluctant enforcement of environmental regulations despite a proactive judiciary and civil society. The question remains whether the outlined state agencies and existing policies are designed to govern the river *and* the city or whether they can only govern the city *or* the river, since inherently sectoral approaches and sectoral mandates forestall any holistic approach for governing Delhi's riverscapes.

A major overarching need for coordinated action on the central level is evident between the MoUD and the MoHUPA dealing with urban issues, while questions of environmental management and pollution are dealt with by the MoEF and the CPCB. Additionally, the sectoral MoWR with the CWC as its subordinate agency plays a major role in the governance of rivers across the country – both in the rural and urban context.

Merging the insights from this chapter, Figure 22 provides a combined overview of the actors involved in governing Delhi riverscapes. While most actors have already been introduced, the remaining actors will be introduced in the course of the case studies (e.g. Yamuna River Development Authority, Yamuna Removal of Encroachment Monitoring Committee).

The multi-level analysis provided in this chapter highlights that the DJB, responsible for cleaning the Yamuna, and the DDA, responsible for the general land-use planning, emerge as key agencies in governing Delhi's riverscapes. While the DJB functions under the Delhi Government, the DDA is under the oversight of the Central Government. The following chapters will outline that these two agencies have largely failed to fulfil their mandates, because of various external (e.g. funding, availability of land) and internal (capacity, knowledge) limitations. These aspects are further superimposed by intergovernmental rivalries and interdepartmental competition.

Urban environmental governance: introducing actors and policy debates 161

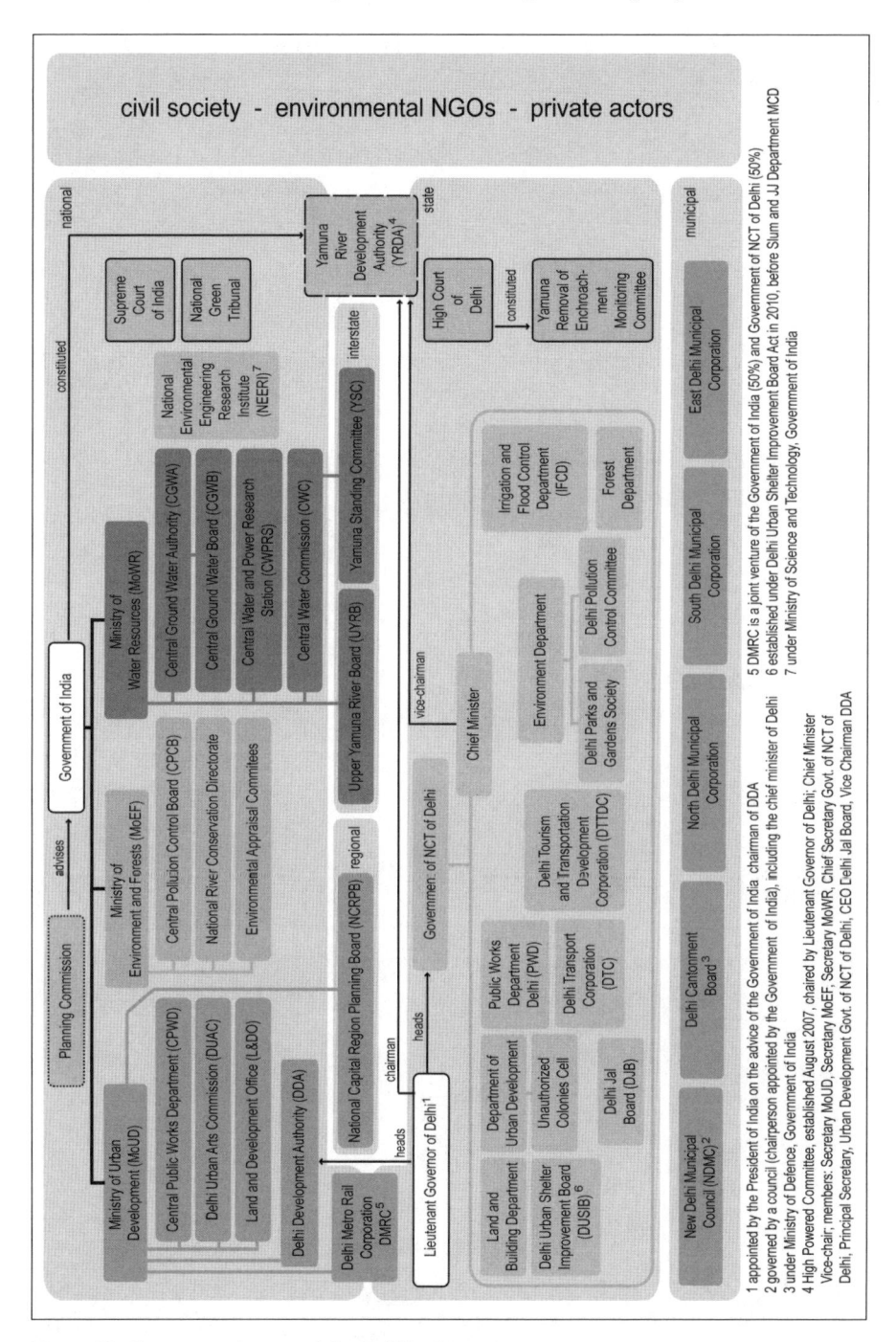

Figure 22: Governance framework for Delhi's riverscapes
Source: Graphic by author based on various sources and information from interviews

V HISTORICAL-GEOGRAPHICAL INSIGHTS INTO THE MATERIAL AND DISCURSIVE REMAKING OF DELHI'S RIVERSCAPES

This Part provides a multi-temporal analysis of both the material and discursive change of Delhi's riverscapes. The urbanization of the Yamuna is outlined from the Mughal period and early colonial period to the last decades of the 20th century. This retrospect provides the basis for the present-day research into Delhi's riverscapes. The central questions addressed in this Part are the following:

- Which processes and practices have produced Delhi's riverscapes?
- Which ideas and imaginaries have dominated the discourses around Delhi's riverscapes?
- What kind of relationships between the river and the city have emerged?

9 THE EARLY REMAKING OF DELHI'S RIVERSCAPES

The history of Delhi is deeply connected to the Yamuna as a lifeline of the city. As described by WILLIAM DALRYMPLE in his book 'The Last Mughal – The Fall of Delhi, 1857' the river was very much part of everyday life for Delhiites:

> "At Raj Ghat Gate, the earlier-rising Hindu faithful – at this time of the day women and their cotton saris far outnumbering the men – were streaming out to perform their pujas and have their morning bathe in the waters of the holy Yamuna before the crowds gathered and the dhobis appeared. Only the pandits kept them company this early in the morning: in small shrines lining the banks of the river up to Nigambodh Ghat […] the bells were ringing now for the morning *Brahma Yagya*, celebrating the creating and re-creating of the world over and over again, morning after morning" (DALRYMPLE 2009: 93).

Hindus worshiped the Yamuna throughout history. When compared to other cities along the Yamuna, especially to the holy cities of Vrindavan or Mathura (HABERMAN 2006), the importance of the river in the daily lives of the majority of the population has, however, shifted from an urban common area used for worship, bathing, washing, and other riverside activities, to a long-neglected backyard of the city (BAVISKAR 2011c). This chapter traces the roots of an ambiguous and consistently contested city-river relationship.

9.1 From the Mughal period to colonial times

While the Yamuna has always been vital for human settlement in the region, the *ancient cities of Delhi* did not depend on the river for drinking water, but rather drew their supply from the streams originating from the Ridge as well as harvested rainfall. A decentralized water supply system of tanks (*hauz*), ponds and stepwells (*baoli*) provided the storing facilities for monsoon rainfall and to some extent tapped the streams running from the Ridge towards the river (ASHRAF 2004, HASHMI 2010, 2013, HOSAGRAHAR 2011). The historic map 'Sketch of the Environs of Delhi' from ca. 1807 illustrates the streams originating from the Ridge and the adjoining settlements (see Figure 23). About 300 *baolis* are known today, and about 120 of these are listed by the Archaeological Survey of India (ASI) (ASHRAF 2004: 210).[48]

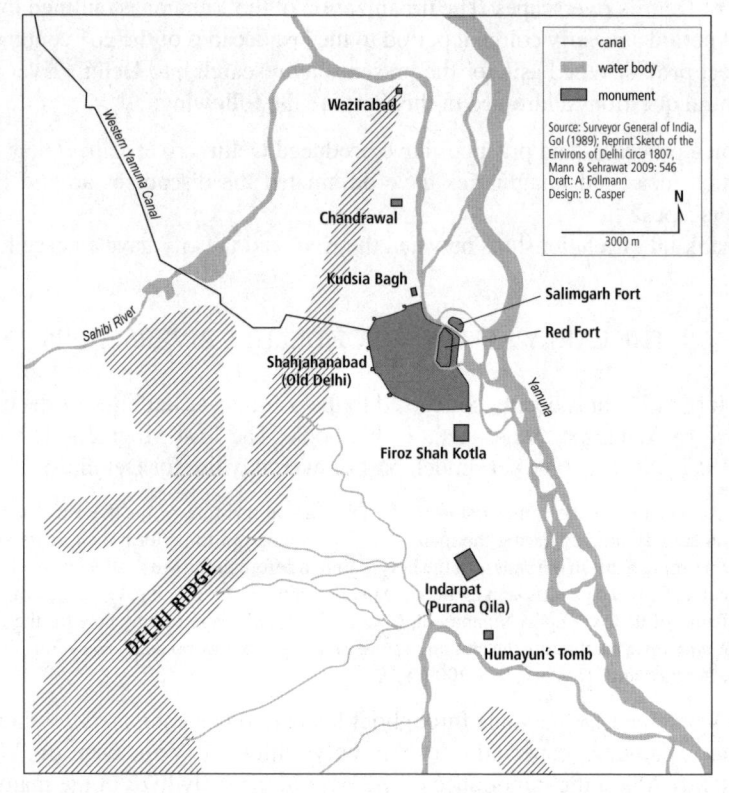

Figure 23: Delhi's riverscapes ca. 1807

---

48 The Delhi-based (water) historian SOHAIL HASHMI (2010, 2013) gives a detailed account on the water sources of the ancient cities of Delhi. He highlights that none of the seven ancient cities from Lal Kot under the Rajputs to Shahjahanabad under the Mughals was dependent on the river for drinking water. The first city (Lal Kot with Mehrauli) and the third city (Tughlaqabad) were located on the Ridge and, thus, had to depend on local wells and rainwater storage. The other ancient cities were located in the plains (Siri and Jahanpanah) and later Firoz Shah Kotla, Purana Qila, and Shajahanabad were not open to the river and did not draw their drinking water directly from the river (HASHMI 2010, 2013).

Under Firoz Shah Tughlaq, the first canal for diverting waters from the Yamuna into the city was constructed in 1355. The Mughal emperors Akbar and Jehangir revived the canal, and for the development of Shahjahanabad (Old Delhi) this canal was redesigned to irrigate the Mughal gardens (HILL 2008: 64).[49] The canal later dried up in the middle decades of the eighteenth century. PRASHAD (2001: 120) explains that "the desert began to claim Delhi". Due to "a drop in the Jamuna's water pressure" and "the gradual shifting of the drainage channels", the city's canals felt dry and "became cesspools" (ibid.). While this might be formulated a bit simplistically, certainly environmental changes in the region in the 18th century including deforestation (e.g. hardwood export by the East India Company) and intensification of agriculture had an influence on the flow regime of the canals and the river itself (HILL 2008). The exact reasons for the hydrological changes are largely unknown (cf. HOSAGRAHAR 2011: 117).

The Yamuna had already shifted its course towards the east from the Sultanate period to the Mughal period. ASHRAF (2004: 215–220) explains that the Red Fort and Humayun's Tomb might actually be located today in an area where during the Sultanate period the river was still flowing. Massive deforestation occurring in the Sultanate period might have contributed to the shifting of the river course due to increased soil erosion and siltation in the region.

In 1817, the British started realigning and reconstructing the Mughal canal as the Western Yamuna Canal (WYC), which supplied water to the city for drinking purposes by 1820 (HOSAGRAHAR 2005, MANN 2007, PRASHAD 2001). The Eastern Yamuna Canal (EYC), which was built under Shah Jahan and known as the *Doab Canal* in the Mughal times, was also refurbished between 1830 and 1854 for irrigation purposes (CWC 2014, HABERMAN 2006: 7). Aside from these new hydroscapes, water availability and water quality became a problem in the growing city in the 19th century as many wells turned brackish (SHARAN 2014: 27). As a result, the British started looking at the river for drinking water. The Sanitary Commissioner of the Punjab Province already in 1869 suggested building a waterworks at the river (GUPTA 1981: 88), but it took another 20 years until Delhi received a loan from the Provincial Government of the Punjab for constructing the first waterworks at Chandrawal in 1889 (ibid. 160).

Thus, while the river's waters were used for irrigation, animal drinking, and washing, the extraction of water from the river for drinking purposes only began during the British colonial times. In the beginning, the bank filtrate was sourced by wells. It was later done by using filter beds, settling tanks, and other technical equipment, until more and more water was directly drawn from the river (SHARAN 2014: 38, 2015: 2). Increasing demand, especially after the decision to shift the capital to Delhi in 1911, led to a "permanent water shortage" in the city (MANN 2007: 18). In 1921, a raw water pumping station was constructed north of all existing major settlements near the village of Wazirabad and in 1925 the Wazirabad waterworks started operation (DOUD 2006: 8-1, SHARAN 2014: 40, TOWN PLANNING

---

49 The canal is known as 'Ali Mardan Canal' named after the military commander in charge of the work under Sha Jahan.

ORGANISATION 1956: 6). By 1935, the raw water was lifted at Wazirabad from a temporary weir across the river (DoUD 2010: 11). The direct physical dependence of the city on the river increased substantially in the colonial period making the Yamuna the most important drinking water resource for Delhi. With the piped water supply now coming from the river, the historic, decentralized water systems became redundant. Many of the tanks and *baolis* felt dry. Today, many of them have been partially filled and built upon.

The sanitation situation changed drastically, too (cf. PRASHAD 2001). Shahjahanabad had already been equipped with drains (so-called *Shahjahani drains*) and the British first planned to refurbish this system. However, preoccupied with the goal to establish "the superiority of the British Empire over the Mughal period", an entirely new hydrological system was put in place by the British with piped water supply and a new underground sewage system (HOSAGRAHAR 2011: 117–118). A first "intramural drainage scheme" was finished in 1909, taking sewage outside the city walls where wastewater irrigation was practiced (PUNJAB GOVERNMENT 1912). In those days, domestic wastewater was still a valued resource for agriculture rather than perceived as a risk to the river or human health; its utilization was even contested among Delhi and the neighboring United Provinces (SHARAN 2015: 2–3).

With the water-based sanitation system, the local streams started turning into drains as a result of the city's increasing load of sewage. While aspects of wastewater with regard to public health were critically discussed in the following decades, the health of the river was no concern in this time (cf. SHARAN 2014). Outbreaks of cholera in 1928 and typhoid in 1930 highlighted the increasing public health problems (ibid.). The first sewage treatment plant was constructed at Okhla, starting operation in 1938 (ibid.). In colonial Delhi, like in many other cities around the world (see among others GANDY 2002, KAÏKA 2005), the new technologies of piped water supply and underground sewage systems were interpreted, and promoted, as the harbingers of modernity. But what did this mean for Delhi's riverscapes?

As the colonial city became more dependent on the river's water, the city's population was "cordoned off from the river […] in the name of 'cleanliness' and 'protection'" of the water resources (JAIN 2011: 36). The population of Delhi was deprived of its common public spaces along the river (BAVISKAR 2011c). Bathing in streams and the river was deemed as unhygienic and uncivilized in colonial discourses, viewing the Indian city as contaminated. The colonial *sanitizing* of the city considered many traditional riverfront activities as a nuisance and unlawful (cf. JAIN 2011). In order to keep the river water clean, bathing, washing and a series of other riverfront activities were prohibited (SHARAN 2014: 40).

The colonial policies alienated the local water resources from the communities, because the municipality centralized the governance of water supply and management, taking over the responsibility from the communities which had earlier been looking after wells, stepwells and ghats (HOSAGRAHAR 2011: 121). It is important to note here that this account is not about romanticizing local traditional knowledge or suggesting any fully harmonious human-nature relationship before the colonial time. Rather, the aim is to highlight that modernity's thoughts inherent to the colonial development agenda completely rejected existing practices and structures and massively transformed the human-nature relationship towards domesticating nature (D'SOUZA 2006b, HOSAGRAHAR 2011).

Further to the direct withdrawal of water from the river, colonial engineering also altered the local watersheds. In the early 19th century the lake-dotted *Sahibi Nala* (or *Sahibi Nadī*, Sahibi River) with the *Najafgarh Jheel* (Najafgarh Lake) used to flow around the offset of the Delhi Ridge and joined a branch of the Yamuna north-east of Wazirabad (see Figure 23). In order to drain out the flood water in the basin during the monsoon period, which "was more than the soil could absorb and was the cause of much sickness and fever", the construction of the Najafgarh drain (also known as *Jheel* Drain) began in 1838 (PUNJAB GOVERNMENT 1912). Colonial water engineering thereby reclaimed the marshy parts of the *dabar* in the Najafgarh basin. Through the use of canals, weirs, dams, and barrages, colonial engineering thereby altered the local drainage pattern of the Delhi region massively in order to assure a perineal water supply for irrigation.

In the 19th century, the spatial character of Delhi's riverscapes remained relatively persistent forming a braided river system with multiple channels. The Delhi District Gazetteer of 1912 describes the Yamuna as a river with a normal depth of "four feet only" in the "cold weather" being "only sufficient to supply the three canals which draw from it [...] and is then fordable in many places" (PUNJAB GOVERNMENT 1912). In the rainy season, the river was known for having a "considerable breadth, swelling in places to several miles, with a maximum depth of some 25 feet" (ibid.). The river in those days was "infested with crocodiles" and rich in aquatic life (ibid.). While the Yamuna used to be "navigable along the whole course" (except for the already existing Okhla Weir), shipping was only of minor importance (ibid.).

In 1866, the *lohā pul* (iron bridge) was constructed by the East India Railway in front of the Salimgarh Fort where a bridge of boats (pontoon bridge) used to exist before. In 1913, the *lohā pul* was converted into a double line bridge by adding two road bridges below the railway tracks (PUNJAB GOVERNMENT 1912). The remaining bridge is referred to as the Old Railway Bridge. Crossing the river was further possible by ferries operating at nine different locations (including Jaitpur, Wazirabad and Okhla) and the Okhla Weir could be used by foot passengers only (ibid.).

In the beginning of the 20th century, the river did not flow freely anymore, but had already been retained by embankments. At the time of the Delhi Durbar in 1911, a continuous embankment about 12 km long existed already along the east bank stretching from south of the railway line to the Okhla Weir (SURVEY OF INDIA 1913b). In 1922, this embankment had a height ranging from 2.5 to 3.7 meters, which is only slightly less than today (SURVEY OF INDIA 1922). Today's Left Marginal Bund (built in 1955–56) and the Pushta Road (also called Noida Link Road) follow this historic embankment up to the later built Hindon Cut Canal. In the north from Jagatpur Village to the offset of the Ridge at Wazirabad, the river was also already controlled by an embankment about 2 km long with a relative height of 8 feet (2.4 meters). In the East, an embankment of 7 feet (2.1 meters) relative height with two spurs[50] controlled the meandering of the river towards the East (ibid.). The low heights of these embankments indicate that they were not primarily designed for flood protection, but rather to ensure that the river in the dry season does not shift and water could be lifted at Wazirabad. On the left bank, embankments were absent north of the railway line and during floods the river expanded all over the Shahdara area up to the alignment of the EYC.

The river was not static in its bed, but rather was shifting towards the East. The structural changes associated with the construction of, and protection work for the Old Railway Bridge resulted in a further shifting of the river. By 1922, the first man-made embankment (called the *Mughal Bund*) was built on the right bank stretching about 400 meters from the boundary wall of the Salimgarh Fort in south-easterly direction (SURVEY OF INDIA 1922). The *Bela* (the swamp between the river and the Red Fort), which was regularly flooded prior to the 1920s, bringing the river's water up to the walls of Old Delhi, was thereby reclaimed. Due to its proximity to the Old City, the Bela area in the following decades emerged as an important location in the remaking of Delhi's riverscapes.

Further south, the construction of the railway line in front of *Purana Qila* (Old Fort) and Humayun's Tomb cut of the ancient monument from the river.[51] Urban development did not exceed over this railway line up until the 1950s. Aside from the development of New Delhi (for details see 9.2), the growth of Old Delhi in the 1920s to 1940s exceeded beyond the city's walls. New urban expansion came up towards the west and the north of the walled city (MANN 2006: 133, MEHRA 2013: 356).

One of these early urban expansions to the north is the area today known as *Yamuna Bazaar*. Located directly north of the Salimgarh Fort, the area used to be part of a large forested river island in the 19th century and became the city's new riverfront when the branch of the river dried up. Development of Yamuna Bazaar started sometime in the late 1920s to early 1930s (cf. SURVEY OF INDIA 1922, SURVEY OF INDIA 1933). Before Independence the area already grew into a major settlement around the cremation ground of *Nigambodh Ghat* (SURVEY OF INDIA 1945). The Nigambodh Ghat is believed to be the site where Lord Brahma took bath in the Mahābhārata and where the Pandavas hold vedic rituals. The ghat is today Delhi's largest Hindu cremation ground and consists of a series of bathing and ceremonial steps leading down to the river.

---

50 Spurs are structures built transversally from the main embankment into the riverbed in order to interrupt water flow limiting the movement of sediment (inducing silting) and directing the flow away from the embankment.
51 Purana Qila (Old Fort) was built at the riverfront by Sher Shah Suri in the 16th century on the site of the former village Indrapat, which is associated with the Indraprastha of the Mahabarata (PECK 2005: 136). Humayun's Tomb is the mausoleum for the Mughal Emperor Humayun built in 1569–70 (PECK 2005: 146–147).

The area along *Kudsia Ghat* (north of Nigambodh Ghat, between Kudsia Garden[52] and the river) was still open to the river without any major built up space in the early 1920s (SURVEY OF INDIA 1922). In the 1930s to 1940s, the area was transformed into a settlement and today's existing ghats and temples pre-date India's Independence (SURVEY OF INDIA 1933, 1945). The riverfront from Kudsia Ghat to the Old Railway Bridge was an early focal point for urban redevelopment schemes as it was in proximity of the British Civil Lines. Already under the DIT, a layout plan was prepared for Nigambodh Ghat and parts of Yamuna Bazaar (DDA 2010b: 24, LEGG 2007: 180). The *Jumna Village*, as the Yamuna Bazaar area was known in those days, was described in the reports of the DIT as an "disreputable and beggar infested area" (Delhi Improvement Trust Three Year Programme 1941–1944 quoted from LEGG 2007: 180). Parts of this area along the Bela Road (the main road stretching from the Old Railway Bridge to Nigambodh Ghat) are Nazul lands and therefore were under control of the DIT. The DDA cleared parts of Yamuna Bazaar in June 1967 as part of the first massive slum clearance drive in the city of Delhi in 1967–1968 and developed a park in this area (TARLO 2000: 57, 2003: 132).

These historic insights highlight that the riverfront along the Western bank has always played an important role in the urbanization of Delhi. The areas were favored locations for urban development and popular residential areas both within and outside the walled city (GUPTA 1981: 58). Through the construction of first embankments on both banks, the meandering of the river was already restricted and the alignments of the major embankments still remain today. The 'natural' riverscapes were thus already massively transformed by the urbanization of the river. The embankments constituted early boundaries of the land belonging to the city and the river.

It is important to highlight that administrative boundaries in those days differed from state boundaries today. When the embankments on the left bank stretching from the Old Railway Bridge to Okhla were constructed in the colonial period, the possession of the land east of the river was with the United Provinces. Therefore the Irrigation Department of the United Provinces (later Uttar Pradesh) was in charge of the embankments. In spite of controversy of the land's location between the governments of Delhi and Uttar Pradesh, the land along the embankment remains in Uttar Pradesh ownership. This land dispute adds to an already complicated interstate relationship between Delhi and Uttar Pradesh, as Uttar Pradesh is not only the upstream and downstream riparian, but also vests powers over land located within the territory of Delhi. This conflict creates a major fault line in the overall governance of the river Yamuna and highlights that political boundaries most often do not correspondent with 'natural' boundaries in environmental governance.

---

52 Kudsia (also Qudsia) Garden is a relic of a palace and garden complex erected by Qudsia Begum, the wife of Emperor Muhammad Shah Rangeela, in 1748. In the colonial period, the Kudsia Gardens were part of the *cordon sanitaire* separating the Civil Lines, exclusive residence for the Europeans, from the 'native' Old Delhi. The Civil Lines were developed with the shift of the British troops to the Ridge Cantonment in 1828.

Land conflicts between the states have a long history. For example, the Delhi Gazetteer described land conflicts between the colonial provinces already in 1921 (PUNJAB GOVERNMENT 1912: 185):

> "The disadvantages of fluctuating boundaries along the Punjab rivers have been recognised and fixed boundaries have been demarcated throughout the Province under the provisions of the Riverain Boundaries Act. The Punjab and United Provinces Governments are now in communication on the subject of creating a fixed juridical boundary along the Jamna [Yamuna]; but, of course, even in the event of a successful issue, alluvion and diluvion [sic] measurements will have to be made for the estimate of changes of land revenue."

The fixing of political boundaries was thus dependent on the domestication of the river and the fixing of the boundaries between land and water. Furthermore, the administrative boundaries also changed with the shift of the capital city to Delhi.

9.2 A new riverfront for the imperial capital

When Delhi was chosen as the new capital of British India, the first plan for developing Yamuna's riverfront was proposed by the Delhi Town Planning Committee in 1913 (DELHI TOWN PLANNING COMMITTEE 1913, cf. SHARAN 2015). The British first planned the new imperial capital in North Delhi as it was close to the river. Based on the recommendation of the Delhi Town Planning Committee, it was decided to locate the new capital on the higher grounds of Raisina Hill south of the walled city. Due to the hazards of mosquito breeding and water-borne diseases commonly associated with floods, as well as standing water in the marshy areas, low-lying extents along the river were rejected (JAIN 1990: 49).

The Delhi Town Planning Committee envisioned a new riverfront for the city with several ghats from Wazirabad in the north to *Purana Qila* in the south. The new capital's central vista, today *Rajpath* (King's Way), was planned to connect Raisina Hill directly to the new riverfront at Purana Qila and a new barrage-cum-bridge was planned across the river (SURVEY OF INDIA 1913b). As SHARAN (2015: 3) outlined recently, the planning committee was of the opinion that the development of the riverfront would "add considerably to the attraction of the new capital". Riverfront development was seen as "an important step towards the complete eventual development and embellishment of Delhi" (DELHI TOWN PLANNING COMMITTEE 1913, cited in SHARAN 2015: 3).

While the planning committee valued the river's special quality for urban development, Lutyens' later planning for New Delhi largely neglected the natural connection between the local watershed of the Ridge and the Yamuna. Many natural streams flowing from the Ridge into the Yamuna were cut off by the development of the new capital city (SURVEY OF INDIA 1922).

In the course of large scale land acquisition for the new capital, the Delhi Town Planning Committee proposed to obtain control over the land along both banks of the river from Wazirabad in the north towards Humayun's Tomb in the south

(SURVEY OF INDIA 1913a).[53] Accordingly, and based on the Land Acquisition Act of 1894, the colonial government took over control of major parts of the land along the river (some parts of the lands were already under the colonial government as *Nazul lands,* e.g. *Bela Estate*). In 1915, the Shahdara area east of the river came under Delhi Province (JAIN 1990: 72).

The colonial plans of the Delhi Town Planning Committee to develop Delhi's riverfront were not realized. In contrast, Delhi's riverscapes were largely neglected and many of the river's tributaries were realigned as storm water drains for the development of the new capital. The land of the floodplain, which was once partly a wooded area and grassland, was more and more used for cultivation under control of the District Board in the northern part and the DIT for the *Nazul lands* south of the Old Railway Bridge (see SURVEY OF INDIA 1922). If the grand vision of the colonial time had been realized, ground realities would undoubtedly be much different today. Despite their lack of materialization, these early ideas have lasted, as the following discussion shows. The planners' wishful dreams continued to influence "urban imaginations" in the following decades (SHARAN 2015: 3).

9.3 The riverfront ideas of the early post-colonial period: the 'recreation plan'

After Independence, the Town Planning Organisation was set up by the Indian government under Jawaharlal Nehru in 1955 to formulate the First Master Plan for Delhi. The Town Planning Organisation presented its Interim General Plan in 1956, which outlined that the riverfront from Wazirabad in the north to Okhla in south should "be developed as a major recreation area" (TOWN PLANNING ORGANISATION 1956: x). Against the backdrop of the later plans for channelization and today's most recent discourses on the river it is worth quoting the planners' vision for the river in the 1950s at length:

"Fortunately not much development has taken place near the banks of the river except for Kashmere Ghat Area [Yamuna Bazaar]. The entire water front should be developed for recreational purposes, and sections [...] should be developed more intensively for public recreation such as playgrounds, swimming pools, fishing areas, bathing ghats, beaches, etc. Some thought should be given to the possibility of developing the flow of the river and building a dam across it which would serve many purposes at one time. The lake behind the dam would maintain an even level of water throughout the year and would also control the river during the monsoon season" (TOWN PLANNING ORGANISATION 1956: 25).

In line with the overall goal to stop the growth of the city in its "haphazard fashion" (TOWN PLANNING ORGANISATION 1956: 57), the committee substantiated its call for planned development along the riverfront development with the fear of unplanned development taking over the valuable land:

53 The committee proposed three different priorities of lands for acquisition: 1) area recommended for purchase for imperial city and cantonments, 2) area recommended for purchase for the extension of the present city and civil lines and for sanitation, and 3) area recommended for purchase or firm control. Major parts of the riverbed were demarcated as second priority lands.

"Jumna [Yamuna] River front has been lying undeveloped. A beautiful park should be developed and incorporated in the proposed national park system all along the river front for public recreation, before this vacant land is infested by any intensive use. This could join in the north with the new planned golf course and major recreation park to form a well planned integrated programme for recreation" (TOWN PLANNING ORGANISATION 1956: 28).

Since the riverfront had at that time "remained unspoiled from intensive land use" (TOWN PLANNING ORGANISATION 1956: 69), the committee aimed to link the river to the city on both banks. The multiple recreational activities were envisioned to be "inter-connected by a parkway and pedestrian walkways" with the monuments of the Red Fort, Jama Masjid, India Gate and Humayun's Tomb (ibid.). The Interim General Plan of the TOWN PLANNING ORGANISATION (1956: iii) further indicated the importance of green spaces in its foreword:

"Green spaces and open recreation areas recede further and further. Unplanned growth in Delhi has caused population to run ahead of water supply and sewerage capacity. [...] A greenbelt must be adopted both for a healthy metropolitan breathing space and as a limit to the city's growth."

These early postcolonial ideas for the riverfront were undoubtedly influenced by western planning concepts. In particular, the European and North American urban legacy of parks and green belts as 'green lungs' ensuring the circulation of fresh air positively contributing to human health and well-being (e.g. see for New York GANDY 2002: 83, for Delhi's green belt JAIN & SIEDENTOP 2014). Therefore, Delhi's riverscapes were considered to be important 'green lungs' for the sprawling city. It is important to note that the water quality of the river in the 1950s was much better than today.

The committee was, however, aware that the implementation of the recreation plan "would not be easy or cheap", but in the view of the committee a policy decision had to be taken since "[d]elay would inevitably result in good land being used up for all purposes except recreation and parks" (TOWN PLANNING ORGANISATION 1956: 70). The committee further proposed a "multi-purpose scheme for flood control" to open up East Delhi for "urbanisation and industrialisation" (ibid.). The expansion of East Delhi, where developments by that time had already taken place along the railway line to Ghaziabad, was an important component of the recommendations of the Town Planning Committee. The committee was for example in favor of locating the new international airport across the Yamuna (TOWN PLANNING ORGANISATION 1956: 72).

After Independence, the government gave land in the floodplain to farmer cooperatives on lease contracts for agricultural production (for details see chapter 13.1). Other parts of the land along the river were still used by the public for various recreational activities. For example, up to the early 1970s, flood fair festivals and 'Flood Olympics' were organized at the riverfront and competitions for crossing the river were held during floods. A senior environmentalist recalls the situation from his memories (NGO-06/2010-09-24):

"Some thousand people [...] gathered on the banks to learn swimming and to enjoy swimming. [...] This scene was regular exercise during summer up to 1960s. I have seen a Yamuna River

Festival in 1973, but it was about to vanish. Very few people came and very few means by their definition some thousand people were there on the banks. But then it [Yamuna] became a drain."

With the development of the Mahatma Gandhi memorial site (today commonly referred to as Raj Ghat) at the site of his cremation in front of the Red Fort[54] a first step was undertaken to realize the ideas of the 'recreation plan'. Later the *Shanti Van* for Jawaharlal Nehru and *Vijay Ghat* for Lal Bahadur Shastri and other samādhis (memorials of Indian political leaders) were also developed between the Red Fort and the Mughal Bund. While the park-like samādhis remain as green and recreational areas, they are all oriented towards the Ring Road (Mahatma Gandhi Road) and the city; a direct link to the river is still lacking.

As the following discussion will illustrate, the development of the riverfront into large-scale recreational areas did not come to fruition as it was envisioned by the planners of the Town Planning Committee. The river itself, by bringing massive floods, and increasing levels of pollution played a major role in this process. The increasing levels of pollution in particular discouraged the population from using the river for recreational purposes, while the city's dependency on the river's water further rose.

9.4 Taming the river for the development of the city

At about the same time when the ambitious recreation plan was outlined by the Town Planning Committee, the need for a safer and secured urban water supply became undeniable. During the floods of 1955 and 1956, the river shifted its course towards the east, making way for the wastewater from the Najafgarh drain to contaminate the water in front of the intake point. This proved to be the cause for a massive outbreak of jaundice in 1956 (SHARAN 2014: 66, 2015: 4). Thus, the growth of the city and the water-health nexus required drastic interventions in the flow regime of the river.

In order to separate the waters of the Najafgarh drain from the fresh water of the river, it was decided to construct a barrage across the river at Wazirabad. The construction of the Wazirabad Barrage was finished in 1957 and since then fresh drinking water for Delhi is taken out from the river above the barrage. Additionally, the Najafgarh drain was diverted slightly to the south to outfall into the river downstream of the barrage.

The construction of the Wazirabad Barrage had serious consequences for the Yamuna. The barrage obstructs the free flow of the river and led to an ever increasing abstraction of water from the river in the following decades.

Besides taming the flow of the river for water supply, the river was further 'domesticated' by additional flood control works, which aimed at restraining the river

---

54 After his assassination, Mahatma Gandhi was cremated in front of the Red Fort on 31st January 1948. The development and maintenance of the memorial site was entrusted to the Rajghat Samādhi Committee under the Rajghat Samādhi Act by the Government of India in 1951.

to its 'bed'. Already in the year of Independence (1947), Delhi had experienced a major flood during which the waters of the Yamuna and the Hindon merged north of Delhi and the low lying areas of East Delhi were seriously flooded (NGO-04/2010-04-01). Drawing on historic newspaper reports from the Times of India, SHARAN (2015: 3) outlined that these floods had washed away houses, damaged crops and even "marooned people and cattle". Serious floods occurred also in 1924 and 1933. During these large scale flood events, the limited heights of the existing embankments were often overtopped and monsoonal floods inundated large areas up to 5–6 miles in width (IFCD 2010, SHARAN 2015: 3).[55] The low-lying areas of East Delhi were virtually inundated every year (IFCD 2010).

As already recommended by the Town Planning Committee at the time, and as a response to the major floods in 1955 and 1956, the *Shahdara Marginal Bund* was constructed north of the Old Railway Bridge in 1957. South of the railway line the *Left Marginal Bund* was gradually strengthened over the following years. Together these two embankments formed the first continuous embankment (today generally referred to as the *Yamuna Pushta Embankment*) on the left bank. This can be interpreted as the beginning of the extensive urbanization of East Delhi since the early 1960s. The *Yamuna Pushta Embankment* defined the active floodplain of the Yamuna towards the east.

In sum, the 'taming' of the Yamuna has been motivated by two aspects: an increased and safe abstraction of water was the reason to dam the river by the Wazirabad Barrage, and flood protection was the aim behind restraining the river to a spatially fixed riverbed. In doing so, the free meandering and shifting of the river to the East was largely constrained. Therefore the technical, engineered boundaries of the river in Delhi were already fundamentally formed well before the first Master Plan for Delhi was adopted in 1962. Additionally, engineered interventions formed the right bank through the alignment works for the Old Railway Bridge and the later construction of smaller embankments (e.g. the Mughal Bund in front of Red Fort). As a result, areas along the west bank dried up and provided open spaces which were envisioned to be reserved for green and recreational uses by the Town Planning Organisation. These recommendations also found their way into the first Master Plan.

---

55 Information confirmed during interview NGO-04/2010-04-01.

9.5 Delhi's riverscapes under the first Master Plan

Delhi's riverscapes, as defined in the late 1950s by the Yamuna Pushta Embankment in the east and the existing urbanized areas in the west, were considered flood prone areas and remained largely outside of the envisaged planned development of the city of Delhi in the first Master Plan for Delhi (MPD) enacted in 1962 (DDA 1962, cf. JAIN 1990: 130–131).

In the west, the *urbanisable limit* (land for development) set in the first Master Plan largely reflected the natural transition from the high grounds of *bangar*, which in some parts extended close to the river, to the lower areas of the *khadar*. Following the request for recreational areas along the river by the Town Planning Committee, the land-use plan of the first Master Plan proposed riverfront parks and greens stretching almost the entire length of both sides of the river (see Appendix V). The first Master Plan explicitly recommended the redevelopment of the Nigambodh Ghat "with more bathing ghats and parks" (DDA 1962: 37) and the relocation of the JJ colonies from the earlier mentioned Yamuna Bazaar area to develop a riverfront park south of the Old Railway Bridge (DDA 1962: 33).

With regard to the pollution of the river, the first Master Plan stated that "proposals to discontinue sewage overflows into the Yamuna will have to be executed soon" (DDA 1962: 37). The proximity of the Rajghat power plant to the national memorial sites (samādhis) was also considered problematic and a shifting of the power plant was proposed (DDA 1962: 37):

> "The existing power house, being too near Rajghat, will continue till such time as the machinery becomes obsolete, so that the Rajghat site will be completely open to the river. A new site has been earmarked for the location of a big thermal plant south of the present site as it will need constant and large supplies of water from the river and also a railway siding for coal."

The power plants came along with the development of large fly-ash ponds. The fly-ash ponds are located east of the samādhis towards the river. These bulky and polluting infrastructure developments dramatically changed the physical appearance of Delhi's riverscapes in this stretch. Furthermore, the construction of the Ring Road along the riverfront began under the first Master Plan. JAIN (1990: 130) noted that the "vast and intensive traffic axis [...] along the urbanisable limits [...] has acted as an effective barrier between men and the river on the western bank". Like in other cities around the world, the river-city interface was dominated from that time onwards by the road transportation corridor. Additionally, the railway line connecting the power plants for coal delivery to the main railway line was another physical barrier separating the city from the river. The Ring Road and the railway lines, together with the power plants and fly-ash ponds, cut off the river from the city and prevent direct access.

While the city at large neglected the river, the flood-prone, increasingly polluted and overall marginalized 'left-over' spaces between the power plants and fly-ash points and along the transportation lines and the river have been settled informally by the poor since the early 1970s (BATRA & MEHRA 2008, BAVISKAR 2011c,

BHAN 2009, FOLLMANN & TRUMPP 2013, SHARAN 2015). The Mughal Bund provided some protection from the yearly flooding and acted as the nucleus for the development of the slums in this stretch (for details see 13.2). The development of the slums was diametrically opposite to the visions for the area outlined by the Town Planning Committee. However, the state agencies – especially the DDA – were not able, and to some extent not willing, to enforce the regulations of the MPD. Most importantly, because the housing provisions of the DDA "failed to meet the demand of the poorest section of the population" (DUPONT 2008: 79) forcing them to develop 'their own city' informally (cf. BATRA & MEHRA 2008, BHAN 2009, 2013, DATTA 2012). The developments along the river confirmed the planners prevision and fear of unplanned, "haphazard" development (TOWN PLANNING ORGANISATION 1956: 57), which is referred to as the *fear of encroachment* storyline in the following chapter.

The river itself was further engineered through the construction of the ITO Barrage (barrage-cum-bridge, generally referred to as the 'ITO Bridge') by the Haryana Irrigation Department to provide water supply to the power plants in 1966–68. Besides a further obstruction of flows, the involvement of the Haryana Irrigation Department further added to the complex water and river governance setting for the Yamuna in Delhi. The next major step in the engineering of the river was the creation of the new Okhla Barrage for water diversion to the Agra Canal in 1986.

Furthermore, besides the longitudinal transportation corridor of the Ring Road, the MPD gave great importance to the transportation connectivity across the river in order to facilitate the development of the Trans-Yamuna area. In the early 1960s the Old Railway Bridge was the only bridge over the Yamuna; the road over the Wazirabad Barrage was under construction; and three more road bridges (including the Nizzamudin Bridge opposite Humayun's Tomb) and the New Railway Bridge (opposite Purana Qila) were proposed (DDA 1962). The first MPD also proposed strengthening of the existing embankments in the east "so that there is no danger of floods from the river" and highlighted the danger of waterlogging in low-lying areas (DDA 1962: 25).

9.6 Interim conclusion

As outlined in this chapter, riverfront development focusing on the remaking of Delhi's riverscapes for recreational uses dominated the urban imaginations and planning discourses in the late colonial period, as well as the early post-colonial period. The envisioned schemes aimed to connect the monsoonal river and the city through riverfront parks and other recreational areas. Yet as the city grew, the floods of the river increased and the development of the city became increasingly endangered. Flood protection measures in form of embankments were constructed and every new embankment contributed to the disconnection of the river from the city; both in terms of visibility and direct access to the river. In contrast to the vision of the Town Planning Committee and the first Master Plan to connect the river to the city through large recreational areas, the actual development of infrastructure provisions (roads, power plants, etc.) and increasing levels of pollution had the opposite effect, disconnecting the river from the city.

The area between the Red Fort and the river along the Mughal Bund is no exception to the separation of the city and the river. The development of the *samādhis* in the area was in line with the 'recreation plan' of the early post-colonial period. However, the park-like areas do not connect the city to the river. Rather, together with the transportation infrastructure, the *samādhis* appear as a physical boundary between the city and the river.

Towards the river, the formation of the slums along the Mughal Bund added another physical barrier between the river and the planned city. The planned city and the river moved away from each other physically (since the river course shifted slightly to the east), as well as visually and mentally during both the colonial as well as post-colonial period. The very existence of the river has been "gradually forgotten" (JAIN 2011: 42) by the majority of the city's population. In the eyes of the planners the riverfront, the former 'vacant land', as foreseen by the TOWN PLANNING ORGANISATION (1956: 28), had been lost to the growing city.

Considered as flood prone area, the planners had evaluated the 'land' along the river as not suitable for planned development. The *khadar* was generally considered as space of the river, and only some stretches were aimed to be transformed into parks. Delhi's riverscapes were thus not integrated in the planned development of the city. Originally they fell outside the spatial urban development areas (DDA 1962), yet, when the city grew faster than forecast in the plan, and urban growth exceed far beyond the urbanisable limits set by the plan, the planners became aware of the 'failure of planning' – or rather non-planning for the river.

The dichotomies of the Master Plan including river/city and urbanisable/non-urbanisable areas were increased by the dichotomies of planned/unplanned, formal/informal, legal/illegal (cf. BHAN 2013: 69). Large parts of Delhi's riverscapes had become fragments of the unplanned city. As the 'recreation plan' was jeopardized by the slums and the increasing levels of pollution, the planners themselves – under the fear of further encroachment – started to reclaim land along the river for 'planned' urban uses (e.g. further power plants), often in an ad hoc way. Thus, with the shifting of the river's course the 'new' land resources emerging on the western side of the river were becoming part of the urban fabric of the growing city; partly planned but principally unplanned and informal.

The historical analysis further indicated that urbanization has affected the river Yamuna remarkably, both within and beyond the city's boundaries. Overall, the 'early' remaking of Delhi's riverscapes suggests that the city-river relationship was originally defined by the plenty resources the riverscapes offered the city (water, fish, agricultural products). While the early history of Delhi's hydrological system was shaped by the rhythm of the monsoon and largely independent from the river, the hydrological connection between the river and the city increased substantially over time. The historical development of Delhi's riverscapes illustrates the effects of human interventions aim to alter nature in order to produce a river according to the ever-increasing needs of the city. As a result, the urban metabolisms intensified with regard to water. But the socio-natural link between the people and the river, based on the religious importance of the river (spatially manifested in the ghats) and the recreational opportunities along the river, weakened as the river was more and more cut off from the city.

The historical perspective further reveals that the boundaries of the 'active' floodplain of the Yamuna, in its basic geography, were already defined by the construction of the embankments on the left bank at the beginning of the 20th century. These boundaries between land and water were still fluid in times of monsoonal floods; especially when the river overtopped the embankments. With the construction of the consisting Yamuna Pushta Embankment in the mid-1950s the physical boundaries between the river and the city were manifested and have largely remained stable. This also had political repercussions beyond Delhi.

Previously, the natural movement of the river had produced shifting boundaries and boundary making between the provinces was a complex and conflicting exercise. With the shift of the capital city state boundaries were realigned which in turn led to conflicts in controlling the boundaries set by the historical embankments.

In sum, the river has long been at the center of a political inter-state conflict, both with regard to its flows as well as the land of the floodplain. These conflicts were further complicated by the shifting of administrative boundaries and the colonial state system being transformed into a democratic setting. Apart from conflicts about water, land conflicts persist as the neighboring state of Uttar Pradesh, in particular the Uttar Pradesh Irrigation Department, remains in control of the land along the embankments in the NCT of Delhi. This situation adds to the complexity of governing Delhi's riverscapes, making not only water, but also land a disputed interstate issue.

Based on these historical insights into the early material and discursive remaking of Delhi's riverscapes, the following chapter outlines the ideas for channelization that started to dominate the discourse on the river in the mid-1970s.

10 CHANNELIZATION AND RIVERFRONT DEVELOPMENT – A PERSISTENT IDEA

Reviewing planning documents of the DDA and other state actors, including the minutes of the YSC, the chapter outlines the emergence, structuration and institutionalization of the channelization discourse for the Yamuna in Delhi from the mid-1970s to the late-1990s.

10.1 The emergence of the channelization idea

The minutes of the meetings of the YSC indicate that the "training of river" (YSC/33/1975-05-23) downstream of Tajewala, and the channelization of the Yamuna in the Delhi stretch in particular, have been discussed since at least the mid-1970s. In the 33rd meeting of the YSC in May 1975 the interstate committee agreed to make an attempt for the channelization of the river Yamuna. In those days, the channelization plans were not restricted to Delhi alone, but channelization was proposed for the entire stretch of about 250 km from Tajewala Barrage to Okhla.

The minutes of the YSC outline the view of the concerned engineers in the mid-1970s (YSC/33/1975-05-23). In May 1975, the river is described in the YSC minutes as "a braiding river" deposing its silt in a rather "irregular fashion sometimes at one bank or the other" (ibid.). As the river meandered "over a very wide Khadar area", its shifting course and eroding forces were "endangering the villages and agricultural land" (ibid.). Thus, hoping to "reduce its destructive effect and permit reclamation of land" the engineers "felt that the river can be tackled in its totality" pressing it into a stable river channel of about 2000 feet (610 m) to carry its discharge (ibid.).

The committee decided that an attempt should be made to channelize the river in its entire segment from Tajewala to the south of Delhi (ibid.): "Though it may not be possible to attempt the channelisation in its entire length, it is, however, desirable to take up this work reach by reach." In this context, the YSC had initially discussed the purchase of dredgers for river control works in the Delhi stretch, but this proposal was rejected and "rather training of the river by inducing siltation and utilising the rivers' [sic] own force in securing out a stable channel for itself" was agreed upon by the members of the committee in May 1975 (YSC/33/1975-05-23).

In the following meeting in December 1976, the 'Channelisation of river Yamuna' was, for the first time, discussed as a separate item in the meetings of the YSC. During the discussion, the Chief Engineer Floods, Delhi (IFCD) made a proposal for the construction of a continuous "forward embankment" on both banks of the river between ITO Barrage and Nizammudin Road Bridge in a phased manner starting with an embankment on the left bank between the Nizamuddin Road Bridge and Nizamuddin Railway Bridge (YSC/34/1976-12-16). This stretch has therefore been at the center of urban development since the mid-1970s. Nevertheless, the development of a forward embankment in this stretch was only adopted about

25 years after the first proposal (for a detailed discussion of the Akshardham Bund see chapter 14).

The following analysis illustrates how the idea of channelization was criticized for various reasons from the beginning. In the meeting of the YSC in April 1978, the representative of the Indian Railways and other committee members requested detailed model studies for the assessment of the effects of the proposed forward embankment on the flow of the river and possible adverse effects upstream and downstream. Responding to the critique, the engineer from IFCD ensured "that the proposal was not at present under active consideration" (YSC/35/1978-04-25). Yet, other materials indicate that the IFCD had already requested the Central Water and Power and Research Station (CWPRS) to conduct model studies for channelization in 1977 (GUPTA 1995: 684). The DDA had also been working on first ideas for channelization in the Delhi stretch in the late 1970s (cf. GUPTA 1995, JAIN 1990, 2009a).[56]

While the minutes of the YSC-meetings in the mid-1970s suggest an overall agreement in the engineering community that channelization of the river Yamuna should be taken up, the members "felt that some sort of guideline may be evolved for adoption in this reach on experimental basis" (YSC/33/1975-05-23) since "channelisation of river in totality" had not been attempted in India. Additionally, the members from the CWC warned that already existing embankments were "cutting of the valley storage" and additional embankments might have a "consequential effect on increase in flood peak discharges and/or rise in flood levels" along the river (YSC/35/1978-04-25). It was thus agreed that the riverine states collect all available data on existing and planned embankments for model studies coordinated by the CWC. The floods of 1978, however, marked a turning point in the discussions at the YSC.

10.2 Learning from the floods of 1978

During the floods of 1978, the highest water level in recorded history was measured at the Old Railway Bridge in Delhi, at a height of 207.49 meters (for details see chapter 7.5). Breaches in the Right Marginal Embankment upstream of Wazirabad resulted in flooding of residential areas causing an estimated damage of "nearly Rs. 10 crores [100 million rupees], eighteen lives were lost and thousands of people were rendered homeless" (IFCD 2010). On the east bank, the Shahdara Marginal

---

56 The authors and former DDA planners, Ashok Kumar (A.K.) Jain and R.G. Gupta, were both involved in the planning for the Yamuna. A.K. Jain was Commissioner (Planning) DDA. R.G. Gupta worked with the DDA from 1965–1994. In their books, 'Shelter for Poor in the Fourth World (Vol. 2).' (GUPTA 1995), 'The Making of a Metropolis' (JAIN 1990), and 'River Pollution' (JAIN 2009a), they provide detailed information on the different schemes for channelization and riverfront development. R.G. Gupta prepared the detailed plans for the channelization schemes between 1988 and 1992, and shared the materials with the author. He was involved with the project from 1977–1994. The author discussed the channelization schemes with both retired DDA planners during meetings in Delhi.

Bund also "reached the point of distress but could be saved by raising its heights in certain reaches with earth filled bags" (IFCD 2010). While the discharge of the river during the floods of 1978 remained distinctly below the calculated hundred year flood mark, major breaches in the upstream between Tajewala and Delhi were considered to have considerably reduced the flood wave in Delhi, saving the city from greater destruction while causing devastation in the upstream areas (YSC/37/1979-04-26, cf. JAIN 2009a: 43).[57]

The YSC evaluated the floods of 1978 to be of "unusual magnitude" (YSC/36/1978-10-28). The committee stipulated that a detailed mapping of the flooded areas should be prepared by the Survey of India "to have a proper documentation of this event" and the breaches in the embankments (ibid.). Additionally, the Ganga Flood Control Commission (GFCC) was requested to review the existing guidelines for the spacing of embankments.

The flood of 1978 moreover resulted in serious tensions between the states with regard to existing and future extension of flood protection measures. For example, the Uttar Pradesh Irrigation member emphasized that while the YSC was set up in 1961 "for safeguarding the interests of Delhi against undue increase of flood levels in Yamuna river due to flood protection works upstream", the situation had changed considerably insofar that "Delhi had been encroaching on the flood plains of the river Yamuna" (YSC/37/1979-04-26). He concluded that the neighboring states "could not ensure safety of those encroachments by flooding the areas of Uttar Pradesh and Haryana" (ibid.).

The discussions further reveal that Delhi's unwillingness to provide any floodplain zoning was critiqued by the neighboring states and the CWC. The CWC had requested the Delhi Government to "enforce flood plain zoning regulations" (ibid.). Additionally, the minutes reveal a certain dissatisfaction of the CWC, which requested the states and the Central Government to give more powers to the YSC, empowering the YSC to be able to force Delhi to obtain floodplain zoning. The matter of floodplain zoning remained a critical topic from this time onwards and yet to be solved. Nevertheless, based on scientific calculations the YSC decided in the meeting of April 1979:

> "that the minimum spacing between future embankments on the banks of the river Yamuna should be 5 km and the embankment should be aligned at a minimum distance of atleast [sic] 600 metres from the active river edge at the time of construction of embankment" (YSC/37/1979-04-26).

Since the existing space between the embankments in Delhi was evidently under the postulated width of 5 kilometers, the development of any new embankment in Delhi would not have been possible if these decisions would have been followed in due course (cf. MISRA 2010b). In the following meeting of the YSC, the item 'Training of river Yamuna downstream of Tajewala and Channelisation' was changed to

---

57 A study by the IIT Delhi, discussed in the meetings of the YSC estimated that due to the breaches in the upstream the maximum discharge at Wazirabad Barrage was reduced by almost by 50% from about 5,007,000 cusecs (14,356 m^3/s) to an actual 2,750,000 cusecs (7,787 m^3/s) (YSC/37/1979-04-26).

'Yamuna River Surveys' since the matter was now restricted "only to the surveys in Yamuna river flood plains" (YSC/38/1979-07-20). The idea of channelization from Tajewalla to Okhla was therefore dismissed by the YSC after the floods. The 1978 floods thus emerge as a critical point in the history of the interstate river governance.

The findings of the flood survey were discussed in the proceeding interstate meetings and the CWC outlined that the strengthening of the breached embankments after the 1978 floods had again resulted in a considerable loss of flood retention area (YSC/42/1982-06-05). It was further observed that "[i]n the event of recurrence of 1978 flood, the flood peak under present condition may be as high as 5.5 lakh [550,000] cusecs at Delhi for which neither the existing flood control works along the river Yamuna in Delhi reach are adequate nor the bridges and barrages" (ibid.). The committee therefore decided that the "status quo in respect of embankments on either side of river Yamuna between Tajewala and Okhla should be maintained and the existing gaps in the embankments should not be closed" (ibid.). Even a physical lowering of existing embankments was discussed, but was it was anticipated by the engineers that it would be "found impractical due to stiff opposition of local population" (YSC/43/1982-10-12).

While the YSC meetings suggest a rethinking among the engineering community, the rejection of new structural measures did not prevail against the modernist river engineering ideology. Certainly the language changed and the idea of channelization as a concept for *controlling the river in its totality* was rejected, but the actions of the different states did not follow the rhetoric and decisions taken in the YSC. None of the states agreed to the floodplain zoning suggested by the CWC (YSC/39/1979-09-26). Rather, in the following years, all three states requested approval after approval for new and strengthened embankments, which generally approved by the YSC were after some discussion. The states remained keen on protecting their lands from floods through embankments.

The states further argued for the construction of upstream dams (including Renuka Dam) for flood moderation (YSC/42/1982-06-05). The Delhi Government emerged as the strongest advocate for new dams in the upstream for flood protection and increased water supply. Therefore, the belief to be able to control the river Yamuna through engineering technologies persisted. In contrast to the clear rejection of channelization for land reclamation in the YSC, the DDA was pushing for channelization and riverfront development in the 1980s and 1990s.

10.3 DDA's plans of the 1980s and 1990s

DDA did not attend the meetings of the YSC until 1988, but in the course of preparing the second Master Plan (MPD-2001), the DDA had requested CWPRS to conduct model studies for its channelization scheme already in 1983 (GUPTA 1995: 684). In September 1984 the Lieutenant Governor further created a High Powered Committee to overlook the "implementation of various projects on the bank of the River Yamuna" (JAIN 2009a: 107).

In this context, the DDA was then requested by the then Ministry of Works and Housing (today MoUD) "to draw a scheme along with a three dimensional wooden model for the planning of Yamuna bed area to show to the Prime Minister of the country" (GUPTA 1995: 684, cf. NEERI 2005: 2.4, see photograph of wooden model in Appendix X). Thus, since the mid-1980s the DDA was actively involved in planning the remaking of Delhi's riverscapes from an urban development perspective.

In the following decade, the channelization project was discussed in different meetings attended among others by the DDA, MCD, Water Supply and Sewage Disposal Undertaking (today DJB), and different ministries (GUPTA 1995: 108, NEERI 2005: 2.4). In a meeting in February 1987, the MoEF (established 1985), offered to fund 50% of the costs of the channelization scheme (GUPTA 1995: 685, JAIN 2009a: 108). A survey of the ghats in the Yamuna Bazaar area around Nigambodh Ghat was prepared by the DDA in 1987 and the Inlands Water Authority of India submitted a report on the navigability of the Yamuna in 1989 (GUPTA 1995: 685, JAIN 2009a: 109). Furthermore, the DUAC postulated in its 'Conceptual Plan-2001' (prepared in 1988) that "the landscape potential of the Yamuna should be explored through proper channelization" suggesting that such a scheme "can yield a sizeable area for recreation activities such as a sports complex [sic], a cultural centre, a bird sanctuary, a botanical garden safari park lakes, water sports facilities" etc. (quoted from DDA 2006d: 6). Thus at this time, the DUAC also had the impression that a "comprehensive river development scheme is [...] essential" (ibid.). These events suggest an expanded support for the channelization occurred in the state agencies – from the environment wing (e.g. the MoEF) to the city planners, architects and engineers. Furthermore, it illustrates the multiplicity of agencies involved in the planning process from the beginning.

Most importantly for the future development of the river however, was the Central Government's action, based on section 12 (1) of the Delhi Development Act, to declare 3,500 hectares of land in the River Zone as 'development area' on 29th March 1989. This development area encompassed the space between the Ring Road in the west and the marginal bund in the east from one kilometer north of Wazirabad Barrage to Okhla Weir in the south. This notification legally authorized the DDA to plan and develop the River Zone like any other urbanisable area of the city (cf. GUPTA 1995: 686, JAIN 2009a: 109). In Latourian terms, the development area declaration by the Central Government was a major step in the purification of Delhi's riverscapes as *land*, attributing it to the pool of the city rather than seeing it as the space of the river not connected to the city development.

The minutes of the YSC from this period reveal that this move and DDA's preparation for riverfront development (including the request of CWPRS for model studies) had not been coordinated with the neighboring states and the IFCD under Delhi Government. Through its direct links to the Central Government, DDA had largely disregarded all other agencies involved in the YSC and governance of the river. Minimum information about their plans was shared by the DDA (YSC/46/1988-09-22). The developments further suggest that the channelization scheme in the 1980s was considerably pushed by the DDA through its links to the Central Government. A former DDA planner confirms that the DDA had "never called" the neighboring states or informed the YSC in this time (DDA-10/2013-02-08). Thus, it was exclusively the Central Government and the DDA behind the initiative of river channelization in Delhi.

Once the YSC was informed of the plans of the DDA, the committee requested both DDA and CWPRS to attend the YSC-meetings to report about the proposed schemes in 1988. Further, CWPRS was requested to conduct two model studies: one for the whole stretch from Tajewala to Okhla and another for channelization in the Delhi stretch (YSC/46/1988-09-22).

By that time the riverfront development scheme had also entered the public debate. In 1988, the DDA revealed information of their development concept to the public by publishing it on the back cover of its quarterly published in-house booklet 'Delhi Vikas Varta'. This was the beginning of the public discourse about channelization and riverfront development in Delhi (ARCH-01/2011-12-09):

> "The DDA used to bring out a small publication called Vikas Varta – a journal of development. In 1988, it had a bright blue cover and at the back of it was an artist impression of riverbed development. [...] That is the first time this idea became public."

Furthermore, challenging the proposed floodplain zoning, the DDA consistently claimed to hold sole planning sovereignty and development responsibility for the River Zone in Delhi. The minutes of the 50th meeting of the YSC clearly depict how DDA was willing to proceed with channelization without any interference by other state authorities or the neighboring states. The DDA engineer-member is quoted to have argued (YSC/50/1990-02-26):

> "The responsibility of development of land for various uses in Delhi including flood plains of river Yamuna lies entirely with Delhi Development Authority and this has been intimated to Ministry of Water Resources and CWC. [...] no separate enactment of legislation or a flood plain zoning bill was required in this regard. [...] channelisation of river Yamuna shall be considered as a flood management scheme [...]. The reclaimed land as a result of channelisation would be put to suitable uses [...] and land use plan in this regard after preparation would be submitted for scrutiny."

Responding to the DDA, the Chairman of the YSC stated "that channelization of river Yamuna should not be compared with other small rivers wherein there were almost constant flows" and "pointed out that extreme caution had to be exercised in restricting the waterway of such a large river" (ibid.). Consequently, the DDA was again requested by the YSC to prepare floodplain zoning and flood risk maps, which DDA eventually agreed to provide.

Based on an interim report by CWPRS, the DDA then informed the YSC in June 1991 that "there is no significant change in water profile due to channelization [...] in the reach from ITO Barrage to Nizammudin Bridge" (YSC/51/1991-06-25). This stretch was supposed to be phase I of the channelization scheme; phase II was proposed for Nizamuddin Bridge to Okhla Weir and phase III the northern stretch from Wazirabad to ITO Barrage (cf. JAIN 2009a: 111). Highlighting that CWPRS so far had only prepared "an interim report", which did "not take into account the full terms of reference suggested for the studies", the YSC-chairman (CWC) pointed out that, as agreed upon in earlier meetings, floodplain zoning and detailed information on land use planning are prerequisites for the YSC to assess DDA's scheme (YSC/51/1991-06-25). DDA again agreed to submit the materials "as soon as the same is ready" (ibid.). While CWPRS submitted the final report already in May 1991 to DDA, the YSC-members and the CWC had still not received any report or draft for floodplain zoning from DDA by 1995 (YSC/54/1995-06-26). This demonstrates how unwilling the DDA was to share detailed information before getting the scheme approved.

CWPRS's final report outlined that channelization in the stretch from Wazirabad Barrage to ITO Barrage is technically possible, but is subject to the renewal of the Wazirabad Barrage and the Old Railway Bridge (CWPRS 1993, cf. DDA 1998: 6, DDA 2010b: 5). This in turn supported the priority given to the stretch south of the ITO Barrage, since in this stretch the guide bunds of the by then already constructed bridges could be joined for channelization.

Based on the model studies by CWPRS, the DDA prepared the report 'Planning of River Yamuna Bed' in 1995. This report envisioned the reclamation of 490 ha of land in three 'pockets' along the river (DDA-07/2011-11-23). The proposal was circulated to the YSC-members in March 1996. While DDA and CWPRS had been working on the study for almost 20 years (1977–1995), the YSC-members were given only one month to issue remarks (YSC/55/1996-07-15). The YSC-members did not meet this deadline set by the DDA. The neighboring states remained skeptical due to unknown upstream and downstream effects for which they requested that Delhi should bear the costs of any measures arising out of channelization (ibid.).

The proposal for the reclamation of three pockets of land (pocket I, II, and III) was discussed by the YSC in the following as "limited channelization of river Yamuna" (YSC/57/1997-11-18). In April 1997, the reclaiming of 260 ha on the east bank between Nizamuddin Railway Bridge and Nizamuddin Road Bridge (referred to as 'pocket III', today site of the Akshardham Temple and the Commonwealth Games Village) through the construction of an embankment was approved in principle by the YSC (YSC/56/1997-04-08). The CWC repeatedly requested a floodplain zoning from the DDA, which DDA failed to provide (ibid.). However, without a detailed study of its upstream and downstream effects, and without a floodplain zoning concept, the YSC approved the reclamation of the first pocket of land only one and a half years after the initial proposal was made by DDA.

The reclamation of 'pocket I' (50 ha on the west bank between Nizamuddin Railway Bridge and Nizamuddin Road Bridge; today Millennium Bus Depot) and 'pocket II' (180 ha on the west bank south of Nizamuddin Bridge) was, however, denied approval (ibid.). DDA proposed development of parking space and "some permanent structures" for these two pockets – again minimal information was provided by DDA on land use details (ibid.).

Overall, the negotiation process between the DDA and the YSC, therefore, came to a compromise allowing the reclamation of a first pocket of land, which was small compared to the large-scale channelization plans proposed earlier. The developments in the 1980s and 1990s suggest that the protest against DDA's channelization scheme was based on unclear technical feasibility and expected negative runoff and flood impacts on the upstream and downstream. These critical issues dominated the discourse. Additionally, the financial viability of the project was unclear. An article by a former member of the CWC (Floods and River Management) from 2000 quoting reports of CWC-meetings supports this interpretation and highlights that river engineering circles were not in favor of DDA's channelization scheme.

> "There is significant intrusion in the river bed and flood plain, much of which seems politically organized and Government condoned. This impermissible occupation is constantly on the increase. There are even some Government sponsored schemes of this nature. The DDA plan for channelization of the Yamuna and commercial use of the reclaimed area, should be viewed in this context. [In a CWC workshop] 'the proposal … was not considered to be sound prima facie on technical considerations. Further it came in direct conflict with the declared policy of flood plain zoning by the government of India for preventing encroachment on flood plains in order to minimize flood losses in terms of money and human lives […]'." [58]

10.4 Channelization in the second Master Plan (MPD-2001)

Within the same period of time, the second Master Plan for Delhi (MPD-2001) was prepared by DDA and approved by the Central Government in August 1990 (DDA 1990). In comparison to the first Master Plan, the land along the river was defined as the River Zone (Zone O). In doing so, Delhi's riverscapes became part of the 'planned' area of Delhi. They were no longer left out as flood prone land in the planning process. The DDA was therefore also responsible for preparing a Zonal Development Plan (ZDP) for the River Zone.

The entire open floodplain limited by the Ring Road in the west and the continuous embankment in the east was designated as 'Agricultural and Water Body'; except the samādhis (up to the Mughal Bund) were designated as recreational, the Delhi Secretariat and the power plants as utility, and the site of today's gas-based Pragati Maidan Thermal Power Plant (commissioned 2002–2003) was still defined as 'Recreational' (see land-use map of MPD-2001 in Appendix VI). Based on the classification 'Agricultural and Water Body' any development project in the River Zone required change of land use in the MPD-2001. It is notable that the land use 'Agricultural and Water Body' is unquestionably a hybrid category corresponding with the amphibian landscape of Delhi's riverscapes.

In contrast to this declaration, the MPD-2001 outlined the possibility of channelization.

> "Rivers in the major metropolitan cities of the world like Thames in London and Seine in Paris have been channelised providing unlimited opportunities to develop the river fronts. After the

---

58 Rangachari, R. 2000. 'Flood Damage. How Prepared are we?' quoted in the petition WP (C) No. 7506/ 2007 against the Commonwealth Games Village.

results of the model studies for the channelisation the river Yamuna become available, development of river front should be taken up. Considering all the ecological and scientific aspects, as a project of special significance for the city" (DDA 1990). [59]

The MPD-2001 further suggested the development of a "special recreational area on the pattern of Disneyland/amusement park" for the land reclaimed by channelization (ibid.). In total, the MPD-2001 envisaged a development of 3,000 ha in the River Zone for "public utilities" (ibid.). Thus, the earlier recreational ideas of the TOWN PLANNING ORGANISATION (1956) were reflected again in the MPD-2001.

A retired DDA planner confirmed during interviews that the large channelization scheme was, however, never approved by the Central Government (DDA-07/2011-11-13). He indicated that he believed "some very preliminary draft plans were prepared that this river can be channelized to 550 meters [...] but ultimately it was not agreed" (ibid.). He further emphasized that during the preparation of the MPD-2001, the channelization scheme had become "a kind of dead proposal" from his point of view and no further details were given in the MPD-2001 (DDA-07/2011-11-13).

> "The whole proposal was almost dead when in 1990 the Master Plan for Delhi 2001 was approved [since it again] stated that the channelization of river Yamuna can only be taken up after very comprehensive environmental and flood management studies are available" (DDA-07/2011-11-13).

Therefore, while other authors have interpreted the MPD-2001 as "the stamp of approval" for the channelization idea (BATRA & MEHRA 2008: 404), the detailed analysis suggests that this was not the case. Another DDA planner engaged with the planning for the River Zone confirmed this (DDA-06/2011-11-22):

> "Before making this plan [ZDP], we had engaged some experts [...] . All of them concluded that we should not go for channelization. The river should not been channelized."

While earlier technical and financial feasibility of the channelization scheme was unclear and the risk of flooding prevented the technical approval by the YSC, later growing ecological concerns were more widely opposed when reclamation of land along the river was discussed. The DDA planners already begun to realize that it was very difficult to reach an agreement on the 'great scheme' for channelization and riverfront development.

Nevertheless, the construction of the bridges since the late-1970s materially and spatially manifested the persistence of the channelization schemes. The CPWD, responsible for the construction of the bridges over the river, designed and constructed all bridges in accordance with the proposed channelization width of

---

[59] This section of the MPD-2001 was changed a couple of times. For example, BATRA & MEHRA (2008: 404) cite the passage: "Rivers in the major metropolitan cities of the world like Thames in London and Seine in Paris have been channelized providing unlimited opportunities to develop the river fronts [sic]. The possibilities in respect of river Yamuna have been studied in depth and indications are that it could be channelized within 550m width and an area of about 3000-4000 ha could become available for riverfront development." See also JAIN (1990: 131) for another slightly different version of the same section.

550 meters (e.g. Nizamuddin Road Bridge 549m). DDA in turn planned to join the guide bunds (see Figure 24) of the bridges later for channelization. However, the 550 m benchmark was established already before the channelization scheme was prepared as also the ITO Barrage was constructed with a width of 552 meters in the late 1960s (see Figure 25).

Figure 24: Guide bund, Geeta Colony Bridge
Source: Follmann, January 2010

Figure 25: ITO Barrage
Source: Follmann, September 2014

Thus, through the construction of the bridges (and barrages) the river was inadvertently channelized, preventing the free meandering of the river and restricting the river to a rather fixed bed. While the scheme was not approved in its totality, the idea and logic of river channelization materialized in the physical design of the bridges. In this way, the technical studies by CWPRS, which recommended the channel width of about 550 meters, supported the de-facto channelization of the river. Based on the CWPRS-studies, the YSC approved the different proposals for bridge construction, which then institutionalized the discourse of channelization. Furthermore, the DDA interpreted the constructed guide bunds of the bridges as the already laid out foundations for the next steps in reclaiming the land between the bridges and approach roads.

In this context, it is further important to discuss that for the 9th Asian Games (Asiad), held in Delhi in 1982, the sports complex with the Indira Gandhi Indoor Stadium and the (proposed) players building[60] was built north of the ITO Bridge between the Ring Road and the railway tracts leading to the Rajghat power station. These developments were never referred to the YSC, most likely because the land behind the railway lines was not considered to be part of the 'active' floodplain anymore.[61] Furthermore, the development of the sports complex at the riverfront

---

60 The players building had not been finished for the Asian Games and was later remodelled as the Delhi Secretariat.
61 The question of 'active' versus 'passive' floodplain, and the logic if boundary making, will be reviewed in more detail in chapter 14.

was very much in line with the schemes for channelization and riverfront development discussed above and a major flagship project of the DDA in the early 1980s (quite similar to the Commonwealth Games in the 2000s).

The Asiad also resulted in large numbers of construction workers migrating to Delhi and many of them settled in the growing slums along the river (cf. BATRA & MEHRA 2008, BAVISKAR 2011c). Furthermore, since the possession of land was a crucial precondition for DDA's riverfront development schemes, the state machinery of land acquisition for the channelization project was in full swing from the late 1980s onwards.

10.5 Land acquisition for channelization

When analyzing the development of Delhi's riverscapes, questions of land ownership and possession of land are of major importance, since unsolved land disputes are omnipresent. The Delhi Government's *Land & Building Department* has been, and continues to be, responsible for large-scale acquisition of land. The land acquisition in public purpose is generally conducted by the department using the Land Acquisition Act 1894 (section 4 and 6). In order to secure a high degree of control over land and thereby check land speculation as well as facilitate the implementation of the MPD, the Central and State governments have deployed large-scale land acquisition and disposal in Delhi (cf. JAIN 1990: 114–115, MAITRA 1991: 344, SRIRANGAN 1997: 95–96).

In connection with the channelization and riverfront development proposal, the DDA started identifying and mapping land ownership in the River Zone in 1994 (DDA - SPECIAL PROJECT CELL 1998: 17). Based on DDA's surveys, including 6,081 ha from Wazirabad to Okhla, the land of Delhi's riverscapes was under different ownership: 3,380 ha was with DDA (including among others *Nazul lands* of Bela Estate); 1,359 ha was private lands; and another 1,342 ha of land was notified by the Delhi Government for acquisition from private land owners and existing villages (DDA - SPECIAL PROJECT CELL 1998: 3).

Planning documents of the DDA from the late 1990s indicate that significant conflicts ensued over land. The report of the DDA - SPECIAL PROJECT CELL (1998: 3) stated, for example, that land of about 1,657 ha was "under unauthorised cultivation" by farmers whose leases had expired, and another about 30 ha of land was "under encroachment" by slums and other unauthorized uses (including also ghats). In total, "more than 200 cases" were pending in the High Court with regard to these land disputes (ibid.). A letter from the DDA Land Management to the Co-ordinating

Officer Damages DDA (the demolition wing of DDA) from November 1987 indicates that the DDA had started trying to evict farmers from along the river already in the late 1980s.[62]

In the northern stretch from Wazirabad to Nizamuddin Railway Bridge and along the eastern embankment south of Nizamuddin Railway Bridge, land ownership was largely under different government authorities including DDA, IFCD, L&DO, Indian Railways, CPWD, DPWD and Irrigation Department of Uttar Pradesh. Even among these government agencies land disputes occurred; especially between the DDA and the Irrigation Department of Uttar Pradesh.

In the southern stretch from Nizamuddin Railway Bridge to the NCT border, land ownership was largely private until the end of the 1980s. Large-scale acquisition of land under public purpose based on Land Acquisition Act, 1894 (Section 4 and 6) for the channelization scheme started in 1989 (cf. PEACE INSTITUTE CHARITABLE TRUST & ENVIRONICS TRUST 2009). Land measuring about 3,500 ha covering 15 villages between Nizamuddin Bridge and Okhla Barrage "was sought to be acquired by the Lieutenant Governor in 1989 for planned development".[63]

Affected land owners along the river had filed a series of petitions in the High Court of Delhi against the acquisition of land. The cases were heard by a Division Bench[64] and the two judges involved delivered contradicting judgements in September 1996; while one judge "quashed" the land acquisition, the other confirmed legality.[65]

The petitioners had reasoned their objection among other aspects with the fact that since "no plans were prepared for development of the area" and "without these plans" the area "could not be acquired for [a] particular purpose".[66] The first judge, Mahinder Narain, gave merit to the arguments of the petitioners and went into the details of the purpose of acquisition. In doing so, justice Narain interrogated what 'channelization' actually meant and how the respondents (DDA and others) were planning to implement channelization.

> "In order to understand what is involved in channelisation of River Yamuna as postulated by the said notification [for land acquisition], it is necessary to know what is channelisation. Whereas a canal is necessarily man made, a natural river channel is not. Channelisation is the work of man, just like making a canal is the work of man. Canal making is resorted to take

---

62 Letter DDA Lands Management S.S. Tyagi to Co-ordinating Officer (Damages), DDA, dated 4th November 1987. Subject: Regarding Eviction Proceedings against Delhi Peasants Cooperative Multi-Purpose Society Limited. Letter accessed by the author through the materials of the Delhi Peasants.
63 16th Report of the YREMC, dated 25 April 2007. Wazirpur Bartan Nirmata Sangh vs. Union of India and ors and Public Interest Litigation (published in DUTTA 2009).
64 In the judicial system in India the term Division Bench refers to a case which is heard and judged by at least two judges. A Division Bench can transfer the case to a larger bench.
65 High Court of Delhi, 11 September 1996, Baldev Singh Dhillon And Ors. vs Union Of India (Uoi) And Ors. Equivalent citations: 64 (1996) DLT 329.
66 Ibid.

away waters of river from its channel, channelisation may mean deepening of the natural river channel or creating a new channel in the river bed, or its flood plains."[67]

Based on his own research on channelization schemes from around the world the judge declared the acquisition of land as illegal as "no scheme of channelization of River Yamuna or maps drawn to scale showing the areas which are going to be channelized ha[d] been shown to us [the court] despite the Delhi Administration and the Delhi Development Authority being asked several times to produce such a scheme or map, or both."[68] This order thus supports the interpretation outlined above that by that time DDA had not yet prepared detailed plans for channelization, and further on DDA was not willing to share any more information on account of their fear of the approval being refused.

The other judge of the Division Bench, J. B. Goel, differed with the above quoted judgement so the case was referred to a third judge, A.K. Sikri. The case remained pending in the court until A.K. Sikri delivered a judgement confirming the legality of the acquisition of land in 2005.[69]

While Mahinder Narain in his order had mixed the question of whether the process of land acquisition was legal with his own interpretation of river channelization, A. K. Sikri finally dismissed the petition in 2005 stating:

"If the Government is of the opinion that it needs land for such a project [...] and the decision does not suffer from any vices known in the administrative law which may be ground for interfering with such administrative action, the Courts would not interfere with such an administrative action. Similarly it is not for the Court to give its own interpretation to the scheme by assigning literal meaning to the phrase 'Channelisation of Yamuna River' [...]."[70]

In the light of the pro-active role of the courts in India in regard to environmental issues (see chapter 8.7), A.K. Sikri's reasoning seems to be rather uncommon and contradictory. Compared to the judgements of the High Court delivered between 2003 and 2006 on the slums along the river (for details see 13.2.2), which show an active involvement of the court in policy-making, A.K. Sikri withhold any interference by the judiciary. In addition, the channelization scheme under which the land was sought to be required was long outdated when the order was delivered in 2005. Therefore A.K. Sikri took the decision under completely different circumstances. The land dispute of the case in south Delhi had become an even more explosive and sensitive issue since the area experienced large-scale growth of mainly Muslim-dominated unauthorized colonies (see 13.3). This put the Kalindi Kunj By-Pass Road, a major road development project in this area, in danger.[71] Additionally, the DDA had envisioned large-scale 'planned' development for about 350 ha of land in the area south of today's DND-Flyway from *Khizrabad* to *Okhla* since 1998

---

67 Ibid.
68 Ibid.
69 High Court of Delhi, 26 May 2005, Baldev Singh Dhillion and Ors. vs Union Of India (Uoi) And Ors.
70 Ibid.
71 The Kalindi-Kunj By-Pass Road was eventually upheld by land disputes between Delhi and Uttar Pradesh.

(DDA - SPECIAL PROJECT CELL 1998: 4). These developments (for details see 11.7 on the proposal of a stadium complex at Kalindi Kunj) were also put under pressure by the growing 'unplanned' residential developments in the area.

The channelization discourse thus constituted a powerful logic used by the government agencies for restraining the informal, unplanned growth of unauthorized colonies on private land and other forms of encroachments on public land as well as for accumulating a land bank for different infrastructure developments along the river.

10.6 Interim conclusion

The 'great scheme' for the channelization of the Yamuna and the development plans for Delhi's riverfront remain urban imaginaries. As discussed, the first Master Plan viewed the land along the river as flood prone area and proposed only recreational uses (as recommended by the Town Planning Organisation already in 1955). The riverine land of the floodplain was perceived to belong to the realm of nature, the planners control focused on the protection of the land from the nature through the existing embankments, and the realm of the planned city was located behind the embankments. Despite this, the unplanned city expanded into the realm of the river and along the embankments.

Through channelization, the DDA aimed to reclaim land for the planned development of the city from both the river and the 'unplanned' city in form of the slums. While mentioned in the MPD-2001 (DDA 1990), channelization was never approved due to technical and financial ambiguities. Yet, as the following analysis of the more recent development schemes show, these schemes have had an impact on the remaking of Delhi's riverscapes. Through the construction of the bridges the foundation stones for channelization had been laid out.

Furthermore, since the idea of the 'great scheme' for channelization and riverfront development haunted planners and politicians from the 1970s to the 1990s, they were reluctant to work on any other alternative land-use concept for the River Zone. Inter-sectoral competition and interstate conflicts regarding land, allocation of flows, pollution amendment, and flood prevention added to the difficulty of the possibility of comprehensive planning of Delhi's riverscapes.

The reduced flow in the river due to increasing water abstraction at Tajewala/Hathnikund and Wazirabad moreover supported the reclamation of the land pockets from the river. The interplay of reduced flow and engineered intervention in the form of the guide bunds and bridges suggested that the land between the approach roads had become 'abandoned' by the river. Thus, the DDA viewed the area as 'wasted land'; neither used by the city nor by the river.

The existing embankments had emerged as the physical boundaries between the city and the river, both materially and discursively out of the "constant negotiations between land and water" (LAHIRI-DUTT 2014: 507) – the monsoonal river and the growing city. The land-use planning of the DDA thereby formalized the ontological separation of nature and society (WHATMORE & BOUCHER 1993). The river remained 'outside' the city, while it was actually 'in-between' the two parts of the city (cf. SHARAN 2015). The Ring Road in the west and the continuous embankments in the east (Yamuna Pushta Embankment) remained the persistent boundaries demarcating the realm of the city and the river.

Under the first Master Plan, the realm of the city and the river was more or less clearly defined by the natural boundary of the *bangar* and *khadar* areas. Only some land of the *khadar* plains was reclaimed for the development of the samādhis, the Ring Road and the power plants. The wide open floodplain between the Yamuna Pushta Embankment in the east and the Ring Road in the west was the realm of the river to spread during floods; only seasonal agricultural uses were permitted (DDA 1962). The land-use classification for the River Zone in the MPD-2001 as 'Agriculture and Water Body' expressed this (seasonal) hybridity of Delhi's riverscapes derived from the metabolism of human practices (agricultural use of the land in the non-monsoon months) and the 'natural' processes (flooding in the monsoon months).

In contrast, the great scheme of channelization and riverfront development aimed to purify more land for the sake of the growth of the city as the hybridized space of the floodplain was perceived by the planners to be neither required by the river nor used by the city. This wet category 'Agriculture and Water Body' was "neither fully land, nor entirely water", but rather land temporally "soaked in water" (LAHIRI-DUTT 2014: 508). Although this accurately represented the ground realities, it was perceived as a residual category and, thus, viewed as 'wasted land'. Urban planners were uncomfortable with this hybridity because land uses remained difficult to be controlled (e.g. slums) and boundaries difficult to be fixed. Hybridity created contradictions when translated into the water-land binary inherent to land-use planning. Channelization, in contrast, was about producing clear boundaries between land and water, the city and the river. Since the planners thought mainly about land, and where facing the pressure of a fast-growing city, they aimed for the reclamation of land for the city. In doing so, they envisioned the purification of the hybrid riverscapes into the spatially stable categories of land (riverfront) and a well-behaved river (water channel).

The following analysis of the preparation of the Zonal Development Plan (ZDP) for the River Zone starting in the mid-1990s will highlight how the 'great scheme' of channelization and riverfront development was reworked to legitimize large-scale projects to be developed along the banks despite growing opposition on ecological grounds. The purification of the ambiguous category 'Agriculture and Water Body' remains central in this analysis.

11 DRAFTING THE ZONAL DEVELOPMENT PLAN

During the time when land acquisition was in full swing and the land of Delhi's riverscapes was increasingly under control of the state, the planning visions for Delhi's riverscapes shifted from the 'great scheme' of channelization and riverfront development to spatially limited project-based approaches. The DDA started preparing the Zonal Development Plan (ZDP) for the River Zone (Zone O) in the mid-1990s.

This section reviews the policy making process for the ZDP, starting with the first draft of the plan prepared by the DDA in 1998 (in the following referred to as 'draft ZDP-1998'). Besides the draft ZDP-1998, two other drafts in 2006 and 2008 are analyzed, which finally resulted in the approval of the ZDP in 2010.[72] In doing so, this chapter aims to outline the negotiation and argumentation process between the DDA, other state agencies, and in the later period, non-state actors.

11.1 Change of land use for pocket III: the starting point for channelization

In spite of the growing opposition both within and externally to government agencies against the reclamation of land in the River Zone through channelization, the DDA issued a public notice inviting objections and suggestions for the proposed change of land use for 'pocket III' (later Akshardham Temple and Games Village) from 'Agriculture and Water Body' to 'Public and Semi-public Use' on 23rd June 1997. While the DDA received only three answers to the public notice[73], the Delhi Urban Arts Commission (DUAC) raised major critique against the proposed developments:

> "The urbanization of the river belt [...] would have serious repercussions on the ecology and environment of Delhi [...]. The area in question forms the natural flood basin of river Yamuna and no attempt should be made in an adhoc [sic] manner to change its use on a piecemeal basis. Any such proposal is likely to create major imbalance of the entire ecology of the water shed [...] and would adversely affect the underground aquifer. [...] DDA should initiate the proposal only after viability has been established with the help of all necessary studies. In any case first of all the clearance of Environmental Impact Assessment Authority be obtained by DDA."[74]

Notably, the DUAC did not refer to the River Zone or riverfront, but rather used the terminology of the 'river belt' drawing on the idea of the green belt outlined by the

---

72 The author is aware that in 2003 another draft for the ZDP was circulated among the state agencies. As the author was unable to access the materials from that time, the analysis concentrates on the three available drafts.
73 The author is unclear to what extent the public notice at this time actually reached the public. The notice has been generally provided in the local daily newspapers. The public notice for the change of land use could not be traced by the author.
74 Letter DUAC, Dr. J.P. Singh (Chairman) to DDA, Vice-Chairman P.K. Ghosh, 19th August 1997. The DUAC issued a copy of this letter also to the Ministry of Urban Affairs and Employment. Material accessed by the author via interview partner.

DDA in both master plans (DDA 1962, 1990). The second Master Plan outlined the rationality of the green belt concept for Delhi (DDA 1990):

> "Green Belt and its other synonyms are a planning tool to restrict the growth of towns and cities to definite limits. Delhi Master Plan prescribed and agricultural green belt around the urbanisable limits [...]. This green belt was also intended to be inviolable and was to restrict physical growth or over spilling of urban development beyond the urbanisable limits."

The DUAC appeared to view the floodplain as defining the 'natural' urbanisable limits for the city, which should be protected and conserved as the green belt. By strategically avoiding the term riverfront, which was coined by the DDA, the architects of the DUAC adopted the language of ecologists and environmentalists referring to Delhi's riverscapes. The aesthetic vision for a new riverfront deployed by the DDA was not adopted by the DUAC – which is again noteworthy as the DUAC consists of a group of architects. The DUAC did not draw on the story-lines of the DDA making it difficult for the DDA to form a discourse-coalition supporting its riverfront development scheme.

The comments of the DUAC were discussed in the Technical Committee Meeting of the DDA held in September 1997 ensuring DUAC that "DDA is taking all necessary clearances [...] before starting any development of the river Yamuna Project".[75] The DDA Authority Meeting then sanctioned "provisional" change of land use for 35 ha based on the condition that DDA will obtain an environmental impact study from the National Environmental Engineering Research Institute (NEERI) on 23rd December 1997. DDA approached NEERI in August 1998 for an 'Environmental Management Plan'. NEERI delivered an interim report in June 1999 and a final report in 2005 (see detailed discussed in section 11.3). Based on the interim report and before NEERI had finalized the study, the change of land use was granted by the Union Ministry of Urban Affairs in September 1999 (MoUAE 1999).[76]

Thus, after DDA failed to obtain approval of the 'great scheme', they started focusing on the development of 'smaller' schemes for different subzones. Proponents and opponents of the change of land use for 'pocket III' viewed its development as the starting point of DDA's larger vision. Simultaneously to the change of land use procedure for 'pocket III', the DDA proceeded in planning for the entire riverfront and circulated a first draft for the ZDP in 1998.

11.2 Seeking approval for channelization – the draft ZDP of 1998

While the above cited project report 'Development of Yamuna Riverfront' prepared by DDA in 1998 stated that the "engineering feasibility" for channelization "is yet to be examined" and the outlined scheme should provide only "a canvas for generating dialogues, discussions and further work" (DDA - SPECIAL PROJECT CELL 1998: 3), the DDA had already requested approval of change of land use for the

---

75 Comments DDA to DUAC, 8th September 1997. Internal material DDA.
76 While the public notice was issued for 35 ha only, later 42.5 ha were approved by the ministry (MoUAE 1999).

three pockets (460 ha) along the river from MoUD (in those days Ministry of Urban Affairs and Employment).

In the 1990s, the DDA considered different land utilization concepts based on the final CWPRS-study (CWPRS 1993), which provided in principle feasibility of channelization of the river for a width of 450 to 800 meters. These included the development of "amusement parks, aqua sports, water parks, stadia and other [...] tourism activities of international standard", a "Time Theme Park and other innovative centres of unique style" as well as a "financial district on pattern of Manhattan in New York and Raffle Square in Singapore" (DDA - SPECIAL PROJECT CELL 1998: 4). In the eyes of the DDA all these land uses were considered to be "ecologically and environmentally viable yet remunerative" (ibid.).

The project plan further highlights that besides the unknown final technical feasibility of channelization three other critical aspects were largely unsolved when those schemes were floated:

1) Availability of land (due to various stay orders by the High Court and encroachments[77]);
2) Financial viability ("including cost-benefit analyzes, sources of investment, likely returns"; and
3) Pollution control (including "shifting of hazardous activities/land uses", development of sewage treatment plants, etc.) (DDA - SPECIAL PROJECT CELL 1998: 5).

Despite these critical aspects, the DDA – backed up by the Central Government's MoUD – tried to pit the three approval bodies (YSC, MoEF, and NCRPB) against each other putting pressure on all of them for timely approval. When a first draft for a ZDP for the River Zone was circulated in March 1998, the DDA had already pressed for approval of their development plans by YSC, MoEF, NCRPB and MoUD (DDA - SPECIAL PROJECT CELL 1998, DDA 1998). DDA envisioned developing about 7,300 hectares of land promoting that most of the land (about 85% or 6205 ha) was planned to be used for recreational activities including parks, golf courses, and other green spaces, but also tourist cottages, camping sites, small shopping plaza, auditoriums, restaurants/cafés, amusement halls, sports centers and other uses (NEERI 2005: 2.10).[78] Further, about 6% (438 ha) was designated for institutional uses, which are generally referred to as 'public and semi-public uses'

---

77 The project report mentions unauthorized colonies as the major encroachment along the entire eastern bank and the project refers to unauthorized colonies south of the Nizamuddin Bridge on the western bank, but it does not refer to the JJ colonies (slums) in various stretches along both banks. A reason for this might be that the JJ clusters in those days came under the MCD Slum and JJ Wing and detailed information was not with DDA. However, it seems more feasible that the highly sensitive issue of slums was kept quiet about in order to let the implementation of the schemes sound more realizable.
78 The here quoted study by NEERI outlined the envisaged riverfront development plans in more detail than the actual material accessed from DDA. The background of the study is discussed in the next section.

– a land-use category which will be discussed later in detail. The public and semi-public uses included a "large institutional buildings of national importance like museums of Indian history and independence movement" (NEERI 2005: 2.9). The expected financial viability was based on the development of high density residential uses (3% or 219 ha) for a population of about 120,000 people, commercial uses (3% or 219 ha) as well as government funding (NEERI 2005: 2.10). Even if the mentioned figures are to be viewed critically, they give some insight into DDA's ideas, and highlight the still large dimension of the proposed scheme. Based on the outlined scheme the DDA prepared a first draft for the ZDP in March 1998 (DDA 1998).

The draft ZDP-1998 acknowledges the "special character" of the River Zone "in terms of being a flood prone natural feature with large stretches of land available beyond the predominant water course and existing bunds" and, moreover, recognized that all "stretches may not [be] used/usable for any kind of development" (DDA 1998: 1). The DDA endorsed the ZDP to "act as a policy framework for formulating action plans for eco friendly [sic] development" (DDA 1998: 2). While the DDA acknowledged the development limitations in the floodplain on the first pages of the draft, DDA promoted the riverfront development projects to provide "a unique opportunity for developing a strong city image on the pattern of River Thames in London and River Seine in Paris" (DDA 1998: 13). In order "to bring the river closer to the citizens", the DDA proposed the development of "bathing ghats", "pedestrian boulevards", and "landscaped avenues" as recreational spaces (DDA 1998: 14). Further "three dimension developments" including a financial district, convention centers, stadiums, theme parks, government uses (secretariat, assembly), a race course, a polo ground, etc. were proposed in the 1998 ZDP (DDA 1998: 13).

A detailed reading of the draft suggests that the DDA aimed to install a legal document sanctioned by the Central Government in order to obtain final approval for the long prepared channelization and riverfront development scheme. In doing so, the DDA made a request for approval of the ZDP to the NCRPB in 1998. Already in 1997, the DDA had made a request for approval of the three above mentioned pockets in the YSC (YSC/56/1997-04-08). However, as interviewed professionals (ARCH-01/2011-12-09, ARCH-02/2011-11-17, DDA-06/2011-11-22, SCIENTIST-02/2011-11-08) and environmentalists (NGO-02/2011-12-09, NGO-04/2011-12-06, NGO-12/2011-11-20) recalled, the DDA's plans were highly criticized in public meetings and talks on the proposed riverfront development scheme already in the late 1990s. The following interview quotes give an account of the overall interpretation by the interviewees.

"[...] a great deal of resistance has been put up which has prevented the DDA from [...] colonizing the floodplain area. If that had not been done, this land which is lying so centre to the city would have been colonized as very valuable land. [...] there were people who were talking about this can become Manhattan. [...] At that time in the 90s the environmentalists had sort of upper hand, because when they spoke [...] they were talking with some greater wisdom, [...] with some moral authority, and it appeared that the planners had not applied their minds. They obviously were on the back foot. And so it went through that 'sorry' river cannot be channelized" (NGO-12/2011-11-12).

"There was too much opposition to the riverfront development right from the beginning [...]. The idea [channelization] was discussed in many public forums and not accepted by the general public at large. So they actually went slow on it" (ARCH-01/2010-09-23).

Moreover, the NCRPB vehemently criticized the proposed developments along the river. Based on the provisions outlined in the *Regional Plan – 2001* (NCRPB 1988), the NCRPB highlighted that the River Zone is "not meant for urban uses and activities" and recommended "to preclude activities that could harm the aquifer underneath or adversely effect [sic] the environment or ecology".[79] Emphasising the risk of flooding, the NCRPB outlined that "it seems inadvisable to tinker with the natural flow of a monsoon river and take-up permanent construction activities of this magnitude".[80] The NCRPB thus concluded that DDA's draft was "not permissible".[81]

Besides NEERI, DDA had also approached the School of Planning and Architecture (SPA) to prepare a study on riverfront development. The SPA study, which analyzed three different development alternatives, observed that the channelization scheme prepared by CWPRS was not financially viable and proposed only partial channelization (cf. DDA 1998, JAIN 2009a).[82]

Correspondences of the NCRPC with the MoEF[83], the SPA and the NRCD evidence that all these agencies did not support DDA's scheme and demanded detailed environmental studies.[84] A letter of the then director of the SPA to the NCRPB highlights that in his eyes the main question regarding the draft ZDP "centres around the wisdom of reclaiming the riverbed for non-riparian land uses".[85] The letter further indicates that "reclaiming the riverbed for any alternate use may prove to be hazardous not only for Delhi but also both upstream and downstream settlements on the river".[86] The comments further highlighted that the River Zone should

---

79 The information and quotes are based on a letter from the NCR Planning Board to the DDA (no date, 1998), which the author received from an interview partner requesting anonymity.
80 Ibid.
81 Ibid.
82 The author was unable to access the study by SPA as of this writing. Due to this, the information about this study are limited and based on secondary sources. The SPA study analysed an "ecosystem based approach", an "integrated development scenario", and a "post channelization development scenario". The latter was based on the recommendation by CWPRS proposing major real estate developments whereas the first alternative considered only recreational uses. The "integrated development scenario" outlined a middle way of both (DDA 2010b, NEERI 2005).
83 The ZDP does not require EIA from the MoEF. Yet, the MoEF was requested by the NCRPB for comments.
84 The information are based on a letters from the different agencies accessed by the author trough interview partners requesting anonymity.
85 Letter SPA, Director to NCRPB, dated 12 October 1998. Letter received from an interview partner requesting anonymity.
86 Ibid.

be considered as a wetland and since India is committed to Ramsar Convention[87] should be protected. Protecting the river as a wetland under the Ramsar Convention would certainly mean purifying the River Zone as the realm of nature.

Another critically discussed aspect at that time was the target to control the growth of Delhi by de-concentrating development in the region towards the satellite cities, which was the overall philosophy of the *Regional Plan – 2001* (NCRPB 1988). It was felt that the proposed riverfront schemes would however "attract migration" to Delhi and could be realized "only [...] at the cost of balanced regional development".[88] As is evident here, DDA's ZDP draft was widely critiqued on various grounds and eventually denied approval.

11.3 Negotiating scientific 'facts' and recommendations

While several expert institutions had refused approval to DDA's scheme for partial channelization and riverfront development, DDA assigned the study to NEERI with the expectation that the government-sponsored environmental expert body would scientifically support DDA's channelization scheme.

NEERI, in the interim report, supported a modified land utilization plan of the DDA focusing only on the first phase of the riverfront development schemes. This was the favored stretch of land between Nizamuddin Railway Bridge and the Nizamuddin Road Bridge (NEERI 1999). When clarifying their vague recommendation made in the interim report, NEERI delivered a 'first version' of its rejuvenation plan for the whole river stretch in April 2000 (NEERI 2000). The 'General Guide Lines for Development of Riverbed' outlined by NEERI in 2000 stated (NEERI 2000: 13):

"The riverbed slope towards the water channel should be maintained. No modification of topography (except minor levelling using soil from the riverbed itself) should be permitted. [...] No residential, commercial, industrial or public/semipublic facility requiring permanent structures should be provided on the riverbed."

The 'first version' of NEERI's report was not made public, and major modifications in the recommendations were made after discussions between DDA and NEERI as well as based on the inputs of various other central and state departments and ministries (cf. NEERI 2008). NEERI submitted the 'final version' of the report in October 2005 – more than five years after the study was done.

This long delay indicates that the findings and recommendations by NEERI likely did not correspond with the interests of its sponsor DDA; and highlight a long

---

87 The UN Convention on Wetlands is commonly known as the Ramsar Convention. Adopted in 1971 in the Iranian city of Ramsar, the convention is "the oldest of the modern global intergovernmental environmental agreements" and came into force already in 1975 (RAMSAR CONVENTION SECRETARIAT 2015).
88 Letter Director SPA to NCRPB, dated 12 October 1998.

and difficult negotiation process between the agencies regarding the report's content. In the production of the final report, the above quoted section of guidelines was modified in the following way (NEERI 2005: 3.33):

> "The riverbed slope towards the water channel should be maintained. Unless the channel dredging and subsequent dressing of river bed as proposed in this study is followed, no modification of topography (except minor levelling using soil from the riverbed itself) should be permitted. [...] No residential or industrial facilities requiring permanent structures should be provided on the riverbed."

Comparing these two version of the same paragraphs of the study, important changes in the wording appear. In 2000, NEERI had clearly rejected the construction of any further embankments, but in the final version of the report NEERI weakened this strong statement and allowed structural interventions if connected to the dredging of the river in principle. Additionally, while originally NEERI had disqualified not only residential and industrial uses but also commercial and public/semi-public uses, the research organization altered this paragraph as a result of pressure from DDA and other state organizations. Accordingly, commercial and public/semi-public uses[89] were not explicitly excluded in NEERI's recommendations.

These changes in the recommendations suggest that the overall assessment and findings of the study had been considerably altered and clearly weakened regarding the proposed development restrictions.[90] The findings by NEERI emerge as (discursively) 'negotiated facts' indicating the 'co-production' of science and policy (FORSYTH 2001, HAJER 1995).

DDA's target remained as reclaiming land for the city, and NEERI's findings did not discourage DDA from pushing the riverfront development scheme further. Despite its negotiated recommendations, DDA withheld the final report by NEERI and did not make the results public. Furthermore, NEERI's recommendations – being a government organization of national importance – gave pronounced credibility to DDA's future planning and development actions.

> "NEERI is a government organization. For the DDA it is also like a consultant, but it is a government consultant. You authenticate your project by putting this report of the high arm technical body, so that you are not asked any questions later. If you go to a private consultant, then people will say this consultant is already biased or you must have given him huge money. But within a government system they may have a levy" (NGO-08/2011-11-12).

The analysis below shows that this was how the DDA proceeded in implementing its plan for Delhi's riverscapes; with the argument that their schemes had already

---

89 DDA's category of public and semi-public uses includes eight land use categories namely Hospital, Education and Research, Social and Cultural, Police Stations, Fire Stations, Communication, Cremation and Burial, as well as Religious Uses.
90 The author was unable to obtain the full report of the 'first' version and could only review this main section, which was circulated by NEERI also to the media after critical newspaper reports were published on NEERI's role in the Games Village case.

been 'sanctioned' by NEERI. Eventually, the relative truths produced by the negotiating process between DDA and NEERI were strategically revealed by the DDA in order to build legitimacy for riverfront development.

As in the Foucaultian sense, truth is an effect of power/knowledge (FOUCAULT 1984), DDA provided planning power through its control over land and expertise; and NEERI provided scientific knowledge and credibility. Thus, a powerful discourse-coalition was born.

During an intense discussion on the role of the DDA in governing the River Zone, a DDA landscape planner captured the self-evaluation of the DDA succinctly (DDA-03/2011-03-03):

> "This land [in the River Zone] as a land bank remains mainly with the DDA, even if certain parts of land are not under DDA [...]. We are the people, who are making the plan for it – the Master Plan. So DDA is very strong in land and power; and has the power to tell what is to be done [and] where."

While the interlinking story-lines deployed by this discourse-coalition are outlined in detail in section 14.2, it is important to highlight that this discourse-coalition, by drawing on certain ideas, scientific concepts, and the categories of land-use planning was in a powerful position to gain credibility and acceptability, and had thus a great (rhetorical) power in structuring the future discourse and push for its institutionalization.

11.4 The ZDP draft of 2006: proposing 'partial' channelization

After the rejection of the draft ZDP-1998 and the change of land use for 'pocket III', DDA focused on developing the land. In 2000, the construction of the Akshardham Temple Complex started on the reclaimed land. In 2003, DDA intended on paving the way for the development of the Games Village in this pocket (for details see section 14.2). Therefore, another draft for the ZDP was circulated among state agencies already in 2003, but a reworked draft ZDP was made public by DDA only in May 2006 inviting objections and suggestions.

The draft ZDP-2006 still proposed large-scale developments, but was reworked especially in order to ensure the post-facto authorization of the already realized developments in the River Zone (e.g. Akshardham Temple Complex). The draft ZDP-2006 outlined the development of a series of projects declared as 'public and semi-public uses' with a focus on the eastern bank (e.g. sports complexes, hotels, shopping plaza, convention center, amphitheater, cultural center) along with recreational facilities including parks, a race course, a polo ground and botanical gardens (DDA 2006d, an insight in the land-use plan of the draft ZDP-2006 is provided in Appendix IX). The draft ZDP-2006 did not differ much in terms of proposed development compared to the earlier version of 1998. On the western bank, a cricket and football stadium complex (85 ha) was proposed by the DDA south of the DND-Flyway (opened in 2001) and a large green area was proposed south of the Nizamuddin

Road Bridge towards the proposed new bridge linking East Delhi to the so-called 'Barapulla elevated corridor' (this area was earlier discussed as pocket I).[91]

In 2002, the DDA had specified the land "utilisation proposal" for the area along the Kalindi Kunj Road (see Appendix IX, page 384) by approaching the YSC for approval of two stadia for cricket and football (YSC/62/2002-07-31). This project also highlighted the linkage between bridge construction and land reclamation, as the DDA and CWPRS argued that due to the guide bunds of the DND-Flyway "there was only stagnant water in the area" and "only during peak discharge [...] some flow" (YSC/67/2004-09-27). The area was thus conceptualized as having been abandoned by the river, neither used by the river nor the city at the time. This description supports the earlier argument made, that parts of the River Zone, secured by bunds, have increasingly been portrayed as 'wasted land'.

The story-line is, however, fundamentally at odds with the fact that the Yamuna floods so frequently. This indicates that story-lines, while deploying simplification, are often not free from contradictions, especially when being used to separate the spheres of land and water in floodplains. In this case it is even more complicated. Is 'stagnant water' (in contrast to flowing water) an indicator that the river had abandoned the land in this area? Or is 'some flow' during floods an indicator that it still belongs to the sphere of the river? Again these questions indicate a process of boundary-making and purification based on the intrinsic logic of land-use planning and zoning (cf. MURDOCH 2006, WHATMORE & BOUCHER 1993).

In 2004, another argument was adopted by the DDA in their ongoing attempt to achieve the riverfront development plan. When the YSC convened a special meeting in September 2004, the DDA claimed urgent need of approval from the YSC since these two stadia were "required" for the Commonwealth Games 2010 and, thus, of "national importance" (YSC/67/2004-09-27). This proposal included "mass filling" of the site and a construction of a bund to protect the stadia complex (ibid.). DDA's scheme was approved "in principle", yet on the condition that no bund would be constructed and only the area of the stadia (27 ha) would be raised about the High Flood Level (HFL) (ibid.). The YSC-members denied clearance for the proposed scheme due to non-conformity with the floodplain zoning norms indicating that in their eyes the land in question belonged to the river.

In 2006 and early 2007, the DDA again approached the YSC for approval of the construction of a bund for the "closure of gap" in the embankment arguing that "filling up of earth on the 27 ha land 1 m above the HFL will be very costly option" (YSC/72/2007-01-08). The YSC denied approval stating "the gap must not be closed as it would help in moderation of floods and contribute to safety of Delhi" (ibid.). The DDA largely neglected the proposal thereafter. However, the plan for constructing one stadium south of the DND-Flyway remained and was finally rejected only in the course of the approval of the ZDP by the MoUD in 2010.

---

91 While the Barapulla elevated corridor was built for the Commonwealth Games in 2010 linking Sarai Kale Khan to Jawaharlal Nehru Stadium, the bridge over the Yamuna (phase III of the project) is still under discussion and plans are not finalized.

On the eastern bank, the draft ZDP-2006 outlined the construction of a consistent forward embankment from the ITO Barrage all the way south to the Chackchilla area (see Appendix IX). From north to south, this embankment was envisioned to reclaim land for a socio-cultural complex (30 ha), the Yamuna Bank Metro Depot (39 ha), a sports complex (175 ha), public and semi-public uses (42.5 ha, including Akshardham Temple Complex), residential uses (11 ha, Games Village), a commercial complex including hotel (5.5 ha) and another sports complex (175 ha). Unmistakably against NEERI's recommendation, DDA still planned large-scale developments including residential and commercial uses for the River Zone in the mid-2000s. With regard to the river, the draft ZDP-2006 projected further de-facto channelization in the central stretch without making this explicit.

DDA certainly plays the most important role in the promotion of the channelization and riverfront development agenda. Yet, it needs to be mentioned that other state agencies, including the Delhi Tourism and Transportation Development Corporation (DTTDC) and the Housing and Development Corporation Ltd. (HUDCO) as well as private developers, also made development proposals for the River Zone in the 2000s (cf. BATRA & MEHRA 2008, GUPTA 1995, JAIN 2009a). The influence of these other state agencies must not be neglected in the planning process. A retired DDA planner even claimed that DDA was never the driving force behind the channelization and riverfront development ideas (DDA-07/2011-11-23):

"Channelization was not an idea of the planners. It came basically from the leading community, from the bureaucratic community, who found that this attractive land in the center of the city is waste land or being encroached upon. I don't know whether there was a mind-set among the planners for channelization of the river anytime."

While other interview partners supported this statement to some extent, almost all highlighted the fundamental role of the DDA as the public planning body in the preparation of the different schemes.

The most controversial and greatest transformation of Delhi's riverscapes was proposed by HUDCO in collaboration with the foreign developer consortium Fairwood in 2005. They proposed a Rs 80,000-crore riverfront development plan from ITO barrage to Kalindi Kunj covering about 2,800 ha (7,000 acres) of land.[92] The plan included constructing upstream storages and a 44 kilometer long link between the rivers Yamuna and Hindon upstream of Delhi. The managing director of HUDCO at that time was cited in newspaper articles to have promoted the scheme by arguing that "the connectivity of both the rivers will be an alternative way of storing 'extra' water", which can be released and used during non-monsoon period and the scheme would provide funding for cleaning the river.[93] The project proposal included large-scale commercial development, 80,000 residential units and sports complex (cf. BATRA & MEHRA 2008: 406, BROSIUS 2010: 139).

---

92 See among others: Hindustan Times, 18th July 2006: 'Riverfront development – carving a city out of Yamuna' by Aruna P. Sharam.
93 See among others: Indian Express, 26th August 2006: 'Years after proposal, Yamuna-Hindon link-up may finally happen' by Teena Thacker. See also: Outlook, 16th July 2006: 'Yamuna river front development to change face of Delhi'.

The DTTDC made a similar development proposal to create more than 2,800 ha (7,000 acres) of recreational space along the river Yamuna by channelizing the river to a width of 400 m. The proposal included 12 stadia and sports complexes along the Yamuna at similar locations as proposed in the draft ZDP-2006 by DDA (see Figure 26).

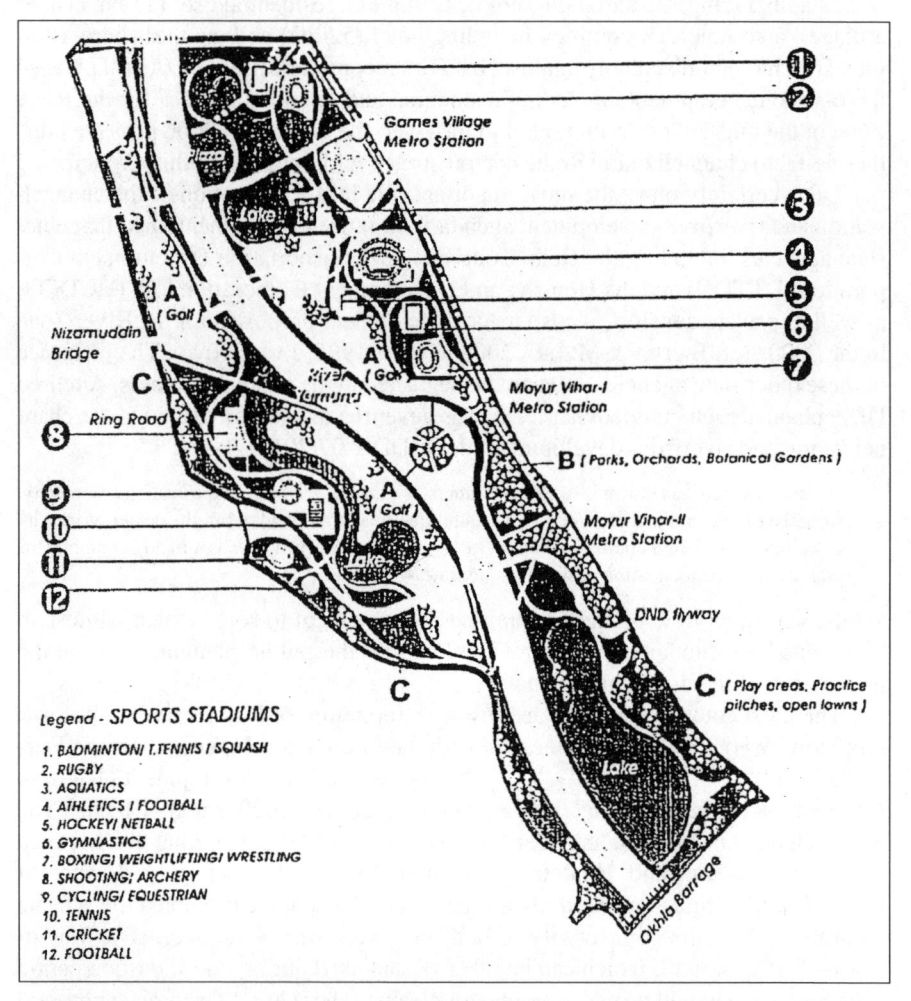

Figure 26: DTTDC's plan for channelization and riverfront development (2006)
Source: DTTDC 2006 reproduced from JAIN (2009a: 150)

While these development proposals have largely remained pipe dreams, they show that the channelization and riverfront idea was fueled by a whole group of state agencies, quasi-state agencies, and public-private partnerships for a long time. The Commonwealth Games further induced a new development 'hysteria' pushing the proposed development schemes in the public debate. In connection with the mega-sports event a powerful discourse-coalition emerged drawing on the channelization discourse and the wasted land and fear of encroachment story-line already deployed by the DDA and other state agencies in the 1980s and 1990s.

11.5 'Wasted land' and the 'fear of encroachment' versus 'ecosystem services'

Picking up the ideas of channelization discourse of the 1980s and 1990s the central story-lines of the discourse-coalition supporting the riverfront development schemes were: the evaluation that 'undeveloped' land in the River Zone means 'wasted land'; and that the land was under a constant threat of 'encroachment'. These two story-lines were closely connected to a city-wide discourse of land scarcity. The land in the River Zone was perceived by bureaucrats, politicians, planners and developers as surplus land 'waiting to be developed':

> "Delhi was growing very fast and everyone used to think that this is vacant land, potential land for urban development [...] the common perception among various administrators, bureaucrats and even sometimes the planners is that whatever land is available should be used to the optimum use, seen as a property without going into the depth of the environmental issues or the issues of ecology" (DDA-07/2011-11-23).

Reasoning DDA's riverfront developments, the planners promoted high-density development along the river that – they argued – would cater to control urban sprawl "and make optimum use of the land" in the region (NEERI 2005: 2.7). The development of recreational uses in the riverbed would further "be a post-independence gift not only for Delhi but also for the entire Delhi Metropolitan Area" (NEERI 2005: 2.7). Further, DDA's plans highlighted that riverfront development will also bring "relief and decentralization of economic activities" and, therefore, would help to decongest Old Delhi (NEERI 2005: 2.8).

The arguments are quoted here from NEERI's report. It is evident where the language of the DDA planners was 'inserted' into the report in contrast to the language used by the technical-engineering body, therefore merging different strands of technical knowledge into simplified story-lines.

The planners believed that the schemes could provide the much needed green and open spaces for East Delhi, which was "extremely deficient in recreational facilities and public facilities" (GUPTA 1995: 689). The scarcity of land discourse, which enabled the *wasted land* story-line to gain integrity, in turn also constituted the fear of encroachment story-line, since remaining land resources needed to be protected – and the best protection was considered to be 'planned' development. By 1998 the DDA had already recommended to take up "areas susceptible to encroachments and / or unauthorised developments" at first priority irrespective of any phased development in the draft ZDP (DDA 1998: 15).

The fear of encroachment story-line is also apparent in the publications of retired DDA planners. For example, A.K. JAIN (1990: 130) observed:

> "The lands which were identified as flood prone [in the first Master Plan] now fall in the midst of urban development. As such these have become prone to encroachment and therefore need to be immediately planned and controlled."

In the same way, R.G. GUPTA (1995: 688) noted:

> "Being centrally situated and considering pressures on the land, the land in river bed is precious, if not developed in a planned manner, it would be encroached upon and built upon as already going on. Prevention and removal of encroachments is always difficult so such situation should be avoided."

It is characteristic for such a simple but powerful story-line constituted by a powerful actor like DDA that such a story-line would also find its way into writings and policy recommendations of other government bodies and adopted by consultants. As an example, the cited paragraph above (GUPTA 1995: 688) was used verbatim in the DDA-sponsored NEERI report of 2005 (cf. NEERI 2005: 2.8). This suggests that the above quotes from the publications of retired DDA planners need to be considered as the baseline approach of the DDA towards the River Zone in these years.

The fear of encroachment story-line was also widely reflected by interview partners within DDA (DDA-01/2009-12-02, DDA-07/2011-11-23). Explaining the development of the continuous Ring Road, one senior DDA-planner acknowledged that the road actually "protects the River Zone from unauthorized encroachments, because any occupation of land on the other site of the road towards the river can be easily spotted and action can be taken" (DDA-01/2009-12-02). The DDA-planner further emphasized that the slums "grow like anything [and] if they are not checked, they will be mushrooming and maybe tomorrow you will find that the whole river zone is squatted upon" (DDA-01/2009-12-02).

The outlined fear of encroachment story-line was therefore institutionalized by the DDA through the provision of the ZDP (cf. DDA 1998, 2006c, 2008, 2010b). As an example, the following paragraph is found in all three drafts for the ZDP and was finally approved in the ZDP-2010:

> "Land under public ownership, with no specific assignment of uses and having good accessibility is highly susceptible to encroachment and unauthorised development and construction. Priority development of such lands will ease stress on the land management system" (DDA 1998: 14, 2006c: 22, 2008: 28, 2010b: 28).

Discussing the fear of encroachment story-line with other interview partners outside of DDA it became clear that this story-line is not only prevalent in the planning of the DDA, but is widely accepted by other actors. According to this argument, encroachment is understood to be prevented only by developing the land before it gets encroached. A senior self-employed architect argued with regard to a proposed development of a theme park close to the river by Delhi Metro that the construction of a

> "Theme Park may not be the best use for that [land], but if you leave it as it is, it will be encroached. You just can't leave a huge of land unattended in Delhi and don't expect the poor people to come and sit there. [Encroachment] is not a fear, it is a reality" (ARCH-02/2011-11-17).

In contrast to these pro-riverfront development and channelization discourses using the fear of encroachment story-line, the same was also developed into an argument opposing channelization. A water conservation expert from a Delhi-based environmental NGO explained this logic (NGO-05/2011-03-05):

> "Obviously, people started thinking, if urbanization cannot happen what will happen in this 90 square kilometre. We said, if you want to avoid encroachment on the floodplain one way is to fill it up with water. People or slums will not come. We had some concrete argument for the Delhi Government, since people were saying if we don't develop it then it will get encroached. We said fill it with water and people will not come."

From this point of view the lands of the floodplain, only temporally flooded and providing marginal spaces for the poor, should be fully given back to the river by creating engineered structures (e.g. barrages) for withholding the monsoon flood water in order to create groundwater recharge ponds. These would then 'naturally' prevent encroachment. This argument is aimed at purifying Delhi's riverscapes into a waterscape – lands temporally soaked in water should become lands covered by water throughout the year. Thus, the story-line also found merit in counter-discourses against riverfront development drawing on the groundwater recharge potential of the floodplain.

Furthermore, other ecological functions and benefits of the river's floodplain gained recognition, when the riverfront developments were discussed in the public domain. A World Bank-funded study, prepared under the aegis of the MoEF by a team of scientists from the Delhi University (including members of CEMDE), is characteristic for the inter-discursive debate at that time and the framing of an *ecosystem services* story-line (BABU et al. 2003: 2):

> "Land is a scarce resource in Delhi, thus strong socio-economic justifications are given to carry out such [channelization] development programmes. These justifications neglect the hidden 'economic value' of the ecological functions and benefits that are provided by the wetlands to the urban society and local inhabitants on a sustainable basis. Moreover, due to characteristic position of these wetlands in the landscape they have a critical role in the urban ecosystem of Delhi particularly with respect to ground water recharge."

While highlighting the various problems and shortcomings of the economic valuations of ecosystem services, the cost-benefit analysis undertaken in the study suggested that the conservation of the floodplain area is economically more viable than the channelization scenario (BABU et al. 2003: 142). As the quote above indicates the study also drew on the scarcity of land discourse to develop a counter-argument. Media coverage[94] and interview partners (NGO-04/2010-09-18, NGO-17/2013-02-11, CEMDE-01/2011-03-09) indicated that the study had an important impact on the overall debate around the Yamuna in Delhi.[95]

---

94 See among others: Times of India, 10th October 2002: 'Yamuna one of the most threatened ecosystems'. The Hindu, 19th September 2007: 'A shot in the arm for 'Save Yamuna' drive' by Smriti Kak Ramachandran. The Hindustan Times, 14th September 2007: 'Report against Yamuna channelisation' by Aruna P. Sharma.
95 The co-author of the study Dr. Pushpam Kumar is today Chief of the Ecosystem Services Economics Unit, Division of Environment Programme Implementation, UNEP.

"What the study did [...] was to put in a value on the function that would be lost by channelization. And the press did pick up the positive elements of it saying that conserving floodplains makes more business sense - nothing beyond that. [...] to get something moving on public terms – something easy to assimilate and understand – valuation studies do play an important role. I think in this case it was used very proactively. [The study] was one of the key studies that blocked [channelization] any further" (NGO-17/2013-02-11).

The counter-discourse had to adopt the rationalities of economic value of the land along the river in some way if it hoped to be heard. The economic value of the groundwater recharge potential of the river emerged as the most credible story-line in the light of the pervasive 'water shortage' discourse in Delhi (cf. MANN 2007, MARIA 2006, SELBACH 2009). The 'water shortage' discourse is a very politically sensitive issue but also easy to communicate to the public by linking it to events of interruption in the water supply system. The groundwater recharge capacity of the floodplain became therefore the most important argument in the ecosystem services story-line.

11.6 Floodplain zoning: institutionalization or co-optation?

In addition to the ecosystem service story-line, a discourse structuration concerning floodplain zoning is also identifiable in the draft ZDP-2006. Based on the NEERI report of 2005, DDA finally prepared a floodplain zoning map. This seemed to indicate a certain step towards institutionalization of flood risk prevention in the land-use plans of the DDA. Yet, a detailed reading of the draft ZDP-2006 makes apparent that the DDA adopted the policy of floodplain zoning rather to please critical voices by showing some goodwill to the CWC. The prepared floodplain zoning clearly contradicts the findings of the in depth study by the IFCD around the 1978 floods, which had also been widely discussed in the YSC meetings (see above section 10.2). Adopting the terminology of 'floodplain zoning', NEERI and DDA strategically located the areas supposedly 'safe' from floods. In the stretch from Wazirabad to Okhla the DDA declared 117 ha safe from any floods, 545 ha safe from 100 year return floods and 2,197 ha safe from 10-year return floods. In total 3,988 ha were envisioned for different kinds of development (DDA 2010b: 39, NEERI 2005: 3.94, see Figure 27). However, even 10-year flood events are known to impact far more area in the River Zone as calculated by NEERI/DDA and in 1978 the entire remaining floodplain between the embankments was inundated (EXPERT COMMITTEE MoEF 2014). The floodplain zoning provided by NEERI/DDA was indisputably faulty; and the overall analysis suggests the zoning was done wrongly by design in order to legitimize the proposed developments.

NEERI/DDA had proposed cut-and-fill methods to actually create these different zones by dredging the river and filling up certain areas. In addition, DDA has interpreted NEERI's recommendation loosely, by stating that:

"[...] three dimensional development is envisaged in the central areas which have good locational potential and are either comparatively free from inundation or can be made free from inundation expeditiously and/or at low cost" (DDA 2006d: 20–21).

PROPOSED LANDUSE PLANNING AND MODIFIED RIVER TOPOGRAPHY FROM WAZIRABAD BARRAGE TO OKHLA BARRAGE

RECOMMENDED LAND USE

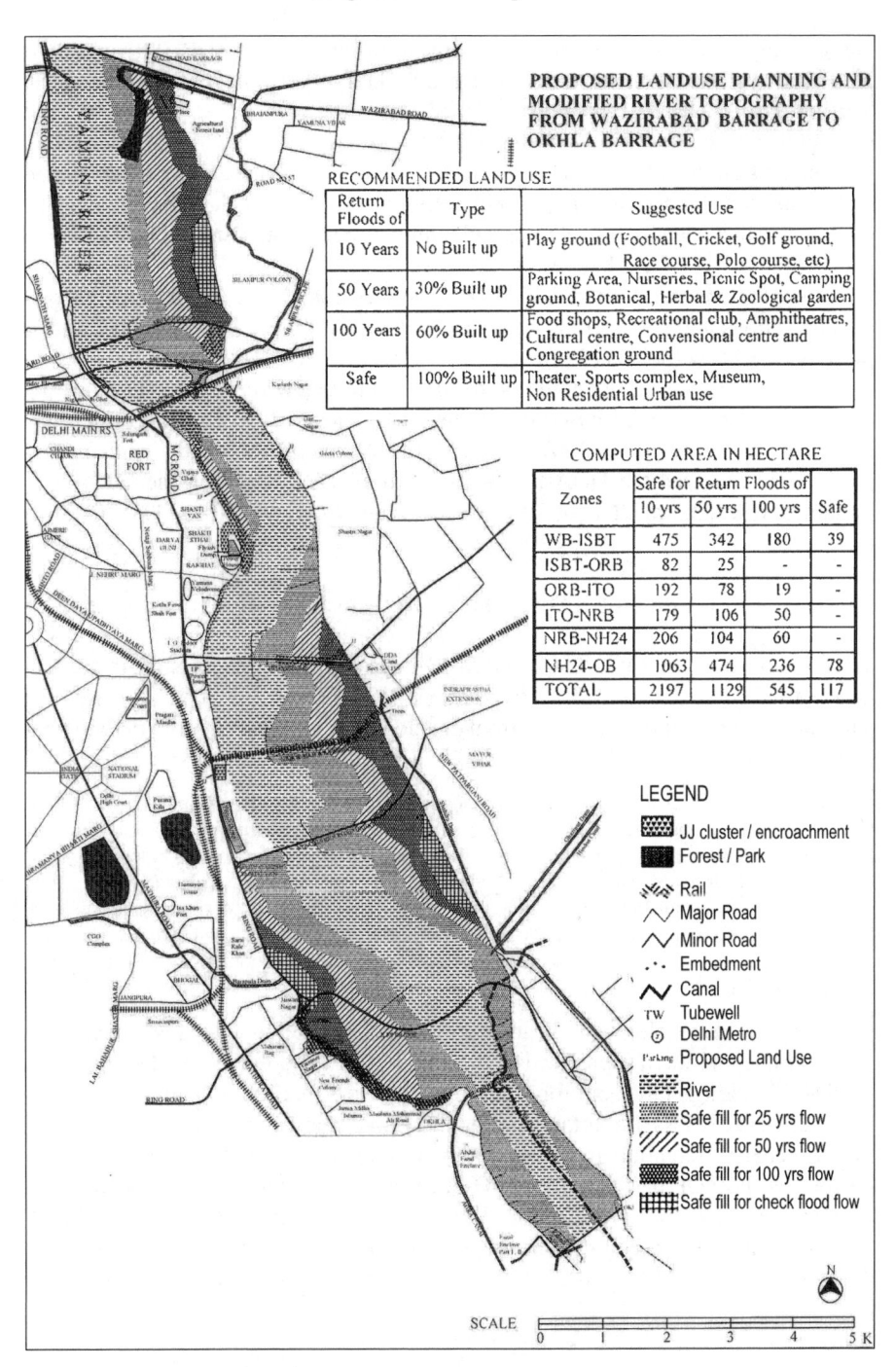

Return Floods of	Type	Suggested Use
10 Years	No Built up	Play ground (Football, Cricket, Golf ground, Race course, Polo course, etc)
50 Years	30% Built up	Parking Area, Nurseries, Picnic Spot, Camping ground, Botanical, Herbal & Zoological garden
100 Years	60% Built up	Food shops, Recreational club, Amphitheatres, Cultural centre, Convensional centre and Congregation ground
Safe	100% Built up	Theater, Sports complex, Museum, Non Residential Urban use

COMPUTED AREA IN HECTARE

Zones	Safe for Return Floods of			
	10 yrs	50 yrs	100 yrs	Safe
WB-ISBT	475	342	180	39
ISBT-ORB	82	25	-	-
ORB-ITO	192	78	19	-
ITO-NRB	179	106	50	-
NRB-NH24	206	104	60	-
NH24-OB	1063	474	236	78
TOTAL	2197	1129	545	117

LEGEND
- JJ cluster / encroachment
- Forest / Park
- Rail
- Major Road
- Minor Road
- Embedment
- Canal
- TW Tubewell
- Delhi Metro
- Parking Proposed Land Use
- River
- Safe fill for 25 yrs flow
- Safe fill for 50 yrs flow
- Safe fill for 100 yrs flow
- Safe fill for check flood flow

Figure 27: Floodplain Zoning (O-Zone) prepared by DDA
Source: DDA (2010b: MAP II) based on NEERI (2005: 3.94), graphic modified by author

Remarkably, the DDA used the exact same wording in draft ZDP-1998 and draft ZDP-2006 (DDA 1998: 14, DDA 2006d: 20–21). Furthermore, even the approved version of the ZDP-2010 contains the cited paragraph verbatim (DDA 2010b: 28). This indicates that the results of the NEERI study, which were critical of channelization, and the overall critique against the proposed riverfront developments had only limited influence on DDA's actual policy framings for the River Zone. The ZDP was rather seen by the DDA as a planning tool for administratively reclaiming land from the river for urban development.

The public notice for the draft ZDP-2006 was issued on 21st August 2006 and 112 objections and suggestions were received (DDA 2010b: 10). These comments were reviewed by the Board of Enquiry and Hearing of DDA in May 2007, and it was decided that the ZDP should be reworked and a new round of public hearing should also be convened based on the new provisions of MPD-2021, which had been approved in February 2007 (ibid.). Moreover, the NCRPB had already in January 2005 requested the DDA to rework the draft ZDP based on the MPD-2021 framework, which the DDA had nonetheless failed to do so (DDA 2008: 10).

11.7 The ZDP draft of 2008: Seeking post-facto approval for developments

In comparison to the MPD-2001, the MPD-2021 outlines a major shift in the language used referring to Delhi's riverscapes. While the MPD-2001 still used the phrases 'channelization' and 'riverfront development', these terminologies are non-existent in the MPD-2021 (cf. DDA 2007). The MPD-2021 rather states that

> "[…] a strategy for the conservation/development of the Yamuna River Bed area needs to be developed and implemented in a systematic manner. This issue is sensitive both in terms of the environment and public perceptions. Any such strategy will need to take into account the cycle of flood occurrences and flood zones, the ground water recharge potentials and requirements, potential for reclamation derived from the foregoing considerations, designation and delineation of appropriate land uses and aesthetics of the River Front which should be more fully integrated with the city and made more accessible-physically, functionally and visually" (DDA [2007] 2010: 94).

The adopted terminologies indicate that the counter-discourse against channelization had become institutionalized in the MPD-2021. The recommendations of the MPD-2021 though left room for interpretation as reclamation of land was still considered. Nevertheless, important arguments against channelization (e.g. the groundwater recharge potential) found their way into the larger policy framework. After reworking the ZDP based on the MPD-2021 provisions, DDA invited public comments on the new draft in July 2008 (DDA 2008). While the language of the MPD-2021 indicated an emerging dominance of the anti-channelization discourse, riverfront development (e.g. Games Village) was more and more 'decoupled' from the 'great scheme' of channelization.[96] Nevertheless, it is insightful to review DDA's most recent 2008 draft, as the draft outlines new discursive framings with regard to the ecological sensitivity of the River Zone in addition to adopted new terminologies for separating the river and the city.

First of all, the draft ZDP-2008 formulates the "basic goal [...] to rejuvenate river Yamuna as a 'CITY LIFELINE' by striking a balance between developmental parameters and ecological aspects" (DDA 2008: 13 [emphasis in original]) and, thus, clearly adopts the language used in the MPD-2021. Yet, by indicating that the ZDP aimed to make "more land available" for development, the DDA still proposed the development of "remunerative purposes" (DDA 2008: 14). DDA aimed to remove "non-conforming and polluting" land uses (e.g. JJ cluster and unauthorized construction) in order to protect the ecological sensitive zone, while defining its own planned, desired developments as "suitable, positive, compatible, and conforming" (DDA 2008: 16, cf. DE MELLO-THÉRY et al. 2013: 230). The detailed analysis of the land-use changes in the river zone (see chapter 12) highlights the spatial effects of this controversial and discriminatory treatment.

The reclamation of the 'riverfront' (now also spatially defined by DDA, for details see below) was overall propagated by the DDA as an economic driver for the city as a whole and as a project to be taken up without extensive spending of public funds (DDA 2008: 16):

> "Reclaiming the riverfront ensures not only its environmental integrity but also stimulates the economy. The aim is to achieve maximum economic benefit with minimum public investment to ensure a self-sustaining set up. Thus, the riverfront development will be based on financial and environmental catalysts. The revenue generated by the various activities will allow the riverfront to further develop parks and trails, commercial sites and infrastructure improvement within the area for an economically and environmentally healthier riverfront. This healthier vision is critical, as it is the foundation for new, long-term economic growth in the city and region."

The draft ZDP-2008 hence indicates a 'balancing act' performed by the DDA between formulating (strong) policies for the conservation of the floodplain and ensuring development potential.

For example, the DDA on the one hand outlines that the "river front vision includes multiple land uses" (DDA 2008: 16) and that the ZDP is not meant to "limit the variety of possible uses" (DDA 2008: 25), while on the other hand it argues that the "eco-sensitive nature" of the River Zone needs to be taken into account.

In contrast, to the NCR Regional Plan-2021 (approved in 2005), which clearly states that the River Zone should be treated as a "natural conservation zone" and that in "the flood prone areas/river beds/banks, no construction or habitation activities [shall] be permitted" (NCRPB 2005: 153–154), the DDA largely avoided the

---

96 In the meantime, the construction of the Games Village on the floodplain had already prompted an intensified critical debate on developing the river zone. Therefore, the 2008-draft ZDP must be interpreted as an approach seeking post facto approval for the development already undertaken in the river zone (for details see 14.2).

term 'conservation' in the draft ZDPs.[97] Instead the DDA, based on its floodplain zoning, claimed that "[c]ertain highland pockets can be identified in the river basin for any kind of proposed development" (DDA 2008: 26).

The controversial floodplain zoning remained the basis for the planning of the River Zone despite the protest of environmental NGOs. The environmental NGOs critically assessed the proposed zoning concept in the public participation process:

> "The flood zonation map [...] is grossly incorrect in the light of the recent floods whereby all such areas in the flood plain which had been shown safe even in 100 yr return flood went under water. [...] The NEERI flood plain zoning map is no longer valid and it should not be used any more. As a matter of fact in view of limited width of the available flood plain there is no flood plain zonation possible for the Zone O as has been proved during the recent floods of 2008."[98]

Despite the critique, the floodplain zoning, including the resulting land utilization concept, was finally sanctioned by the MoUD with the approval of the ZDP in 2010 – about 12 years after the first draft was circulated in 1998. Overall, the draft ZDP-2008 became the legal planning document for the River Zone with only minor changes (cf. DDA 2008, 2010b).

In comparison to the earlier drafts, the approved ZDP (based on the draft ZDP-2008) introduced a differentiation between "three distinct morphologies" for the River Zone: the area under river water defined as "river-bed", the area between the river water course and the embankments defined as "river flood plain" and the "river front" demarcating the area outside the embankments (DDA 2008: 25, cf. DE MELLO-THÉRY et al. 2013: 230). The DDA gave the following development guidelines for the three morphologies:

> "The area under water course is part of the river hydrology. The area between water course and embankment is to be conserved as flood plain. Any development in this areas [sic] should be taken up only after the detailed hydrological studies and with approval of Yamuna Standing Committee/Central Water Commission. The area out side the flood plain ie out side [sic] embankments (River front) should be conserved and developed considering the eco-sensitive nature of the river zone and based on comprehensive scheme" (DDA 2008: 25).

The draft ZDP-2008 indicated that the DDA through this invention of the three morphology terminologies aimed to redefine the scope of the YSC, while at the same time ensuring approval from the NCRPB by differentiating between riverbed/floodplain inside the embankments and riverfront outside the embankments. The draft ZDP-2008 (almost verbatim with the draft ZDP-2006 and the approved ZDP-2010) states that any development outside the embankments (river front) "will not require technical clearances" from the YSC. In doing so, DDA claimed the absolute planning autonomy for the land protected by the embankments (riverfront) clearly aiming at developing the riverfront without taking approvals from the YSC

---

97 The NCR Regional Plan-2021 permits only agriculture, horticulture, aquaculture, social forestry and plantations including afforestation, and regional recreational activities with no construction exceeding 0.5% of the area (NCRPB 2005: 156–157). The draft ZDP-2008 uses the term 'conservation' only in regard to built heritage and water (DDA 2008)
98 Letter Yamuna Jiye Abhiyaan to DDA: Comments / Suggestions on draft zonal plan for Zone O (River Yamuna), 14 October 2008.

or any other agency involved in river governance. Thus, through the three morphologies the DDA clearly *purified* the hybrid space along the river into the domains of land and water plus a remaining, hybrid intermediary space of the so-called 'active' floodplain.

In addition, by indicating that "development is a continuous process" (DDA 2008: 25), the DDA left the door ajar for reclaiming more land from the river through the construction of new embankments, which will then result in the 'hydrological fact' that these area are abandoned by the river and are thus no more part of the 'active' floodplain.[99] The case of the Akshardham Bund will outline how the DDA in the following relied heavily on the three morphologies.

Comparing the draft ZDP-2006 (see Appendix IX) with the approved land-use plan of the ZDP-2010 (see Appendix IV) illustrates that the DDA simply changed the color of the former blue marked areas for public and semi-public facilities (e.g. sports complexes) into green for recreational uses (parks). The extent of these green patches along the eastern bank is largely congruent with the former pockets earmarked for large-scale development. While 'River & Water Body' (ca. 68%) and 'Recreational' (ca. 21%) are the dominant land-use categories, any clear logic explaining the boundaries of these green areas is nonetheless missing in the ZDP. The change to 'River and Water Body' (white) remains unclear. Surprisingly, the ZDP land-use plan does not provide any information on the existing embankments. Thus, the spatial demarcation of the three morphologies also remains unknown.

11.8 Interim conclusion

The detailed analysis of the policy-making process of the ZDP for the River Zone indicates that while the reclamation of land from the river for urban development was highly criticized based on various grounds for a long period of time, the institutional powers vested with the DDA allowed the central urban planning institution to address these increasing critiques half-heartedly, making only minor changes in the drafts for the ZDP. When the draft-ZDP 1998 was prepared, channelization was still the dominant discourse and it was vehemently opposed.

In the early 2000s counter-discourses started emerging. The 'great scheme' of channelization was finally dismissed, but project-based approaches remained on the agenda of the DDA. The DDA aimed to legitimize its prepared schemes and obtain post-facto approval for its developments. The construction of the Akshardham Temple Complex and Commonwealth Games Village only encouraged the DDA to push riverfront development, which was increasingly decoupled from river channelization.

---

99 The author likes to thank Veronique Dupont for highlighting that the three morphologies were already introduced in the draft ZDP-2008 (personal communication) and the inspiring analysis of the ZDP-draft in her co-authored article 'Public Policies, Environment and Social Exclusion' (DE MELLO-THÉRY et al. 2013).

The floodplain zoning conflict between DDA and CWC evolved in so far that a certain institutionalization of the concept is traceable in the ZDP. Yet, the implementation of floodplain zoning by DDA differs substantially from the original purpose and idea of floodplain zoning. Based on the 'negotiated facts' of the NEERI study, the DDA was able to co-opt the concept of floodplain zoning. A concept, which was originally intended to keep the floodplain free from development, was reinterpreted and reformulated as an urban zoning concept. In this process, the idea of an optimized land utilization established through the wasted land story-line emerged as the baseline rationality for creating a "modified river topology" by engineered interventions in form of embankments and cut and fill methods (DDA 2010b: 39).

The negotiation and reformulation of scientific findings and its implementation through the floodplain zoning concept developed by DDA/NEERI highlight the 'co-production' of science and policy (KEELEY & SCOONES 2003). The scientific 'facts' – whether negotiated, correct or incorrect – simplified the hydrological system of the river and the separation between water (river) and land (city) to a matter of topography. By comparing the level of the land and the calculated level of water during floods, the DDA defined the cut and fill measures. All other ecological connections between the land, temporally "soaked in water" (LAHIRI-DUTT 2014: 508) and the river are not covered in the applied concept of floodplain zoning.

Furthermore, this 'topological simplification' allowed that, through cut and fill and dredging of the river, the hydrological system could be altered and new topographies could be produced. The scientific 'facts' provided by NEERI and CWPRS further reduced the earlier felt uncertainty and, thus, legitimized the reclamation and development of Delhi's riverscapes. Based on the technical studies, the DDA aimed to install a scientific document, which by its data would delegitimize the opposition against riverfront development and establish the ecological feasibility for urban development in the floodplain. In this context, the title of the study 'Environmental Management Plan for Rejuvenation' (NEERI 2005) is remarkably.

The ZDP drafting further demonstrates the process of boundary-making and separation of the city from the river by imposing "a site-based, rather than system-based" logic (WHATMORE & BOUCHER 1993: 169). The three outlined morphologies of 'riverbed', 'floodplain' and 'riverfront' illustrate this. Its implications and legitimizing powers will become more evident when reviewed in the context of the implementation of the 'suitable, positive, compatible, and conforming projects' developed by DDA. The following chapters review the details of the actual land-use changes along the river.

VI THE RECLAMATION OF DELHI'S RIVERSCAPES FOR A WORLD-CLASS CITY IN THE MAKING

12 MULTI-TEMPORAL ANALYSIS OF LAND-USE CHANGE (2001–2014)

In the 1980s and 1990s, various schemes were developed and discussed for developing Delhi's riverscapes. None of these were approved and aside from the bridges, not much 'planned' development happened on the ground. With the start of the new millennium large-scale development projects started and slum demolitions (long envisioned in the channelization schemes) reclaimed the riverfront for the city's makeover to become world-class. Based on multi-temporal mappings and information from qualitative interviews this chapter presents the results of the analysis of the land-use changes, focusing on the stretch from Wazirabad Barrage to Okhla Barrage (research area) between 2001 and 2014. This chapter highlights the dynamic physical remaking of Delhi's riverscapes in the new millennium.

12.1 Selection of research area for land-use change analysis

The research area covers about 5,200 ha and is bounded by the Wazirabad Barrage in the north and the Okhla Barrage in the south, the Ring Road in the west and the existing embankments in the east. The eastern and western boundary of the research area corresponded with the boundary of the River Zone (Zone O) defined by the DDA (see Figure 1, page 30).

While this study focuses on the urban environmental governance in the NCT of Delhi, the southern part of the research area encompasses areas belonging to Uttar Pradesh. The territory of Uttar Pradesh is included in the land-use change analysis for the sake of coherence with the riverscapes physical characteristics. In total about 500 ha of the research area belong to Uttar Pradesh, of which large parts (ca. 400 ha) are protected as the Okhla Bird Sanctuary, a wildlife sanctuary notified by the Government of Uttar Pradesh in 1990 under the Wildlife Protection Act 1972.[100]

The width of the research area varies from about 800 m at the Old Railway Bridge to more than 3.5 km south of the Nizamuddin Bridge. During the non-monsoon period, the waterway of the river in this stretch of the river is at certain places less than 150 meters. The slope of the ground level of the floodplain in this

---

[100] Following the construction of the new Okhla Barrage in 1986, a new year-round lake with low water levels was created attracting a large number of birds. The protection of the Okhla Bird Sanctuary has also been a matter of the courts for many years.

stretch of the river is approximately eight meters from about 202 m ASL at the Wazirabad Barrage to about 194 m ASL at the Okhla Barrage (VIJAY et al. 2007: 384).

The selected research area is justified by the central location, the dynamic recent changes and the fact that this segment was the main focus of the channelization and riverfront development schemes in the 1980s to 1990s.

12.2 Land-use changes 2001–2014

Delhi's riverscapes were massively transformed within the last two decades through the demolition of slums and the development of urban mega-projects. However, agricultural uses still dominate large stretches along the banks of the river. Nevertheless, new roads and other transportation infrastructure (metro and bus depots) have altered the physical appearance of the banks. Table 8 provides detailed insights in the land-use changes for different stretches of the river.

Stretch of the river	West existing uses pre-2001	West new developments post-2001	East existing uses pre-2001	East new developments post-2001
1) NCT of Delhi boundary – Wazirabad Barrage	unauthorized colonies, Wazirabad Village, Wazirabad Water Treatment Plant	Yamuna Biodiversity Park	unauthorized colonies, Sonia Vihar Water Treatment Plant, Delhi Police Firing Range	ongoing construction in various parts
2) Wazirabad Barrage – ISBT Bridge	Tibetan Colony (unauthorized), Majnu Ka Tila Gurudwara, Chandrawal Water Treatment Plant, Tibetan Market	Gas station (CNG) Signature Bridge	Usmanpur Village, Old Garhi Mendu Village, restricted forest	electrical substation, Signature Bridge with approach roads
3) ISBT Bridge – Old Railway Bridge	bathing ghats, Nigambodh Ghat (cremation ground), residential colony Yamuna Bazaar (unauthorized extensions and religious structures), JJ cluster***	-	unauthorized colonies, JJ cluster***	Metro Depot Shastri Park*, Metro Station Shastri Park, Metro IT-Park, Residential Project Delhi Metro
4) Old Railway Bridge – ITO Bridge/Yamuna Barrage	electric crematorium, samādhis (Vijay Ghat, Shanti Van, Raj Ghat, etc.), Rajghat Power Station, fly-ash ponds, Indira Gandhi Indoor Stadium cum Sports Complex, Delhi Secretariat, Ring Road By-Pass Road [jj-cluster]	Golden Jubilee Park, Rajghat Bus Depot, fly-ash brick plant	cremation ground	-

	West		East	
Stretch of the river	existing uses pre-2001	new developments post-2001		existing uses pre-2001
5) ITO Bridge/ Yamuna Barrage – Nizamuddin Railway Bridge	Indraprastha Power Station, Pragati Maidan Power Station, JJ cluster***	Metro Station Indraprastha	restricted forest	Yamuna Bank Metro Depot, Yamuna Bank Residential Project Delhi Metro
6) Nizamuddin Railway Bridge – Nizamuddin Road Bridge	fly-ash ponds, fly-ash brick plant, JJ cluster***	Millennium Bus Depot (on former fly-ash pond)	restricted forest	Akshardham Temple Complex**, Akshardham Metro Station, Games Village
7) Nizamuddin Road Bridge – DND Flyway	Rajiv Gandhi Smriti Van, construction and temporary soil disposal Delhi Metro, JJ cluster***	electrical substation, electric crematorium, site of Times Global Village Festival (temporary in 2007)		-
8) DND Flyway – Okhla Barrage	unauthorized colonies (Batla House, Abul Fazal Encalve I & II, Dawat Nagar, Shaheen Bagh), Canal Colony (by UP Irrigation), Delhi Rites (amusement park), Okhla Bird Sanctuary, JJ cluster***		NOIDA area (Uttar Pradesh)	
			Okhla Bird Sanctuary (notified 1990)	approach road to DND-Flyway, Rashtriya Dalit Prerna Sthal and Green Garden (Dalit Inspiration Place and Green Garden)
9) south of Okhla Barrage	Madhanpur Khadar Resettlement Colony, fly-ash ponds, Indian Oil Depot, LPG bottling plant, unauthorized colonies (Jaitpur Extension), Jaitpur, Meethapur	Kalindi Kunj Metro Depot (under construction)	NOIDA area (Uttar Pradesh)	
			ongoing residential developments	

Table 8: Land use in the different stretches of the Yamuna in Delhi
Source: Compilation by the author from different sources and own mappings
*Remarks: Agricultural uses are found in all stretches, * construction started 1998, ** construction started 2000, *** jj cluster demolished 2004–2006*

The compilation of data illustrates the dynamic changes which have occurred, particularly in the central region of the research area. Based on the multi-temporal mappings for the years 2001, 2006, 2010 and 2014 (see Map 4, pages 388–389), Map 2 and Map 3 (pages 386–387) illustrate the combined land-use changes. In order to visualize the spatial land-use changes, the areas cleared from the slums have been provided as an overlay to the land uses in 2014. The overlay of the different mappings highlights major new developments which have been realized either within the areas of the former JJ clusters (slums) or in close proximity. For example, the Ring Road By-Pass Road in front of the Red Fort following the historic Mughal Bund has been developed on land earlier occupied by JJ clusters and the approach roads to the ITO Bridge today cover land earlier used by JJ clusters.

Overall, the built-up area increased from about 567 ha in 2001 to 582 ha in 2014 in the study area, making up about 10% of the total area in 2014 (for details see Table 9 and Figure 28).

	in ha				in %				Change in ha 2001 to 2014
	2001	2006	2010	2014	2001	2006	2010	2014	
Built Up Area	567	560	647	582	9.8	9.1	9.1	10.3	+ 15
Institutional	86	122	112	158	1.7	2.3	2.2	3.0	+ 72
Commercial, industrial and power plants	71	83	96	96	1.4	1.6	1.8	1.8	+ 25
Residential	302	215	214	229	5.8	4.1	4.1	4.4	- 73
Slums demolished (2004-2006)	97	-	-	-	1.9	-	-	-	
Transportation**	323	446	545	600	6.2	8.6	10.5	11.5	+ 277
Roads**	244	263	313	365	4.7	5.1	6.0	7.0	+ 121
Delhi Metro**	66	125	128	143	1.3	2.4	2.5	2.8	+ 77
Bus Depots	0	0	0	30	0.0	0.0	0.0	0.6	+ 30
Other (including embankments, fly-ash ponds, waste disposal, construction zones)	188	231	314	199	3.6	4.4	6.0	3.8	+ 11
Green	3341	3359	3070	2891	64.2	64.6	59.0	55.6	- 450
Agriculture* (including nurseries)	2236	1925	1792	1609	43.0	37.0	34.5	30.9	- 627
Forests	200	211	193	271	3.8	4.1	5.2	5.2	+ 71
Grasland, reed, fallow land*	776	1089	923	833	14.9	20.9	17.7	16.0	+ 57
Parks (including samādhis)	130	134	162	177	2.5	2.6	3.1	3.4	+ 47
River*	838	690	796	974	16.1	13.3	15.3	18.7	+ 136
Total	5200	5200	5200	5200	100.0	100.0	100.0	100.0	

Table 9: Land-use change in the research area 2001–2014
Sources: own calculation based on multi-temporal mappings, Eicher maps (different years), GoogleEarth (different years). * land-use changes influenced by seasonal variations in area under inundation, ** including under construction

In 2001, slum settlements along the river still covered about 97 ha. Their demolitions (2003–2006) resulted in a decrease of residential area in the research area from about 300 ha in 2001 to about 214 ha in 2010. The growth of the unauthorized colonies in the southern stretch (Kalindi Kunj area), the development of the Commonwealth Games Village into a residential colony, and residential projects by Delhi Metro have led to a slow growth of the residential areas up to 2014. The development of new institutional, public and semi-public areas (+72 ha) as well as an increase of commercial and industrial uses (including the development of the Metro IT Park and two electrical substations) resulted in overall growth of the built-up area.

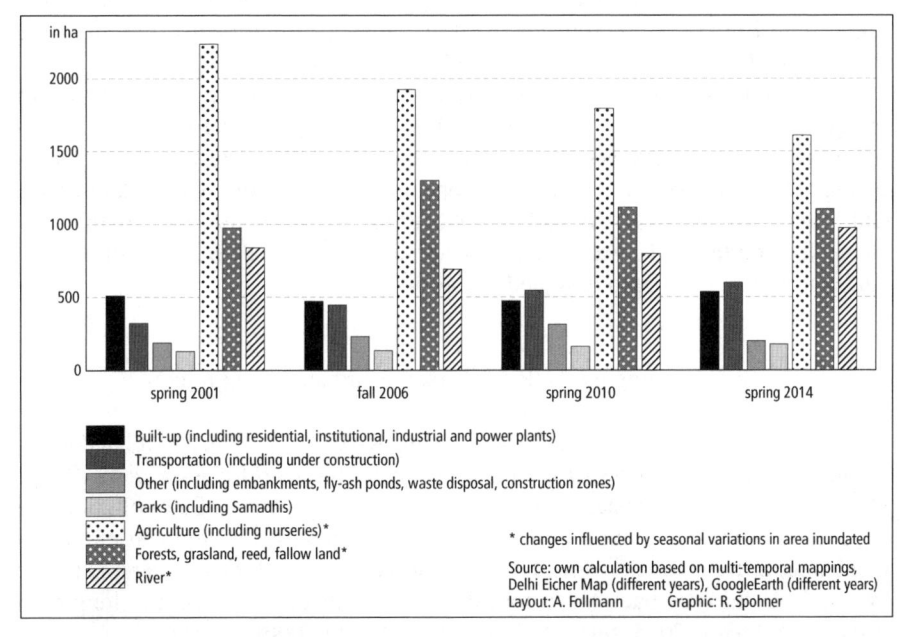

Figure 28: Land-use change in the research area by main categories (2001–2014)

In terms of land-use change, the development of new transportation facilities emerges as the most important change in land use in the study period. The area covered by transportation facilities increased from about 320 ha to about 600 ha between 2001 and 2014. Despite the fact that the construction of Delhi Metro in the research area only began in 1998, the utilities of Delhi Metro (two depots, four stations and different lines) today cover an area of more than 140 ha. Widening of roads (especially along the embankments) as well as new bridges and fly-overs resulted in a considerable growth of the area covered by roads (almost +50%). In total about 11.5% of the research area in 2014 was covered by transportation facilities.

Agriculture is the dominant land use in the research area. However, agricultural land considerably declined in 2014 from more than 2.200 ha in 2001 to about 1.600 ha (-28 %). Alongside seasonal variations, the decrease in agricultural land was caused by a slight increase in built-up area (including institutional uses, commercial and industrial uses, power plants, etc.) and a substantial increase of transportation utilities.

The construction of new road and metro bridges increased connectivity across the river. In 2014, a total number of nine bridges crossed the Yamuna between Wazirabad and Okhla including three barrage-cum-road bridges (Wazirabad, ITO and Okhla), four road bridges (ISBT, Geeta Colony, Nizamuddin, DND-Flyway), two railway bridges (Old Railway Bridge, Nizamuddin Railway Bridge), and two Metro bridges leading to the Metro depots at Shastri Park and Yamuna Bank. Two further Metro bridges are under construction: one south of the Nizamuddin Road Bridge linking central Delhi to Mayur Vihar and the other south of the Okhla Barrage linking Kalindi Kunj to Noida (outside the research area). Additionally, the Signature Bridge is under construction south of the Wazirabad Barrage.

The approach roads and guide bunds of the bridges (except the new metro bridge south of the Nizamuddin Road Bridge, which spans over the whole floodplain) have further narrowed the waterway considerably and restricted the meandering of the river. The narrowest bridge opening is found at the Wazirabad Barrage and is only 455 m wide. As outlined above, the technical recommendations for channelization favored a water channel width of 550 m and all bridges, constructed since the 1970s, have been designed accordingly.

12.3 Opening up the River Zone: the urban-mega projects

The expansion of transportation facilities in the River Zone is closely related to the urban restructuring process of the megacity under the world-class agenda. Struggling with increasing private automobile traffic and high levels of air pollution, the population of Delhi had long been desperate for a smooth development of a high-quality public transport network. The recognition of the need for a mass rapid transit system in Delhi goes back about four decades and its development has long been discussed (cf. BON 2015, MOHAN 2008, SIEMIATYCKI 2006).

The Delhi Metro Rail Cooperation Limited (DMRC) under the leadership of its managing director E. Sreedharan started construction of the Delhi Metro network in 1998.[101] DMRC was created as a joint-venture of the Government of India and the Government of the National Capital Territory of Delhi (GNCTD) in 1995. The Indian government equipped DMRC with tremendous powers for land acquisition and allowed DMRC to generate parts of the initial project costs through property development. Therefore, land especially around the metro stations and lines has been transferred to DMRC on a 99 years lease and the DMRC uses "the instrument of land value capture borrowed from Hong Kong" to fund its development (BON 2015: 224).

With regard to environmental concerns, the DMRC was supported by clear exceptions from existing environmental laws and regulations. First of all, the Metro Railways (Construction of Works) Act from 1978 allows DMRC to "divert or alter

---

101 As DMRC denied multiple requests by the author for an interview, DMRC's view of these developments could not be explored by the author. The analysis concentrates here on official documents and newspaper reports.

as well temporarily as permanently, the course of any rivers" for the purpose of constructing any metro railway line or any other work connected therewith. Furthermore, since considered to be railway projects, which are by law exempted from environmental clearance, DMRC does not require environmental clearance for its lines, station and operational areas. DMRC's property developments are, however, obliged to obtain environmental clearance and EIA notification.

SIEMIATYCKI (2006: 277) observed that "officials in Delhi have moved mountains in making the metro a reality". Indeed, this can be taken literally, since DMRC created two metro depots at Shastri Park (see Figure 29) and Yamuna Bank by filling up the floodplain with the excavation materials from the tunneling of underground lines as well as layers of fly-ash from nearby power plants (ARCH-02/2011-11-17). Furthermore, by developing an Information and Technology Park (see Figure 30) and residential quarters at Shastri Park as well as residential quarters at Yamuna Bank, DMRC has acted against orders by the YSC and the DUAC (cf. FOLLMANN 2014: 132).[102]

Figure 29: Metro depot (DMRC) at Shastri Park, East Delhi
Source: Follmann, February 2011

Figure 30: Information and Technology Park (DMRC) at Shastri Park, East Delhi
Source: Follmann, February 2011

By developing large-scale non-metro projects the DMRC has been clearly invading into a field originally under the monopoly of the DDA. Therefore, the creation of a new powerful parastatal agency being equipped with exceptional powers to develop large-scale urban projects has resulted in conflicts and controversies especially between the DMRC and the DDA (cf. BON 2015: 225).

During interviews, high-ranked DDA planners emphasized that DDA was by and large not in favor of the location of the metro depots in the River Zone, but they have been pressured by the DMRC and the MoUD to comply with the development needs of the DMRC (DDA-06/2011-11-22, DDA-07/2011-11-23). Across the board, DDA officials complained about the high-handiness of DMRC in acquiring land across the city, particularly in the research area. Nevertheless, DDA was aware

---

102 The IT Park at Shastri Park is one of the flagship projects of DMRC. For the IT-Park an EIA-study was prepared by RITES, a Government of India Enterprise; for details see BON (2015).

that based on DMRC's requirements "no land was available in other areas of this dimension" (DDA-01/2010-09-21). Thus, the agency ultimately had to agree to DMRC's plans, but tried to reduce the land allotments in the River Zone to DMRC to a minimum "considering its ecological value" (DDA-01/2010-09-21). When irregularities were proven in terms of non-compliance of DMRC with land-use regulations, the DDA "could have cancelled the allotment" made to the DMRC, but the DDA wanted "to maintain a good relationship with the Ministry of Urban Development" (DDA-07/2011-11-23). "If it would have been [sic] a private developer", as a high-ranked DDA-planner outlined, DDA "would have cancelled the allotment and also sealed all the buildings" (DDA-07/2011-11-23).

DMRC enjoyed wide public support given the existing pressure on the transport system. Furthermore, Delhi Metro has been, and still is, viewed as very efficient, professional and free from any corruption. The managing director E. Sreedharan is widely admired for the timely construction of the mass-rapid transit in Delhi. Criticizing DMRC's developments along the river therefore appeared extremely difficult: "Metro is the holy cow. You cannot touch it. If you say anything against the Metro, you are an anti-national" (ARCH-01/2010-09-23). Discursively DMRC's powers rested therefore on its efficient performance and the tremendous need of the city for public transport.

In the eyes of the environmental NGOs the DMRC has been "the first offender against the river" and "the number one enemy of the river in the city" (NGO-04/2010-09-18). The developments of Delhi Metro clearly opened up the River Zone for large-scale construction activities. Equipped with exceptional powers to acquire land and by-pass environmental rules, existing land-use policies (e.g. MPD-2001) and environmental regulations (e.g. EIA) had little to no effect on the developments of Delhi Metro. The development projects by Delhi Metro were challenged by environmental activists in the court as part of the petitions against the Commonwealth Games Village. However, the courts never critically scrutinized the developments of Delhi Metro (see details in chapter 14.2).

13 'CLEANSING' AND 'RECLAIMING' DELHI'S RIVERSCAPES

The following sections review the dealings of the DDA and other state agencies with the above mentioned "non-conforming and polluting" land uses, which inter alia refer to the JJ cluster (slums) and agricultural uses along the river (DDA 2010b: 15). The widespread agricultural uses of the floodplains had long been under suspicious tolerance of the DDA. In the late 1980s context of land acquisition for channelization, the DDA had already attempted to expel farmers from the River Zone, deeming them as unauthorized cultivators and encroachers. Several cases were pending in the courts already at that time (see section 10.5).

Nevertheless, urban agriculture has remained the dominant land use in the research area, covering about 30% of the land from Wazirabad to Okhla. The land under cultivation was reduced significantly in the course of the developments along the river in the last decade – even when acknowledging seasonal variations and changing areas under inundation. A certain displacement of the traditional use of the floodplain is therefore traceable. The following analysis will connect these material changes to the discursive realities faced by the farmers.

The slum population was displaced from Delhi's riverscapes. The slums and their inhabitants were seen as a major nuisance for the city and deemed to be spoiling the river (GHERTNER 2011b: 146, SHARAN 2015: 5). A significant number of researchers (see among others BATRA & MEHRA 2008, BHAN 2009, DUPONT & RAMANATHAN 2008, GHERTNER 2008, 2010, MENON-SEN & BHAN 2008, RAMANATHAN 2006) have reviewed the slum demolitions along the Yamuna, commonly linking them to larger social dislocations and the diminishing right to the city of the urban poor in India. While incorporating these aspects in the analysis, the discussion of the slum evictions in this study aims to provide a slightly different reading of the demolitions in the broader context of the remaking of Delhi's riverscapes. In doing so, the following sections first outline the role of the farmers and then engage with the slum demolitions before merging the insights.

13.1 Agriculture in Delhi's riverscapes

"The cultivation of melons, cucumbers and squashes is really quite a specialty of Jamna [Yamuna] river bed, the plants being rooted in the strong soil below the layer of sand." (PUNJAB GOVERNMENT 1912: 116).

As the quote above from the Delhi Gazetteer indicates, the vegetables from Delhi's riverscapes used to be of superior quality. Like in other parts of the country, the farmers grow different vegetables with two major harvest periods; the post-monsoon *kharif* (October – January) and the *rabi* (April – May). The cultivated crops include, but are not limited to, tomatoes, eggplants, cauliflower, pumpkins, melons, cucumber, carrots, onions, radish, spinach, coriander and other leafy vegetables. Maize is also common as *kharif* crop. Additionally, flower cultivation (e.g. roses and marigold) is found in some locations.[103] The farmers predominantly live in temporary houses (*kuccha houses*) clustered as small 'villages' (see Figure 54, page 393); some houses made of brick exist (*pucca houses*).[104] Some farmers also keep buffaloes and goats for personal milk consumption (cf. COOK et al. 2014).

Observation, multi-temporal mapping and interviews have indicated that the farmers of Delhi's riverscapes are adaptable to the seasonal variations in Yamuna's flow regime. As soon as the sand banks transform into relatively stable banks or islands, cultivation starts in the post-monsoon period, transforming the water soaked land into agricultural fields. Not only do new islands emerge almost every year, but erosion and accumulation also results in shifting the water's edge in many stretches annually by several meters. In years of heavy floods (like 2010 and 2013), new lands of several tens of meters are accumulated, while other lands disappear. Spatially fixed boundaries between sandbanks, natural vegetation and agricultural lands and the river's water body do not exist, but rather shift annually. Whatever land is available is used for cultivation and the fields are newly laid out every year.

More constant patterns of agricultural fields are observable only in the areas on slightly higher grounds and areas protected by the guide bunds of the bridges (e.g. around the DMRC Depot Yamuna Bank). Being less frequently flooded and inundated by flood waters (and for shorter periods) these areas are preferably transformed into nurseries or multi-year flower production. Nurseries along the roads sell various different types of plants on site (see Figure 55, page 393). Green mesh covers protect the nurseries' samplings from the heat of the sun. The farmers' cattle graze along the river's edge and take bath in Yamuna's water. These insights demonstrate that much of Delhi's riverscapes are not unused, vacant or wasted land as indicated in the discourse on channelization. Historic accounts and narratives told by different interview partners suggest that farming and grazing have always been major activities along the banks.

13.1.1 The governance of farming Delhi's riverscapes

Before the shifting of the colonial capital to Delhi, hundreds of small agricultural villages dominated the Delhi region. The majority of the population depended on agriculture. Systems of land ownership and possession were complex. Large scale acquisition of land for the new capital completely changed the land ownership patterns bringing large tracts of land under public domain (JOHNSON 2010). In turn, population growth and the urbanization of former agricultural lands resulted in pres-

---

103 Flowers and flower garlands are important features of religious ceremonies (*pooja*) and also offered to the river Goddess Yamuna.
104 The term *pucca houses* (also *pukkā*) refers to solid and permanent houses made of brick, stone or concrete (from the Hindi word *pucca*, which means literally ripe). In Delhi, in colloquial language and in official documents the term *pucca* is used in contrast to *kuccha* (also *kacha or* kaccā for raw, unripe) referring to houses made of wood, mud and other organic materials (ZIMMER 2012b: 119).

sure on the remaining agricultural land for food production. In the early 1920s, major parts of the *khadar* were already under cultivation around existing villages (SURVEY OF INDIA 1922). The remaining *khadar* was covered by natural vegetation including grasslands (grazing grounds), shrubs and forest patches. These historic developments are important, because they have persisted in different land tenure ship systems.

Early land acquisitions by the government resulted in the situation where upstream of Wazirabad, agricultural land is predominantly privately owned, whereas between Wazirabad and the Old Okhla Barrage land is generally with the DDA and other state agencies. Since this study focuses on the river segment south of Wazirabad, the farming practices and related discourses are reviewed here in more detail with regard to the lands under 'formal ownership' of the state.

As part of the *Grow More Food Campaign* of the Indian Government in the post war period (1949) (cf. SHERMAN 2013), the DIT, the predecessor of the DDA, gave about 750 ha (8902 *bigha* and 53 *biswa*) of land between Wazirabad and Okhla to the *Delhi Peasants Co-operative Multipurpose Society Ltd.* (herein referred to as the 'Delhi Peasants').[105] The lands leased to the Delhi Peasants between Wazirabad and Okhla belong to four Revenue Estates (Bela, Interpat, Chiragah North and Chiragah South) being part of the *Nazul* lands. Another lease was issued to the *Jheel Kurenja Co-op Milk Producers Society* for grazing purposes. While in the 1990s, the DDA had also initiated eviction of the Milk Producers and their situation has developed with some parallels, the analysis of this study focuses on the Delhi Peasants.

Today, the farmers' society of the Delhi Peasants consists of about 1,000 members. A leading member of the society (FARM-01/2011-03-09) estimated that about 10,000 families cultivate the land along the banks. The members of the society either sublease the land to these families, or the families work for them (information based on interviews with different farmers). COOK et al. (2014: 7) estimate about 2,500 small-scale farms are located on the Yamuna floodplain in the NCT of Delhi (including the area upstream of Wazirabad and downstream of Okhla), which seems to be realistic on the visual assessment by the author, but no official data exists.

The conditions of the lease agreement between the Delhi Peasants and the DIT specified that the Delhi Peasants were allowed to use the land only for agricultural purposes prohibiting any construction on the land.[106] The lease agreement between the DIT and the Delhi Peasants was very detailed. For example, it contains detailed regulations on the types of crops not allowed to be grown on the floodplain in different stretches (e.g. maize, sugarcane, or melon). The Delhi Peasants were further made responsible of preventing any solid waste or other material being dumped on

---

105 The following information are based on different materials provided by the Delhi Peasants Co-operative Multipurpose Society Limited (e.g. letters, payment bills, maps), as well as various documents of different court cases and interviews with the farmers.
106 Information based on lease agreement (Five Year Lease for Cultivation, Grazing and Grass Cutting between the DIT and the Delhi Peasants Co-operative Multipurpose Society Ltd.) accessed from the Delhi Peasants.

the land, which should also not be allowed "to become overgrown with jungle and shrubs or to become swampy".[107] In doing so, the state instituted the Delhi Peasants (and concurrently the Milk Producers) as the caretaker and conservator of the land. In some stretches the farmers cleared the land from forest and shrubs or drained swamps to reclaim the land for farming:

> "The land was not made for farming, there were trees and jungle. So our forefathers [...] put in great labor making this land usable for agriculture" (FARM-01/2011-03-09).

While producing food for the growing city, the farmers were further required to protect the land both from human encroachment of any kind and 'nature' taking over the land again. Institutionalized farming on the floodplain was thus established by the state to ensure the growth of the city through food production, but also to check uncontrolled and unplanned growth in terms of encroachment on the fertile lands of the floodplain. The farmer societies thus became a key agent in the city-river relationship.

The farmers were appointed only as temporary 'green keepers' of Delhi's riverscapes. The lease agreement ensured that the DIT (and later the DDA) had the "liberty to resume and take possession of the land" whenever "the premises are required" in public interest.[108] The original lease agreement was issued for five years, with the possibility of being reneweal on condition of rent payment. The DIT (and the later the DDA) was free to repossess the land without paying any compensation to the farmers with only six months notice. The state aimed to guarantee the accessibility of the land and the flexibility to transfer the land to other uses. The lease was renewed officially only once in 1955, extending the contract term until 1966. The legitimacy of the Delhi Peasants to cultivate the land has been contested since 1966. Court cases between the Delhi Peasants and different state authorities with regard to the possession of land date back to the late 1960s.[109]

13.1.2 Reclaiming the land for urban development

From the perspective of the DDA, as stated in several court proceedings and interviews, the lease of the Delhi Peasants expired in 1966. In the eyes of the DDA, the Delhi Peasants have been illegally occupying and using the land since then. The farmers, in contrast, claim that they have paid the lease up to 1991. However, as they admit, payments were made on a non-regular basis. They argue that since the DDA stopped issuing payment notifications, the farmers did not know, and still do not know, how much to pay (FARM-01/2011-03-09).

---

107 Ibid.
108 Ibid.
109 See High Court of Delhi, 22 September, 1970: The Delhi Peasants Co-operative Multipurpose Society Ltd. vs The Collector And Ors. Equivalent citations: 7 (1971) DLT 399. The case deals with lands owned by the Uttar Pradesh Government in Delhi, which were also leased to the Delhi Peasants. The order relates to earlier orders from 1967, which could not been accessed by the author.

A letter from DDA (Lands Management) to the Delhi Peasants of the year 1977 requesting payment of the lease money for the years 1973 to 1977 (Rs. 112,547) as well as a receipt over 25,000 Rs. from 1978 paid by the Delhi Peasants to the DDA indicate that payments were made after 1966.[110] Furthermore, the author accessed documents which contained information that the Delhi Peasants still made payments to DDA in the early 1990s.

More recently, the Delhi Peasants have tried to make payments to DDA, which were not accepted. Therefore, at least, informally the lease money was still demanded by DDA officials from the farmers' society. Farmers from Bela Estate (members of the Delhi Peasants) explained that they pay money directly to "DDA people" (FARM-03/2011-12-04). A leading member of the society argued that these payments were a common practice of bribery collected from the farmers, who are supposed to pay their lease only to the farmers' society (FARM-01/2011-12-06). While the lease amount paid by the farmers to the farmers' society is about 500 Indian rupees (about 7 Euro) per hectare per annum (FARM-01/2011-12-06), individual farmers reported to pay regular bribes of 100 Indian rupees or more to 'DDA officials' almost every month (FARM-03/2011-12-04).

In the context of land acquisition for channelization, internal communication within the DDA described how the state agency already tried to evict the Delhi Peasants in the late 1980s.[111] In 1991, the DDA then initiated eviction proceedings against certain farmers under the Public Premises (Eviction of Unauthorized Occupants) Act.[112] The Delhi Peasants challenged the eviction proceedings in the court arguing that they had not been heard as defined by the Act. Further, the standpoint of the Delhi Peasants was that the DDA must take action against the society and not the individual farmers, since the lease agreement was settled with the farmers' society (FARM-01/2011-03-09). Thereafter, the DDA issued notices to the farmers' society to vacate the land. These were challenged by the farmers' society again in the High Court of Delhi. In its decision of 2004, the High Court partly followed the argumentation of the Delhi Peasants and directed "DDA not to resume possession of land from any member […] except after following the due process of law".[113] The case between the DDA and the Delhi Peasants is still pending in the court (FARM-01/2011-03-09).

---

110 Materials accessed from Delhi Peasants. These materials were also the matter of several court cases.
111 Internal issues led the Land Management Wing request the Demolition Wing to keep demolition activities in abeyance until further notice. Letter DDA Lands Management S.S. Tyagi to Co-ordinating Officer (Damages), DDA, dated 4th November 1987. Subject: Regarding Eviction Proceedings against Delhi Peasants Cooperative Multi-Purpose Society Limited. Letter accessed by the author through the files of the Delhi Peasants.
112 The Public Premises (Eviction of Unauthorized Occupants) Act was enacted in 1958 and has been amended a several times. The latest amendment was in 2014. The act allows the state eviction of unauthorized occupants on public premises.
113 High Court of Delhi, 19th April 2004: W.P. (C) 14260/2004 Delhi Peasants Cooperative Multi-Purpose Society Limited vs. DDA.

Reviewing the accessed materials from the Delhi Peasants is not meant to produce any final truth, but rather aims to highlight that the current state of the farming activities in the River Zone is to be considered as being highly informal. From the point of view of the DDA, the farmers illegally occupy government land without authorization. The farmers have also been accused of having supported the development of the JJ clusters (for details see 13.2). While being aware of these developments, the farmers' society in many cases remained largely reluctant or unable to force individual farmers, who clearly acted in contradiction to the lease agreement. A member of the Delhi Peasants explains the society's action against not-complying members (FARM-01/2011-12-06):

> "We sent notices to such farmers [subletting the land] asking them why are you doing this? You are given land for cultivation and not [for] subletting it to the *jhuggi* developers or other persons. We gave notices to them and asked them to vacate that land. In some cases we have also sublet that land to another farmer, who wants to cultivate the land. [...] But the farmers are also responsible. I said it openly in the general house of the society. It is the mistake of the farmers to sublet the land. Who does not cultivate the land, but give it to other people on rent – that is wrong. Some farmers are of such a nature that they don't want to work. They want to earn through rent or subletting that land to *jhuggi* dwellers. [...] I feel that farmers are also responsible [...] If a farmer does not allow anybody to erect a *jhuggi* on their land [sic], then naturally it will not develop" (FARM-01/2011-12-06).

Farmers from Bela Estate, the area where large JJ clusters used to exist in front of the Red Fort, explained that they were often forced to give up their land by the 'slum mafia' emphasizing that none of the farmers had sold their land, but rather the land was taken over by migrant labors who outnumbered the farmers. Due to fear of violence and "to avoid everyday fighting" (FARM-02/2012-12-02 [translated]), the farmers left the land. Nevertheless, individual members of the society have also sublet their lands to "nursery people" (FARM-01/2011-12-06). These developments are well-known to the DDA and the courts (NGO-04/2011-12-06, DDA-08/2011-12-07) and clearly weaken the Delhi Peasants' arguments to remain in possession of land. Furthermore, the Delhi Peasants are not considered to be the poorest of the poor, who depend on the land for their livelihood.

Additionally, on the eastern bank migrant laborers employed by the farmers are the ones actually working and staying on the land in temporary huts (*kuccha*), while the lease holders themselves have other accommodation outside the River Zone. In the broader discourse, these migrants (especially from Bihar and Uttar Pradesh) are clearly not welcomed in the city (even on temporary basis) and are associated with the informal growth of the slums across the city. They are seen as the manifestation of the fear of encroachment story-line in the River Zone.

Ongoing litigation cases and internal organizational problems of the society (including evidence of corruption, which contributed to irregular payments) resulted in a weakening of their arguments and stand. The failures of the farmers' society in the past were rooted in a missing foresight that the land of Delhi's riverscapes might actually come into the focus of urban development and become high-valuable land. The below quoted society member explains the behavior of the farmers' society in the past (FARM-01/2011-12-06):

"The society did not function properly, because they did not care for the lease of the society. [...] They were generally illiterate people. The president and more over other management [people] were not so interested in the lease of the society. They said: Ok, you continue to cultivate this land, don't worry for the lease. We shall see. We are in power. Some were political people of Congress Party. That is why they did not care for the lease of money. [...] If the lease had been continued, there would be no problem. The main problem is that the lease has been disconnected or terminated. [...] One difficulty is also that the society for a long time [about 15 years] has been under the administrator of the government. He also did not bother for the lease".

The society initially relied upon strong support from early Congress leaders, who set up the society to cultivate the land (most importantly Delhi's first Chief Minister Chaudhary Brahm Prakash[114]) and as such they did not actively review the lease status. Based on political patronage, the farmers took things for granted for some time and this eventually gave rise to a laissez-faire attitude of the farmers' society towards keeping proper records and paying attention to tenure ship security. The link to the Congress Party remains intact today and the famers have written letters to different senior leaders of the party (e.g. Anil Shastri, the son of Lal Bahadur Shastri) requesting support during the last few decades. The Delhi Peasants also wrote to the Lieutenant Governor requesting a meeting, which remained unanswered.

The missing direct link to the land by some, the carelessness and corruption of leading figures as well as limited knowledge and ability to act, has prevented the farmers from making their voices heard beyond the court in a broader discourse on the river. On the other hand, the dominant channelization discourse and the willingness of the DDA to resume possession of land might have resulted in an implicit understanding between the DDA and the majority of the Delhi Peasants that 'their' land one day will be taken for urban development, as has been gradually occurring since the foundation of the farmers' society. From the original 750 ha of land (approximately) leased by the Delhi Peasants, at least around 290 ha has been surrendered and handed over to different government agencies in the last two decades for the development of various projects along the banks (including around 111 ha to DMRC, 98 ha to DDA and 59 ha to PWD as well as 22 ha for the construction of the Akshardham Temple Complex and the Games Village).[115]

When the channelization plans were stalled, farming in the River Zone continued informally. While in the wake of the slum demolitions some farmers were also affected, by and large the DDA remained reluctant to enforce eviction. The DDA was further upheld by the courts not to take over the land without following the due

---

114 Chaudhary Brahm Prakash (1918–1993) was part of Mahatma Gandhi's Satyagraha Movement and was originally from a village in North-West Delhi. He was the first Chief Minister of Delhi from 1952–1955 and was later also Union Cabinet Minister for Food, Agriculture, Irrigation and Cooperatives.
115 Calculation by the author based on different materials, including Order of the District Judge VI (East), Karkardooma Court, Delhi, dated 1st September 2009 '(order accessed through Delhi Peasants).

process of law. Connected to the biodiversity zone concept of the DDA, new eviction notices have been filed against the Delhi Peasants, covering the whole area of land leased by the farmers' society in 2007 and 2009. The more recent debate and the detailed developments of the planning for the biodiversity zone are outlined in chapter 16.4.

13.2 Cleansing Delhi's riverscapes from the slums

In contrast to the long pending legal battles between the DDA and the farmers, the reclamation of Delhi's riverscapes from the slums was not directly initiated by the DDA, but rather by PILs in the courts.

The first slums along the river emerged in the area of Yamuna Baazar in the 1920s to 1930s and they faced evictions as early as the 1960s (TARLO 2000, 2003). In 1967, large parts of Yamuna Baazar were demolished and the inhabitants were relocated to Welcome Colony (Seelampur) in East Delhi (TARLO 2000: 56). After the demolition drives of the Emergency Period, the slums along the river grew up until the turn of the century without facing any large scale demolition.

The slums along the river were generally known as the *Yamuna Pushta slums* comprising of a chain of at least 22 slums with different names. The Hindi term 'pushta' means bund or embankment, but is also referred to as riverbank (BHAN 2009: 127). Some authors (BHAN 2009, GHERTNER 2011b, HAZARDS CENTRE 2004b, MENON-SEN & BHAN 2008) refer to the Yamuna Pushta slums as located on the right bank of the river. Interview partners during this study explained that while the existence of an embankment is the precondition to speak of *pushta*, and a continuous embankment is only found in the east, the slums on both banks were called Yamuna Pushta slums. Notably, the Yamuna Pushta slums were never notified as 'slums' under the Slum Areas (Improvement and Clearance) Act.

The slum dwellers, generally working-class people (day laborers, rickshaw drivers, rack pickers, domestic workers, etc.) either 'bought' (albeit illegitimately) the land from farmers, cleared it from shrubs and trees, or reclaimed swampland in order to build their homes. They had to bribe government officials to get access to the land and to be tolerated (BATRA & MEHRA 2008: 398). As outlined above, the farmers were not legitimized to sell off or sublet their lands based on the lease agreement. Therefore, the slum dwellers had no legal title to the land in either of the cases.

The embankments along the river were the nucleus of many of the slums, since they provided some protection from the yearly flooding. On the right bank a chain of slums stretched from the Old Railway Bridge all along the samādhis to the Indira Gandhi Indoor Stadium, following the course of the Mughal Bund. The slums encircled the Raj Ghat power station and the attached fly-ash ponds. These slums had formed since about the early 1970s. BHAN (2009: 128) refers to a local tale which tells of how the first house was constructed there in 1972. On the left bank, the main slums used to exist along the Shahdara Marginal Bund and on both flanks of the approach road to the ITO Bridge.

While the slums had generally formed as simple *kuccha* houses made of wood, mud and other organic materials, large parts of the slums over time developed into single to double story *pucca* houses made of brick, stone or concrete. The proximity to the employment opportunities of Old Delhi allowed them to flourish and they experienced major growth during the preparation for the Asiad (Asian Games, in 1982) when a large number of construction workers came into the city (cf. BATRA & MEHRA 2008, BAVISKAR 2011c, BHAN 2009). Originally, about 70% of the population was Muslims and migrants from Uttar Pradesh, Bihar and West Bengal (MENON-SEN & BHAN 2008: 2).

For a long time the slums along the Yamuna seemed to suggest a certain degree of security to their inhabitants and public authorities largely turned a blind eye toward the river, which let the slums prosper (cf. FOLLMANN & TRUMPP 2013). As a result, the slums dominated large parts of Delhi's riverscape for several decades before they were demolished in 2004–2006.

13.2.1 The slum demolitions (2004–2006)

In 2004, the MCD and DDA initiated the clearing of the slums on the right bank in the stretch from Old Railway Bridge to ITO Barrage. BHARUCHA (2006: 7), BHAN (2009: 127) and MENON-SEN & BHAN (2008: 2) speak of 35,000 to 40,000 families, making up more than 150,000 people being evicted from the banks of the river between late February and early May 2004. The Delhi based civil society organization Hazard Centre even reported that about 300,000 people experienced eviction (cf. DUPONT 2011, WOAT 2004).

Overall, it seems to be almost impossible to estimate the exact number of all people who have been affected by the evictions, since the statistics remain only rough figures. When it comes to slum population and resettlement action in Delhi, no undisputed numbers exist. Official information on resettlement in Delhi only includes people who were entitled to rehabilitation and resettlement (DUPONT 2008, DUPONT & RAMANATHAN 2008). Therefore, these figures do not correspond with the total number of evicted people, which is likely to be much higher. Official information on the slums along the river provided by the MCD Slum Wing dates back to 1990/1994 and is also fraught with unreliability. Nonetheless, these figures show the dimension of evictions and demolitions within just a few months (DUPONT 2011: 14).

Only a minority of the displaced families were allotted plots. They were located up to 40 km away from their former residence on the very outskirts of the city in the resettlement colonies of Bawana (people from Sanjay Amar Colony, Indira Camp), Bhalswa (from Gautampuri), Holambi Kalan (from Gautampuri),

Madanpur Khadar (from Gautampuri), Narela (from Buland Masjid Shastri Park), Pappankalan (from Shamshan Ghat), and Savda Ghevra (from Nangla Machi).[116]

Keeping in mind that these large slums were important vote banks, it is remarkable that the evictions took place just before the Indian General Elections. Newspaper reports and other sources reveal that leaders of the Bharatiya Janata Party (BJP) were in favor of the evictions and pressed the MCD and DDA to carry out the eviction immediately. Congress leaders including the Chief Minister Sheila Dikshit were rejecting the evictions.[117] After the elections in which the Congress party won the demolition was initially suspended before a second major demolition drive along the Yamuna took place in early 2006 (for details on the eviction and resettlement process see among others BATRA & MEHRA 2008, BHAN 2009, BHARUCHA 2006, FOLLMANN & TRUMPP 2013, MENON-SEN & BHAN 2008).

13.2.2 'Green' evictions: discursive reasoning and the role of the courts

Delhi has a long history of slum evictions and resettlement (DUPONT 2008). However, the slum policies in Delhi adopted in the 1990s provided a three-tier strategy to deal with slums (cf. DUPONT & RAMANATHAN 2008: 318):

– In-situ upgradation (if land not required for another 15–20 year);
– Relocation (if land required in public interest); and
– Temporary improvement of the living conditions regardless the status of land.

Demolition and relocation was thus only one possible way to deal with the slums in the River Zone. Generally the process of slum demolition stretched over years or even decades, since correctly surveying slums and identifying eligible households for resettlement was very difficult and often not even desired by the landowning agencies because of the associated costs of acquiring land for resettlement (DUPONT 2008, GHERTNER 2011b). Landowning agencies often remained reluctant to carry out evictions and demolitions. The case of the Yamuna Pushta slums was no different.

Yet, in the early 2000s, as argued by a series of authors (see among others BHAN 2009, DUPONT 2008, DUPONT & RAMANATHAN 2008, GHERTNER 2008, 2011d, RAMANATHAN 2006), a broader paradigm change is traceable within the jurisprudence of India's courts from 'stipulating the right to livelihood' as an 'important facet of the right to life' to depreciative judgements against slums and slum dwellers in the early 2000s. Most striking in this context is the judgment by the Supreme Court in the case Almitra Patel vs. the Union of India on 15th February 2000.

---

116 Information based on different materials accessed from the Slum and JJ Department of the MCD. The author likes to thank Anna Zimmer for sharing a map and other information on the slums.
117 See among others newspaper coverage: Times of India, 10th January 2004: 'Demolition man cometh to jazz up Yamuna bank'; The Hindu, 23rd January 2004: 'Yamuna Pushta dwellers willing to shift: Jagmohan'; The Hindu, 12th March 2004: 'Shifting of Yamuna Pushta slums approved'; Times of India, 23rd March 2004: 'Jagmohan set on Yamuna Pushta demolitions'.

"Rewarding an encroacher on public land with free alternate site is like giving a reward to a pickpocket. The department of slum clearance does not seem to have cleared any slum despite its being in existence for decades. In fact more and more slums are coming into existence. [...] The authorities must realise that there is a limit to which the population of a city can be increased, without enlarging its size. In other words the density of population per square kilometre cannot be allowed to increase beyond the sustainable limit."[118]

In this case, which was not primarily concerned with slums but rather solid waste management, the Supreme Court equated slum dwellers "with a somewhat Malthusian slant" as encroachers (NEGI 2011: 184). Their perceived criminality was indicated by the pickpocket metaphor, which has also been widely cited in the literature as a major shift in the way the courts dealt with slums. Furthermore, the understanding of slum dwellers as illegal encroachers – or literally pickpockets – led to the cancellation of the existing slum and relocation policies and the courts demanded direct action of governmental authorities to carry out evictions (RAMANATHAN 2006: 3197). While earlier slum demolitions and relocation had been framed as an improvement of the living conditions of the poor, slum residents were now deemed as 'encroachers' (RAMANATHAN 2006) and 'polluters' (GHERTNER 2008, 2011c, 2013). This broader discursive framing provided the legitimation for the government agencies to demolish the slums along the river, which the remaining section will outline in more detail.

Evictions along the banks of the Yamuna in 2004 to 2006 were therefore not the outcome of planning schemes for the long desired remaking of Delhi's riverscapes, but rather they were driven by judicial activism. Several PILs were filed by Resident Welfare Associations (RWAs) and Factory Owners' Associations complaining that slums in their neighborhoods were occupying public land, leading to local environmental problems including among others blocked drainage systems, open sewage disposal, open defecation, congested and occupied streets, and overall unhygienic conditions (BHAN 2009, DUPONT & RAMANATHAN 2008, GHERTNER 2008, 2011b, RAMANATHAN 2006).

In this context, two PILs emerge as particularly important. The first PIL was filed in 1994 by the Okhla Factory Owners' Association based in an industry and trading cluster in south Delhi. The second PIL was filed in 2002 by an industrial manufacturers association (Wazirpur Bartan Nirmata Sangh) from the Wazirpur Industrial Area in north Delhi. Both petitions had sought the relocation of slums from government land close to their businesses arguing that the slums express crime, dirt and unhygienic living. Further, as a response to the several litigations against slums, slum dwellers protesting against evictions filed petitions in the courts. In total about 65 petitions, half of them petitions by middle-class resident welfare associations and the other half by slum dwellers, were dealt with by the court.[119] In none of the cases were the slum dwellers heard, but in both cases the court ordered the eviction of the slums (NGO-1/2010-02-16). Later these cases

---

118 Supreme Court in the case Almitra Patel vs. the Union of India on 15th February 2000.
119 Due to the large amount of PILs and orders by the court, the author was unable to figure out the exact number of PILs; see also HAZARDS CENTRE (2004a).

were heard together by the High Court of Delhi (herein referred to as the 'Okhla/Wazirpur Case').[120]

It is important to highlight that none of these PILs were originally filed against the slums in the River Zone or for the protection of the Yamuna. In contrast, the original petitions had neither mentioned the river nor the Yamuna Pushta slums. It was exclusively the court that shifted its focus within the proceedings of the case from unauthorized occupation of public land in other parts of the city, to pollution caused by the slums along the river Yamuna (DUTTA 2009). In particular, the High Court – in line with the earlier order of the Supreme Court in the Almitra Patel Case cited above – struck down the relocation policy, stating that "[n]o alternative sites are to be provided in future for removal of persons who are squatting on public land"[121], after the Slum and JJ Department of the MCD had argued that under the existing resettlement policy it would not be possible to evict the slums from the challenged neighborhoods. Affecting hundreds of thousand slum dwellers across the city this order resulted in public tumult and a Special Leave Petition was filed by the Union Government against this order in the Supreme Court (HAZARDS CENTRE 2004b: 6). The Supreme Court on 19th February 2003 stayed the earlier order of the High Court.

Acknowledging the stay by the Supreme Court, the High Court by chance turned its attention to the pollution of the Yamuna. The High Court on 3rd March 2003 ordered:

"River Yamuna which is a major source of water has been polluted like never before. Yamuna Bed and **both the sides of the river have been encroached by unscrupulous persons with the connivance of the authorities**. Yamuna Bed as well as its embankment has to be cleared from such encroachments. Rivers are perennial source of life and throughout the civilised world, rivers, its water and its surroundings have not only been preserved, beautified but special efforts have been made to see that the river flow is free from pollution and environmental degradation. The Yamuna River has been polluted not only on account of dumping of waste, including industrial waste, medical waste as well as discharge of unhygienic material but the Yamuna Bed and its embankment have been unauthorisedly [sic] and **illegally encroached by construction of pucca houses, jhuggies and places for religious worship, which cannot be permitted any more.** As a matter of fact, under the garb of reallocation, encroachers are paid premium for further encroachment. **Delhi with its present population of twenty million people can take no more.** In view of the encroachment and construction of jhuggies/pucca structure in the Yamuna Bed and its embankment with no drainage facility, sewerage water and other filth is discharged in Yamuna water [sic]. The citizens of Delhi are silent spectator [sic] to this state of affairs. No efforts have been made by the authorities to remove such unauthorized habitation from Yamuna Bed and its embankment. We, therefore, direct all authorities […] to forthwith remove all the unauthorised structures, jhuggies, places of worship and/or any

---

120 High Court of Delhi, 03 March 2003, WP (C) 2112/2002 Wazirpur Bartan Nirmata Sangh vs Union of India and Others, Okhla Factory Owners' Association vs. Government of NCT of Delhi.
121 High Court of Delhi, 29 November 2002, WP (C) 2112/2002 Wazirpur Bartan Nirmata Sangh vs Union of India and Others.

other structure which are unauthorisedly put in Yamuna Bed and its embankment, **within two months** from today" [emphasis added].[122]

After this order was handed down, the Okhla/Wazirpur Case became a landmark in the jurisprudence related to the Yamuna being heard by the court on a regular basis (DUTTA 2009: 54).[123] The (neo-)malthusian slant remained a discursive element of the order indicating that the city "can take no more".[124]

Moreover, struggling to get around the existing relocation policy the High Court in the further proceedings of the case made use of a dichotomous understanding of land and water:

"It is required to be noted that the policy of DDA for relocation would not apply to the river bed. It is not an encroachment on land. It is a water body and it is required to be maintained as a water body by the DDA and all other authorities. The water body is not maintained as such and construction illegally goes on unhindered [...]. We fail to understand as to why the officers of DDA are not taking any action in the matter."[125]

In this particular order, the court purified the space under 'encroachment' to belong to the realm of the river being a water body. Simultaneously, the court requested the agency dedicated to the management and development of land to take care of the "mess [which] has been created by these officers".[126]

The inherent contradictions in the orders are apparent. In the following hearings of the cases, the court aimed to clearly define the fuzzy boundaries between land and water in order to press the 'land' owning agencies for action.[127]

---

122 High Court of Delhi, 03 March 2003, WP (C) 2112/2002 Wazirpur Bartan Nirmata Sangh vs Union of India and Others, and W.P.(C). NO. 689/2004 Okhla Factory Owners' Association vs. Government of NCT of Delhi. Order quoted from later judgement in the same case: CM Nos.11672-73/2006 in WP(C) No. 2112/2002, Wazirpur Bartan Nirmata Sangh vs. Union of India and other, judgment reserved on 19 September 2006. See also DUTTA (2009: 54–55) and GHERTNER (2011b: 145–146).
123 The Okhla Case W.P. (C) 689/2004 was disposed of by the High Court on 4th February 2009. The Wazirpur Case W.P. (C) 2112/2002 was disposed of by the High Court on 18th April 2012.
124 Ibid.
125 High Court of Delhi, 3rd May 2005. WP (C) No. 689/2004.
126 Ibid.
127 The author sticks to the common terminus of 'land owning agency' to restrain confusion. Yet, following the argument of the court purifying the space as water body, the terminus is actually contradictory here.

13.2.3 Fuzzy boundaries, legal categories: 300 meters for the river

After a series of evictions in 2004 demolitions had stalled, the court set up a court appointed committee in November 2005, the *Yamuna Removal of Encroachments Monitoring Committee* (YREMC, also referred to as Encroachments Monitoring Committee), to oversee the demolitions of the slums along the river.[128] The judiciary thereby encroached into the sphere of the executive (BHAN 2009: 134, DUPONT & RAMANATHAN 2008: 337, RAJAMANI 2007: 315) and pressed the agencies to act. The Encroachments Monitoring Committee made regular inspections of the riverfront areas deemed for demolition and reported to the High Court on a monthly basis (DUTTA 2009: 56). When landowning agencies still showed inaction and disputes arose over which areas along the river were supposed to be cleared from the slums, the High Court clarified the tasks of the Encroachments Monitoring Committee:

> "No effective steps have been taken to make Yamuna free from encroachments and pollution of all kinds. [...] We direct the Committee to take up in right earnestness and on day-to-day basis the task of removing encroachments upto [sic] **300 meters from both sides of River Yamuna in the first instance**. No encroachment either in the form of jhuggi jhopri clusters or in any other manner by any person or organization shall be permitted. Yamuna has to be redeveloped in such a manner that it becomes the habitat for trees, forests and center for recreation. We are making it clear that **no structure** whether it pertains to religious, residential or commercial or any other purpose shall be allowed to exist. We are also making it clear that **no sullage, no sewers, no industry, no factory** shall be permitted on both sides of the embankment of Yamuna so as to prevent pollution of the river" [emphasis added].[129]

The High Court identified the 300 meter zones on both sites of the river without any scientific rationality. The court did not even specify from where to measure the 300 meters. Therefore, the court order left the door open to interpretation. By including different uses and other sources of pollution besides the slums, the court clearly defined that 'no structure' should remain and the land should be 'preserved' and 'beautified' (see cited orders above).

Furthermore, the phrase 'in the first instance' indicates that the court had envisioned a phased manner and setup the 300 meter benchmark only as a starting point to proceed action (cf. DUTTA 2009: 59). Yet, the landowning agencies were interested in a selective application of the orders:

> "DDA has claimed that if the distance of 300 meters is measured from the river bank (actual flow of river water) about 300 kuccha and pucca residential structures and three religious places

---

128 The committee included Ms. Justice Usha Mehra (retired High Court judge) as chairperson, Shri V N Singh, a former Commissioner of Police, Vice-Chairman DDA, Commissioner-MCD, Commissioner of Police, Chief Engineer, Uttar Pradesh State Irrigation Department at Okhla Barrage and Shri A S Chandhiok (Senior Advocate at the High Court). Shri SM Aggarwal, a retired Additional District and Sessions Judge, was appointed as convener of the committee.
129 High Court of Delhi, 08 December 2005, WP (C) 2112/2002 Wazirpur Bartan Nirmata Sangh vs Union of India and Others, and W.P.(C). NO. 689/2004 Okhla Factory Owners' Association vs. Government of NCT of Delhi, emphasis added.

[...] would come within the purview of the demolition. However according to DDA if the distance of 300 meters is measured from the river middle point of the river flow about 100 residential structures and one religious place would be covered."[130]

The landowning agencies, including DDA, demolished only the slums which existed right on the water's edge and in some stretches of the river enclosed the river with a fence up to 5 meter height. Landowning agencies and affected 'owners' (including residential, commercial units but also religious structures like ghats or temples) approached the Encroachment Monitoring Committee and the High Court pleading not to remove their structure, since their structures had been in one way or another approved by government authorities in the past. In response, the High Court again clarified its earlier orders in June 2006:

"We make it clear that **no structure** which comes on the River bed or within 300 meters of the edge of the water of river Yamuna **can be regularized** even if it finds mention somewhere else and that would be playing not only with River Yamuna but it will be making mockery of the various orders passed by the Supreme Court, the plan made by the Central Government called 'Channelisation of River Yamuna' and tax payers' money amounting to thousands of crores of rupees which have already been spent for this purpose without getting any desired result. Apart from that **such regularisation will have a massive ecological and environmental imbalance and degradation due to large scale unplanned construction on both sides of River Yamuna and its embankment**. We must not forget that river Yamuna is the lifeline for the citizens of Delhi and **if we allow encroachment or unauthorised construction on the river bed or its embankment** it will convert Yamuna into a huge sewage drain causing irreparable damage to the vast majority of the citizens of Delhi. [...] The right of people of Delhi to have clean potable water from river Yamuna and health and friendly environment from its bed and embankment is a constitutional right" [emphasis added].[131]

Expanding on its earlier order, the High Court aimed to limit the scope of the landowning agencies in selectively applying its orders, highlighting that any regularization of any construction in the 300 meter zone is impossible. In the second part of quote, the court explicitly used the clause 'on the river bed or its embankments' to be kept free of 'encroachment or unauthorised construction'. By including the embankment, the court widened the area to be cleared in many stretches. On the other hand by referring only to 'encroachments' and 'unauthorised construction', the court enabled the landowning agencies to define these terms. Significantly, the court also referred to the channelization scheme and the Yamuna Action Plan in its order.

Despite these clarifications by the court, the demolitions affected by in large only the slums (DUTTA 2009: 59). The High Court simplified the complexity of the city-river relationship by defining the 300 m zone as a new legal category, guiding the demolitions and the protection of the river. The problems associated with the fuzziness of the order of the High Court and the flexibility it gave to the concerned

---

130 Minutes of the 9th meeting (15th May 2006) YREMC, for reference see also DUTTA (2009: 81).
131 High Court of Delhi, 1st June 2006, WP (C) 2112/2002 and CMs. 4715/2006 and 6310/2006 and WP (C) No.689/2004: Wazirpur Bartan Nirmata Sangh vs Union of India and Others Okhla Factory Owners' Association vs. Government of NCT of Delhi.

state authorities play a major role in the course of the later developments along the river.

Besides the critique of the brutality and arbitrariness of the orders of the court, it needs to be highlighted here that the former slums had certainly violated planning norms and they were, indeed, contributing to the environmental degradation of the river. However, they were by no way the biggest polluter (HAZARDS CENTRE 2004a, 2004b). A coordinated, planned resettlement process – certainly a long-term exercise and connected to a series of financial and political difficulties (cf. DUPONT 2008, GHERTNER 2011b) – would have been needed since in-situ upgradation of the slums was by and large impossible. However, ad hoc solutions continued to be favored over any long-term development plan, which in turn did neither improve the water quality in the river nor the living conditions of the poor.

In sum, the discursive reasoning of the slum eviction was based on two major arguments: unauthorized occupation generally defined as 'encroachment'; and direct (visible) pollution of the Yamuna through open defecation, disposal of untreated sewage, polluting activities and industries like waste recycling, etc. In short, as SHARAN (2014: 164) noted, "the issue of encroachment got entangled with the question of pollution." Together with the larger paradigm shifts in the jurisprudence with regard to slums, these discursive logic of bourgeois environmentalism located the blame and the responsibility for the pollution of the river on the poor (cf. BATRA & MEHRA 2008, BAVISKAR 2011c, BHAN 2009, DUPONT & RAMANATHAN 2008, GHERTNER 2008, 2011b, 2011e, RAMANATHAN 2006, SHARAN 2014, 2015). Therefore, GHERTNER (2011b) rightly argued that the demolitions along the Yamuna were "Green Evictions" and part of the "bourgeois spatial imaginary" of the slum-free, world-class city discourse. In particular, the courts have used the dominant, hegemonic story-line deeming the slums (and their inhabitants) as "nuisance" negatively affecting the quality of life in the city by polluting the river (GHERTNER 2008, 2011b, 2011e). While the court to some extent acknowledged that the reasons of river pollution are complex and far from limited to the slums and the poor, they concurrently simplified the complex problem of river pollution "into the visible presence of a degraded population living on the banks of the river" and came to the conclusion "that slums had destroyed the natural beauty and ecology of the river" (GHERTNER 2011b: 146). The other reasons for the pollution of the river remained unexplored in the case in the High Court.

The reinterpretation of the nuisance law and the fear of encroachment story-line also highlight that the court did not base its verdict on any scientific rationality (neither with regard to the main sources of pollution nor in the process of framing the 300 m zone). Rather, the court used aesthetic criteria, which identified the slum dwellers as polluters of the river based on their visible appearance (for a detailed analysis of the rule of aesthetic see GHERTNER 2011b, 2011e). 'Slums as polluters' emerged as a powerful story-line.

Reports by civil society groups like the Delhi based Hazards Centre argued that the contribution of slums to the pollution of the river Yamuna was very nominal (HAZARDS CENTRE 2004a, 2004b). In addition, scientific evidence based on the data of the CPCB found hardly any recognition in the public discourse and the court (where the slum dwellers were not even heard). Reports by the CPCB (2004) proving that the major reasons for the pollution of the river Yamuna were inadequate sewage treatment capacity and non-functional sewage treatment plants were ignored. This is noteworthy given the CPCB's data is often equated with the *scientific truth* about river pollution (KARPOUZOGLOU 2011). However, in the case of the slum demolitions, scientific evidence – or truth – was never an aim for neither by the court nor the state agencies involved. Otherwise it would have been an easy task for the High Court to let this matter to be dealt with by the Supreme Court, which had been dealing with the pollution of the Yamuna since the early 1990s.

The slums as polluters story-line was produced first and foremost by a powerful discourse coalition, comprising of the landowning agencies (which not in all cases had a genuine interest in removing the slums, but nevertheless often appreciated the action by the courts), the RWAs, the factory owner associations (which could only be entertained by the court in a limited way, since the demolitions only partially, if at all, affected their neighborhoods, but at least they saw that action is taken), and the court (which could show its efficiency and power). Even if Congress leaders were against the evictions in light of the elections, the slum demolitions were based on a "consensus across all wings of the government and all political parties" (BATRA & MEHRA 2008: 403).

The slums as polluters story-line was further anchored in the city-wide discourse on environmental upgradation and beautification, promoted by the Delhi Government's slogan of a 'Clean Delhi, Green Delhi' and the world-class city agenda. The story-line clearly predates these recent discursive framings and was already evident in the city beautification efforts in the Emergency of 1975–77 shifting the question of risks faced by the poor to "the environmental burdens posed by the poor" (SHARAN 2014: 163, TARLO 2003); a typical framing of the brown versus the green agenda. It is further important to note that after removing the slums – considered as the main task – was conducted 'successfully', the Okhla/Wazirpur case could not gain further ground with regard to the river and the Encroachment Monitoring Committee even vanished.

13.2.4 Demolitions, religious structures and the channelization scheme

While actions were taken against the slums, the state agencies remained reluctant regarding the demolition of religious structures, as they were protested against more violently than the demolition of the residential buildings. While the High Court later excluded the bathing ghats from demolition and even allowed residential facilities for the families of the priest (*pandits* or *pandas*) (DUTTA 2009: 57), the messy ground realities made a differentiation between religious structures, residential uses or commercial activities (e.g. dhobi ghats) very difficult. Many of these areas were used for different purposes simultaneously, rather than being clearly defined. As a result, the bathing ghats in the Yamuna Bazaar area and Nigambodh Ghat remained largely untouched by the demolition, but in other areas (e.g. Kudsia Ghat) ghats were demolished.

An order from September 2006, dealing with a petition by a resident from Kudsia Ghat against dispossession of land and demolition, also revealed the connection of ambiguous land ownership, evictions and the channelization scheme.[132] The applicant claimed that the site was "sanctioned" to him in the 1920s and "no unauthorized construction has been raised by him at the site".[133]

Under the riverfront scheme, the Land & Development Office had transferred the land of the bathing ghats in 1972/1973 in various steps to the DDA. While the land was not repossessed by the DDA, the state agencies argued that the lease of the applicant (similar to the lease of the farmers outlined in chapter 13.1.1) had been terminated in this process. Both the Land & Development Office and the DDA had also denied having accepted payments "sent by the applicant from time to time".[134] The High Court dismissed the application referring again to the importance of the Yamuna as the lifeline for the city:

> "Yamuna is one of the holiest river [sic] of India. This river Yamuna is a part of our National ethos and it is mentioned not only in our Rig Vedas but it is proudly mentioned in our National Anthem as well. It is shocking that this sacred, holy and majestic river is now totally polluted with all kinds of domestic waste, industrial waste, silt and other pollutants. [...] It is, thus, bounden duty of all citizens of Delhi to lend a hand of cooperation, particularly, those living along the banks of Yamuna so that once again this holy river is brought back to its original shape. The pure, clean and clear water of river Yamuna and its encroachment free embankments will certainly help not to few but all the citizens of Delhi as river Yamuna is the lifeline for the Capital."[135]

Besides these various changes in the way the state dealt with the slums and other land uses deemed as 'encroachments' in post-liberalization urban India, the demolitions along the river were also connected to more concrete ideas for transforming Delhi's riverscapes – a riverfront promenade in front of the Red Fort.

13.2.5 Background rationalities: plans for a riverfront promenade

In the early 2000s, a riverfront promenade along the Red Fort was pushed to the fore by the Ministry of Tourism and Culture under the then leadership of Jagmohan.[136] With the vision "From Rajpath to the River Front and from Red Fort through Green Walkways and Green Corridors" Jagmohan aspired to achieve both the creation of a new tourist destinations and the clearance of the riverfront from slums

---

132 High Court of Delhi, 19 September 2006, CM Nos.11672-73/2006 in WP(C) No. 2112/2002, Wazirpur Bartan Nirmata Sangh vs. Union of India and other.
133 Ibid.
134 Ibid.
135 Ibid.
136 Jag Mohan Malhotra, known as Jagmohan, Indian politician (Bharatiya Janata Party) was Union Cabinet Minister for Urban Development and Poverty Alleviation from 1999 to 2001 and Union Cabinet Minister for Tourism and Culture from 2001 to 2004. As vice-chairman of the DDA Jagmohan was earlier a major figure behind the slum demolitions during Emergency Period (1975–77).

(BATRA & MEHRA 2008, BHAN 2009, BHARUCHA 2006, DUPONT 2008, FOLLMANN 2014, MENON-SEN & BHAN 2008).

The example of the riverfront promenade mirrors the earlier discussed riverfront development schemes which were also envisioned to connect the historical monuments and the memorials of the national leaders to the river (see chapter 9.3). The Union Minister for Tourism directly influenced the landowning agencies and filed affidavits in the High Court of Delhi.[137] JAGMOHAN (2005, 2009) describes his passion for Delhi's riverfront in his books. He outlines that he had long been waiting for an opportunity to demolish the slums along the river (JAGMOHAN 2005: 466–468):

> "I had all along been planning to rid the Yamuna River Front of all its slums and squatters and create on it a great New National Hub [...]. This plan had been on my mind, no matter where I was and what job I held. [...] I pressed the authorities to comply with the High Court orders."

The reclamation of the riverfront from the slums for 'formal' developments (e.g. the promenade) was supported by instrumentalizing the slums as polluters storyline and the nuisance discourse. The developments along the river therefore highlight the distinct anti-poor character of bourgeois environmentalism. After the demolitions of the slums the plans for the promenade failed to materialize, but in line with the earlier schemes, the idea remained as an urban imaginary influencing the development of the area up to today. As section 16.3 below reveals, today's development of the Golden Jubilee Park is closely connected to the ideas briefly outlined here.

13.3 Remaining residential spaces in Delhi's riverscapes

In August 2006, when the slum demolitions within the 300 m zone were largely completed, the Encroachment Monitoring Committee started studying historic maps of existing revenue records to clarify the extent of additional demolitions. The studied maps, as indicated in the committee's minutes, showed that "the river Yamuna has changed its course drastically in the last 80–100 years" and therefore "the distance of 300 meters from the existing edge of Yamuna falls short" of many unauthorized colonies in the southern part along Kalindi Kunj and Okhla.[138]

These proceedings indicate how complex the implementation of this seemingly straightforward order by the High Court was. The simplification of the city-river relationship therefore remained complicated (cf. GHERTNER 2011b). Purification in the literal sense of the word was declared as the main task:

> "The Yamuna and its water has to be purified. The steps being taken for ensuring purity of the water of Yamuna may cause some inconvenience at present to some but 20 million people

---

137 For a detailed assessment of the situation along the river in the early 2000s by Jagmohan see interview with Jagmohan in BHARUCHA (2006: 164–171).
138 Minutes of the 12th meeting YREMC (18th August 2006).

living in the city cannot be made to suffer for inconvenience of a few, who have no legal right to stay in the bed of river Yamuna."[139]

As outlined above, the demolitions under the High Court orders concentrated almost entirely on the slums. Yet, all other residential areas along the river (today about 200 ha, except the Commonwealth Games Village) are also not connected to any sewage system and discharge their wastewater through open drains directly into the river. Almost all of these settlements are defined by the DDA as 'encroachments' and their existence violates the provisions of the Delhi Master Plan. In total about 60 unauthorized colonies exist within the River Zone covering more than 980 ha of land (DDA 2010b).

These unauthorized colonies include the Tibetan refugees' colony 'Majnu-Ka-Tila' in north Delhi, the area of 'Yamuna Bazaar' around the 'Nigambodh Ghat' in central Delhi as well as Jogabai Extension, Batla House Extension, and Abul Fazal Enclave (Part-I & Part II) along Kalindi Kunj Road in south Delhi. The two villages Old Garhi Mandu Village and Usamanpur Village, located on the east bank south of the Wazirabad Barrage, are today also considered as unauthorized colonies, since the land of the villages has been acquired by the government under the channelization schemes (see 10.5). The transition from urban villages into unauthorized colonies is often blurred and contested. Extensions of former villages are generally defined as unauthorized colonies, while the land of the villages (Village Albadi or Lal Dora area) remains under special land-use regulations (for details see KUMAR 2014).

In addition to the unauthorized colonies and urban villages, the resettlement colony Madanpur Khadar (about 51 ha) has been developed in the River Zone south of the Okhla Barrage since the 1995. It is noteworthy that parts of the slum population from the centrally located slums were resettled in Madanpur Khadar – a location likewise on the floodplain of the Yamuna. Informal settlements have emerged around the resettlement colony of Madanpur Khadar.

Like the slums, these unauthorized colonies discharge untreated sewage directly into the river. Illegal solid waste dumping and informal waste recycling are common practices at the edge of these colonies towards the river and contribute to the environmental degradation of the river (see Figure 58, page 393). Furthermore, many of the unauthorized colonies are located within the 300 meter zone, but they were not touched during the demolition drive in the mid-2000s, as they were supported by a range of different actors. For example, the Tibetan refugees' colony, *Majnu Ka Tila*[140] (see Figure 59, page 394) which came into existence in 1960 and grew especially after the Sino-Indian War in 1962, was patronized by the Ministry of External Affairs and Delhi's Chief Minister Sheila Dikshit.[141] While the colony between the Outer Ring Road (NH-1) and the river was endangered to be demol-

---

139 High Court of Delhi, 20th July, 2006, CM No. 8740/2006 , CMs 4715/2006 and 6310/2006 in WP(C) No. 2112/2002 and WP(C) No. 689/2004.
140 The area is officially called New Aruna Nagar, but generally referred to as Majnu Ka Tila.
141 See among others Times of India, 22nd July 2006: 'Govt bats for 'Little Tibet'.

ished (a court notice was issued in 2006), the colony was in the process of regularization at the time of writing. The areas of Yamuna Bazaar, which stretch up to the ghats and for which a first redevelopment scheme had been prepared in the time of the DIT, were not demolished due to the religious importance.

The unauthorized colonies along the Kalindi Kunj Road also came under special protection of the Encroachments Monitoring Committee due to their location directly at the edge of the Okhla Bird Sanctuary. In 2006, the DDA identified several colonies to be within the 300 meter distance from the river.[142] However, no major action was taken against the Muslim-dominated unauthorized colonies; only some temporary huts at the margins of the colonies were demolished. The fear of public uproar and violence was too great. JOLLY (2010: 132-133) outlines that the Batla House area north of Okhla in particular has been expanding towards the river with the development of multi-story buildings on filled-up land since the early 1990s. DDA field staff were unable to protect the land at that time from further encroachment and demolition operations by the DDA in the 1990s failed due to resistance and violence (ibid.).

Today the entire area form Batla House (south of the DND-Flyway) to the Okhla Barrage along the Agra Canal is a densely built-up residential area. The state agency ultimately focused on checking the spatial growth of the colonies through large scale land acquisition justified by the channelization and riverfront development scheme (see 10.5). Additionally, a new road (Kalindi Kunj-By-Pass) was planned along the water's edge linking the DND-Flyway to Faridabad. Road construction has stalled, due to land conflicts between Delhi and Uttar Pradesh. Further, a fence has been constructed along the Kalindi Kunj Road in order to protect the adjoining wetlands of the Okhla Bird Sanctuary (see Figure 56 and Figure 57, page 393). The fencing though remains largely ineffective to protect the sanctuary from contamination caused by solid waste dumping. Despite all of these actions, field surveys conducted by the author documented an increasing densification of the colonies and construction activities right up to the water's edge in study period.

In the course of the most recent wave of regularization of unauthorized colonies in Delhi, 32 unauthorized colonies[143] were regularized in the River Zone in 2009 (UC 2014) and the Central Government hold out the prospect of regularization also for the remaining colonies before the Delhi Legislative Assembly election in 2015.

Therefore, a modification of the Zonal Development Plan based on the regularizations will need to follow incorporating these changes, also in the land-use plan for the River Zone. Most of these unauthorized colonies (except Yamuna Bazaar) are located towards the outskirts of the city and have, therefore, not been considered

---

142 Jogabai Extension, Batla House Extension, Okhla Batla House, Abul Fazal Enclave Part I & II. See minutes of the 8th meeting (15 May 2006) YREMC.
143 Own survey based on list of unauthorized colonies (DDA 2010b: 33) and information given by the Unauthorized Colonies Cell (UC 2014). The counting of unauthorized colonies is prone to inconsistencies and failures, due to the different naming and spatial demarcation of unauthorized colonies. Different blocks of the colonies have separately applied for regularization. This makes the exercise of counting and comparing them difficult. In total 1,639 colonies in Delhi applied for regularisation and 733 of these colonies were regularised in 2009 (UC 2014).

in redevelopment schemes. The demolished JJ clusters were in contrast located in the heart of the city, clearly visible from the roads, and close to the historic monuments and tourist attractions.

Many of the unauthorized colonies are highly vulnerable to annual flooding and water logging. Along Kalindi Kunj, the riverscapes remain an especially marginalized space of the poor. These areas still represent the river as the backyard and dumping ground of the city. The cleansing and reclamation of the riverfront has so far not transformed the urban environment in this stretches and large parts of Delhi's riverscapes beyond the city center therefore still appear as neglected areas constituting the living environments of poor, who are pushed further to the margins of the megacity.

13.4 The effects of simplification: permitting construction beyond 300 meters

As already indicated in the discussion on the unauthorized colonies, beyond the slums, the 300 meter no-development-zone has been implemented in a very flexible manner by the state authorities. Yet, what did this order mean for the 'rest' of the River Zone beyond the 300 meter from the water's edge? The following analysis outlines that the orders were interpreted as a *carte blanche* for developing the riverfront beyond the 300 meter. In retrospect, a former high-ranked DDA official interpreted the judgement as being clearly counterproductive to protect the floodplain from urban development (DDA-07/2011-11-23):

> "[The Committee] directed that 300 meters along the river should be reserved for no construction and remaining area can be taken up for development. […] Sometimes the courts directions are very good for the conservation of the river and sometimes they have been very negative. If this [order] is implemented […] a lot of land can be retrieved for urban use. […] You get about 50% of additional land or say a 400 meter stretch on either side for urban development. It is a big dangerous kind of judgement".

The court had reduced the liminal space of the floodplain which should remain untouched to 300 meters but beyond this limit development could occur. Reviewing the discussion about the river in the months after the High Court judgements, media reports and minutes of the YSC underpin this interpretation. The High Court orders were in contempt of the guideline decided upon by the YSC already in 1978 (minimum distance between the embankments of 5 kilometers, see 10.2), but the court order directly influenced the deliberations of the YSC. Since different agencies had requested the YSC for a series of development projects (including Delhi Metro, cricket stadium and several power substation and sewage treatment plants), the YSC was again engaged with the question of "how far we could encroach on the flood plains of river Yamuna" in the mid-2000s (YSC/68/2005-03-16). It is remarkable that in the following discussion, the YSC adopted the 300 meter benchmark as the "active river edge" (YSC/72/2007-01-08).

Calculating different parameters to define the minimum width between the embankments, the committee agreed that 1,150 meters should be reserved as the minimum width between the embankments in Delhi. The stipulated width was clearly politically motivated. Technical calculations based on scientific parameters demanded actually that "1650m width must be reserved for the river to play" (YSC/72/2007-01-08). Yet, as the YSC minutes indicate, applying the scientific width of 1,650m was not in the interest of the development lobby and would have gone beyond the High Court order (YSC/72/2007-01-08):

> "[Adopting 1,650m as the minimum width] would mean in most of the river reach within NCT of Delhi [that] it would not be permissible to take any developmental works. After due deliberation, the Committee agreed that a minimum width of 1150 m should be available for the river to flow and play."

At this point the supposed channel width of 550m, calculated already for the channelization of the river, was again applied to define the width of the water channel and by adding two times 300m the YSC defined the 1,150m spacing of the embankments.

Besides diluting the larger policy framework stipulated by the YSC, the High Court orders also directly influenced the project clearance of the YSC. For example, in March 2008 the YSC rejected a proposal for a sewage treatment plant at the Barapulla Nallah in Delhi, because the proposed construction was less than 300 meter away from the river and would had violated the orders of the court (YSC/73/2008-03-18). In contradiction to the importance of the construction of sewage treatment plants along the drains for the overall cleaning of the river Yamuna, the YSC denied approval (YSC/73/2008-03-18).

These examples illustrate the effects of the court's 300 meter benchmark on the statutory interstate body. The earlier guidelines set by the YSC based on scientific knowledge and technical studies were largely ignored through the court's direction. It seems that the members of the interstate body were also not interested in any clarification or tightening of the court's orders, since the court actually left enough scope for all the involved agencies to gain acceptance of their projects already in the pipeline. In retrospect, while the orders stipulated a strong mandate for the demolition of the slums and also some minor projects were rejected in the YSC, other existing structures like power plants or the upcoming Metro Depots were not challenged by the courts.

13.5 Interim conclusion

Within the clean/green discourse of the world-class city, the slums as polluters story-line in line with the 'green agenda' emerged as an extremely powerful discursive framing adopted by a large discourse-coalition including the landowning agencies, the supposed developers of the new 'world-class' city (e.g. Jagmohan) and the courts. The actions against the poor were further backed by a larger consensus shared by the majority of the urban middle-class and across the English-speaking media, which reproduced the story-line. The demolitions were the outcome of bourgeois environmentalism in its purest and most brutal form.

As this section outlined, the aesthetic rationality was clearly played in the case of the slums, while the discourse on the farmers is much more nuanced and subtextual. Yet, by deploying the *fear of encroachment* story-line and denunciating the farmers for supporting the upcoming of the slums, the farmers' standing as the green keepers of the floodplain have long been weakened. Furthermore, similar to the slums, which did not fit into the world-class imaginary of Delhi, the poor farmers were seen as a clear indicator for backwardness and underdevelopment. Their practices were considered to be 'rural' – not appropriate in the 'world-class' city. The dichotomy of the 'urban' versus the 'rural' is characterized by a "sharp polarization" in the historic imaginations of what the city of Delhi would look like (SHARAN 2014: 161).

Even more important, and rooted in Delhi's history as a 'city of migrants', the migrant workers on the fields embodied the *rural, village life* elements which have never found legitimate space in the planned vision for Delhi (cf. SHARAN 2014: 160). The planners considered them to become the 'next generation' of slum residents. While the slums constituted a spatially-limited 'problem of encroachment', which was of top priority also due to its central location, the farmers still operated in the backyard of the city. Additionally, the state actors – after the rejection of channelization – had no real development vision to put the whole land of the farmers to 'appropriate uses'. This resulted in the further toleration of the farmers in the River Zone.

The court orders have certainly come handy for the landowning agencies to evict the slums from prime lands in the center of the city while ensuring an institutionalized flexibility in dealing with 'suitable uses'. The 'cleansing' of Delhi's riverscapes from the slums emerges as a rather logical consequence of the rejection of the channelization scheme. Earlier, the slums were perceived to be dealt with when channelization starts and the banks were supposedly be put to 'appropriate' use in the course of riverfront development. When channelization was dismissed, the temporal tolerance of the slums was challenged too, since a solution for the 'problem' was out of reach for the government agencies. The court orders revealed a solution for the 'problem'.

The simultaneous development of urban-mega-projects in the River Zone proves that the environmental concerns were instrumentalized under the 'green agenda'. The land in the River Zone was reclaimed from the 'encroachers' for the sake of a 'planned encroachment' which was legitimized by the slum-free, world-class city vision rather than for giving back the land to the river.

14 THE POLITICAL ECOLOGY OF EMBANKMENTS AND URBAN MEGA-PROJECTS

Urban mega-projects "provide an unusually revealing window on patterns of influence in urban development policies", since powerful actors "whose activities are normally camouflaged" are visible (ALTSHULER & LUBEROFF 2003: 4). Therefore, this section analyzes the planning and implementation process of two urban mega-projects; the Akshardham Temple Complex and the Commonwealth Games Village. Both projects are located in the precinct of land reclaimed through the Akshardham Bund earlier discussed as 'pocket III' (see Map 5, page 390).

14.1 The Akshardham Temple Complex

The Akshardham Temple Complex covers about 28 ha of land and was built between 2000 and 2005 by a religious trust known as the Bochasanwasi Shri Akshar Purushottam Swaminarayan Sansthan (in short: BAPS trust). The BAPS trust is a subgroup of the global Swaminarayan movement with temples around the world. It is one of the major subsects within the Swaminarayan sect of Hinduism (DWYER 2004).

The construction period of the temple complex occurred concurrently with the major slum demolitions which were taking place only a few hundred meters away. Besides the main temple, the complex comprises among other features of a large parking area, gardens, a musical fountain, exhibition halls, a movie theatre, and a food court (see Figure 60, page 394). SRIVASTAVA (2009) outlined that the planning of the complex was inspired by visits made by the *Swamys* (members of the trust) to Disneyland and Universal studios and, therefore, it emerges as a religious theme park with a sense of "Disney-divinity" for the new middle classes.

The idea of building the Akshardham Temple dates back to 1968, when the religious leader of the BAPS trust wished to build a temple "on the banks of river Yamuna" in Delhi (BAPS 2005).[144] In 1969 the DDA offered land in North Delhi, which the trust did not accept. In the early 1980s the trust again approached DDA to allot land for a temple and DDA offered the trust land (2,000 square meters) in East Delhi across the embankment of today's site. The trust revoked this offer, deeming the site to be too small and not suitable as it was not located on the riverbed (ibid.).

In the following years the trust reapplied for land on the riverbed. In 1994 DDA discussed whether the area between the two bridges on the eastern bank ('pocket III') could be developed "for cultural, institutional, and allied activities" and could

---

144 The BAPS trust never replied to any of the interview requests by the author.

be allotted to the trust.[145] DDA concluded that about 15 to 18 ha could be given to the trust, if a comprehensive proposal for reclamation of land was accepted by the YSC and other authorities.[146]

The 'Making of Akshardham' timeline on the trust's website states that today's site of the temple is an outcome of the riverfront development schemes worked out by the DDA in the 1980s and 1990s (BAPS 2005).

14.1.1 An offspring of the channelization scheme

In April 1997, the YSC in principle approved the reclamation of the pocket III (260 ha of land on the east bank between Nizamuddin Railway Bridge and Nizamuddin Road Bridge, where Akshardham was built), as requested by DDA. In the mid-1990s the DDA planned a convention area, car parking, and a camping site for this pocket (see Figure 31-A).

It remains unknown to what extent consensus was already reached within DDA to allot the land to the BAPS trust at this time, but as an involved planner outlined during the interviews, the scheme for pocket III was prepared before the Akshardham Temple was considered (DDA-01/2010-09-21):

> "In fact, it was a scheme and that Akshardham came later on. The embankment was not constructed for Akshardham. [...] Akshardham people requested for allotment of land. So land was given to them."

In August 1998, the DDA assigned NEERI to prepare an 'Environmental Management Plan' and NEERI delivered an interim report in 1999. In 1998–1999, NEERI evaluated two alternative proposals for the reclamation of land in this river segment either through constructing a forward embankment 1,300 meters (Alternative I) or 800 meters (Alternative II) away from the existing Left Marginal Bund (NEERI 1999: 17–18). The interim report of NEERI concluded that "after close review" Alternative I, originally proposed by DDA, was revised and Alternative II was agreed upon by a group of experts from DDA and NEERI as this smaller scheme did not require dredging of the river (NEERI 1999: 14).

While the YSC had approved the reclamation of the pocket only in principle and "repeated reminders" were issued by the CWC to the DDA that the final clearance of the YSC "is mandatory before any final shape is given to the proposal" (YSC/60/2000-06-27), the DDA did not present the exact alignment of the embankment to the YSC. In contrast, the DDA in the meantime had already obtained the approval for change of land use of 42.5 ha from 'Agriculture and Water Body' to 'Public and Semi-Public Use' in September 1999 (MoUAE 1999).

---

145 Internal material DDA accessed via RTI by YJA (DDA Technical Committee, Item No 85/96, Sub: Allotment of land to BAPS). File Noting, DDA, 15.09.1995, Sub: Allotment of land to BPs, Internal documents DDA, File No. F3(73)2003MP. Annexure to DDA's counter affidavit in the High Court WP (C) No. 7506 of 2007 and WP (C) No. 6729 of 2007.
146 Ibid.

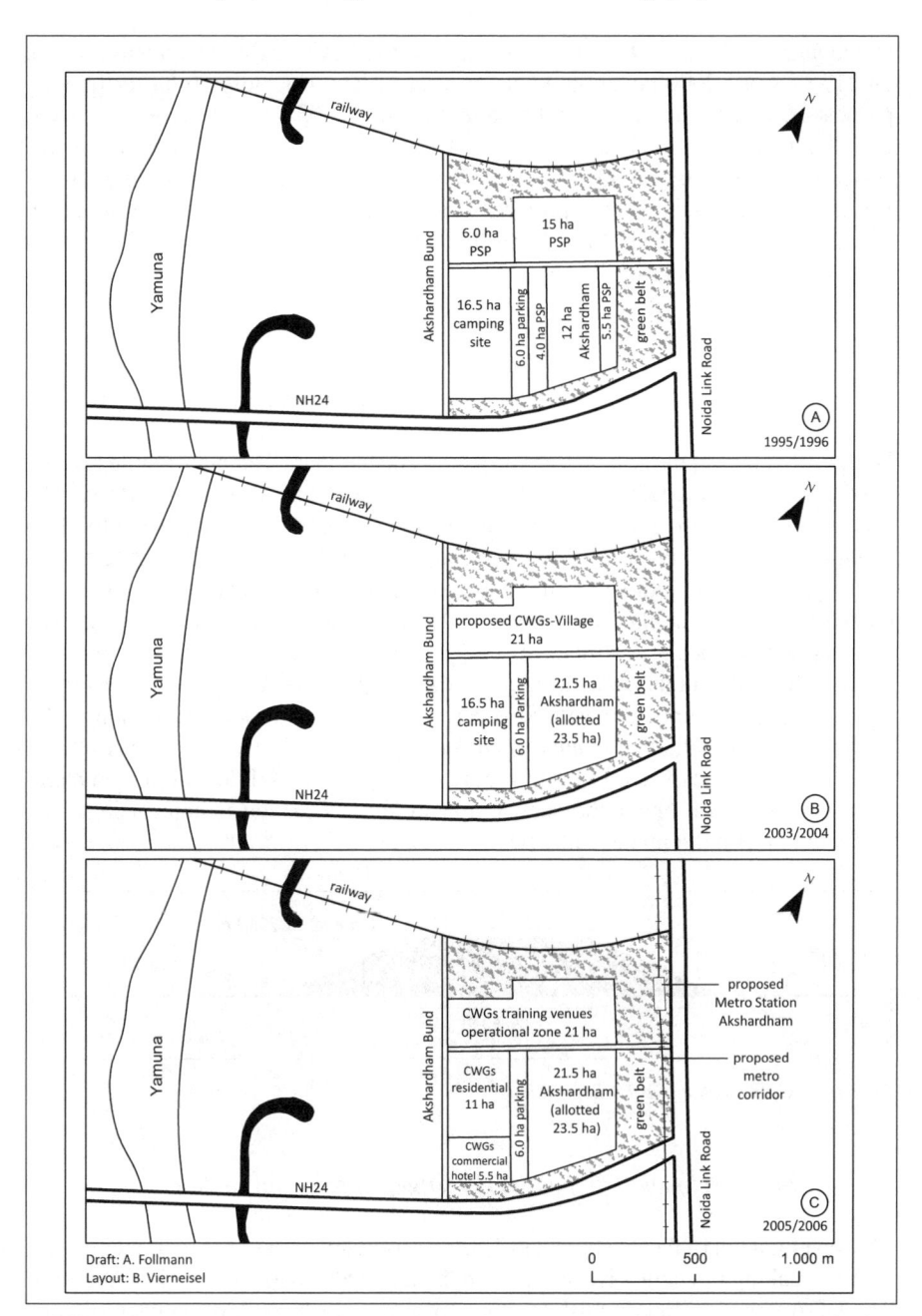

Figure 31: Planning schemes for 'pocket III'
Source: Adopted from internal DDA materials

In January 2000, the DDA offered the land to the BAPS trust, which was allotted on 21st of April 2000 covering about 23.5 ha of land (including 6 ha for parking purposes) close to the eastern embankment (BAPS 2005, DUTTA 2009: 27). Additionally, an area of 4 ha belonging to the Uttar Pradesh Irrigation Department was leased to the trust by the Government of Uttar Pradesh (ibid., see Figure 31-A). Despite the large scale of the project, which would have required environmental clearance from the MoEF, the Akshardham Temple Complex never underwent an environmental impact assessment.

14.1.2 The construction of the Akshardham Bund

With the beginning of the construction of the Akshardham Temple Complex in late 2000/early 2001, the DDA started the construction of the Akshardham Bund. The final alignment of the Akshardham Bund is about 1,050 meter away from the Left Marginal Bund (Noida Link Road), reclaiming about 120 ha of land (CWPRS 2007). The final alignment of the embankment thus lies between the above discussed alternatives studied by NEERI and was never approved by the YSC. The embankment has been constructed by using earth and fly-ash from the nearby power plants. The embankment was designed for a 100 year flood frequency (see Figure 32). On top of the embankment a road was constructed and for the Commonwealth Games a barrier wall of concrete elements of ca. 3 m height with barbed wire fencing was erected (see Figure 62 and Figure 63, page 394). While all other embankments in Delhi are under the supervision of the IFCD, the maintenance of the Akshardham Bund remains with DDA.

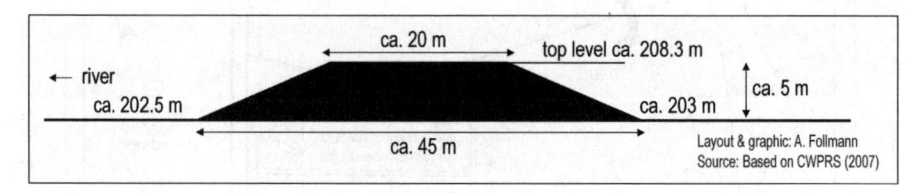

Figure 32: Cross section Akshardham Bund

14.1.3 Religious and political prestige – a powerful coalition

The development of the Akshardham Temple Complex marks an important moment in the remaking of Delhi's riverscapes. In the late 1990s two interests joined forces: The DDA aimed to finally push the long-awaited riverfront development scheme; and the BAPS trust, which had long-awaited the allocation of land on the riverbed, was keen to develop a flagship temple project. Thus, while the DDA earlier envisaged developing a financial district and a convention center at the site (DDA - SPECIAL PROJECT CELL 1998), the construction of the temple fell on sympathetic ears within the DDA as an involved DDA planner explained (DDA-01/2010-09-21):

> "[In India] all the temples are facing the riverfronts, because the river has always been a sacred place. [...] Coming up of temples along the river bank is as for the tradition [sic]. Moreover, we also wanted to develop the River Zone for recreational use, where people can come and enjoy. Religious [use] is one of the functions, which is appropriate for [the] River Zone and that is why Akshardham was considered [to be allocated] land there."

Furthermore, the trust had strong political networks. Lal Krishna Advani, the then Home Minister from the Bharatiya Janata Party (BJP), is known to be a friend of the patron of the temple. Interview partners and media analysis confirmed that he played an important role in the construction of the Akshardham Temple (cf. BAVISKAR 2011c: 51, GILL 2014: 72). The ruling National Democratic Alliance (NDA) – and especially its dominant coalition partner Bharatiya Janata Party – were in favor of the idea of constructing a spectacular Hindu monument in India's capital at that time. Being India's 'Hindu party' the BJP made the construction of Akshardham possible to satisfy its voters (SRIVASTAVA 2009: 339).

Thus, the development of Akshardham must be seen as a symbol of political and religious prestige. A powerful coalition of interests emerged. Additionally, after the terrorist attacks on the Akshardham Temple in Gandhinagar, Gujarat on 24th September 2002 killing 33 people, the project emerged to be even more sensitive on political and religious grounds.

14.1.4 Objection against the Akshardham Temple

The construction of the temple did not proceed entirely without public protest. Right from the beginning the allotment of land and the construction of the temple was criticized by the farmers, who had to give up 'their' land (cf. BAVISKAR 2011c, SRIVASTAVA 2009). The location of the temple was a matter of debate also in the Council of States (Rajya Sabha) and media coverage indicates a heated debate in 2000 about the location and size of the land allotted to the BAPS trust. The Delhi Urban Arts Commission (DUAC) opposed the construction of the temple at the site on ecological grounds (SONI 2006). Despite protest, the project was backed by the powerful coalition and construction started in 2000.

In 2004 the project was challenged in the Supreme Court by the Uttar Pradesh Irrigation Department through a petition made by the Uttar Pradesh State Employees Confederation.[147] The petitioners argued that the construction would adversely affect the recharge of groundwater, would lead to narrowing the river resulting in

---

147 The author was unable to figure out the reasons why the Employees Confederation filed the petition. Interview partners indicated that the Government of Uttar Pradesh might have instrumentalized the Employees Confederation in this matter.

floods and the allotment of land to the trust would be contrary to the existing development plans.[148] Furthermore, the PIL raised doubts whether the land allotment to the temple trust by the DDA was lawful, since the land on which the temple is constructed might actually belong to the Uttar Pradesh Irrigation Department. In court, both the counsel for the state of Uttar Pradesh and the DDA stated that there has been "no violation for the terms of allotment".[149]

Referring indirectly to the 300 m benchmark set by the High Court of Delhi, the DDA further stated that the Akshardham Temple "is nearly 1700 meters away from the river bank".[150] Against the earlier outlined dispute between the DDA and the YSC/CWC about the final approval of the Akshardham Bund, the DDA claimed that all required clearances were obtained. The petition was dismissed by the Supreme Court on 12th January 2005, only eight months before the temple opened. The very short order by the Supreme Court (not even two pages) states that the petition was dismissed because the petitioners were "unable to rebut the statement made by the respondents [DDA, Uttar Pradesh]" and, therefore, the construction of the temple complex is "lawful".[151]

In the eyes of interviewed environmentalists and lawyers it might have been possible to stop the construction of Akshardham if the petition had been better researched and advocated in the court (NGO-04/2010-09-08, LAW-01/2013-02-14). Conversations with environmental activists, lawyers and planners on the Akshardham case further posed the question to what extent the petitioners were really interested in stopping the construction of Akshardham or whether their petition was part of the political dispute about the land along the eastern embankment. Doubts were further raised about why, after being given an opportunity to provide additional documents, the petitioner did not provide any further documents to the court (LAW-01/2013-02-14). While the deeper reasons remain unknown, the high sensitivity of the case might have led to the inaction of the petitioners. None of the interview partners could explain the deeper logic of the case, but some environmentalists regretted their inaction in those days (NGO-04/2010-09-08):

> "[…] the petition was dismissed and no one went into review. No one really bothered about it. And we also blame ourselves, because we were very much here and kept sleeping like everybody else in Delhi."

The construction of the Akshardham Temple Complex paved the way for further developments in the pocket which was reclaimed by the new embankment. In the eyes of the DDA, this land was 'ripe' for development and the Commonwealth Games Village provided the opportunity to capitalize major profits for DDA.

---

148 Supreme Court of India 12 January 2005, U.P. State Employees Confederation & Anr. Writ Petition No. 353 of 2004 and see newspaper coverage Times of India, Bhatnagar, R. (11 October 2004): Akshardham temple to adorn river Yamuna bed and Outlook (11 October 2004): SC declines stay on Akshardham complex construction.
149 Ibid.
150 Ibid.
151 Supreme Court of India 12 January 2005, U.P State Employees Confederation & Anr. Writ Petition No. 353 of 2004.

14.2 The Commonwealth Games Village

Being planned as the first ever green games in the history of the Commonwealth Games[152] and declared as a matter of national prestige and pride, the international sports event was envisioned to boost Delhi on its way to become a 'world-class city' (DUPONT 2011, FOLLMANN 2015, GILL 2014). The athletes' village (ca. 59.5 ha) was deemed to be the flagship project of the games. It was developed in a public private partnership by the DDA and the Dubai-based real estate developer Emaar MGF Construction Private Limited (in short: Emaar MGF). The Games Village residential zone (11 ha) comprises of 34 residential towers with 1,168 apartments and was built to accommodate about 8,000 athletes and officials during the games. After the games the residential zone was to be transformed into a gated community (see Figure 61 and Figure 63, page 394).

The analysis of the development of the Games Village builds upon the development of the Akshardham Temple Complex. The Games Village is reviewed with regard to the practice of EIAs and environmental clearances; the emergence of an environmental activist network against the development of the riverbed (and the Games Village, in particular); and the role of the courts. It is important to note that the protests against the Games Village were organized by a rather small environmental activist network – not a large environmental movement. Other environmental activists had been working on the river before, but with the Games Village they joined their forces.

The information provided in the following sections is based on a series of court documents including several affidavits by the different parties, selected interviews and the report issued by the (High Level) Shunglu Committee setup immediately after the Commonwealth Games by the Prime Minister of India to investigate the irregularities around the mega-sports event and the Games Village in particular (HLC 2011).[153]

14.2.1 The secret 'approval' of the site

After winning the bid for the Commonwealth Games in November 2003, Delhiites awaited the construction of the athletes' village, but for years very little happened before the public's eyes. The report of the Shunglu Committee, however, clearly reveals that the site for the Games Village was already chosen by the Planning Department of the DDA in consultation with the then Lieutenant Governor, Vijai Kapoor, and was approved by the Union Cabinet in September 2003 (HLC 2011: 14–15).

In November 2003, DDA installed a notice board indicating that the site adjoining the Akshardham Temple Complex was to become the Games Village. The

---

152 See for example: The Hindu; 2009/06/15: Delhi getting ready to host first ever 'green' Games.
153 The High Level Committee (HLC) is known as the Shunglu Committee, after its chairman V.K. Shunglu.

selection of the site, as stated by the Shunglu Committee "was a priori and no exercise was undertaken to compare the selected site with alternate sites available with DDA" (HLC 2011: 75). DDA, thus, did not explore other possible sites "because of an implicit desire to construct next to Akshardham" (ibid.). These decisions were made secretly, as several interview partners pointed out.

> "Till 2008 there was a board where the Games Village is coming up, and that board in Hindi said: 'it's a possible site'. It did not say 'it is the only site'. It was one of the possible sites. That was the situation [until 2008]" (NGO-04/2010-04-01).

Planning documents indicate that the DDA originally planned to develop the Games Village along the railway line opposite the Akshardham Temple Complex (see Figure 31B). A site of approximately 47 ha was envisioned to be developed using 21 ha for 'Public and Semi-public Use' (PSP) while reserving the remaining 26 ha as green area (green belt). Most of this area was already in possession of DDA, but about 16 ha were part of the land belonging to Uttar Pradesh Government. The price discussed for acquisition of this land was "upwards of Rs. 5 crores" (ca. 750,000 Euro).[154]

As the internal planning documents of DDA[155] further outline, DDA envisioned to develop the Games Village as a high-class residential project with facilities "comparable if not better" than the Olympics in Beijing. DDA favored a public-private partnership for the Games Village based on the model of Melbourne (2006), which would refund the project by selling off the residential units after the event (in the bidding process DDA had proposed university accommodation). Yet, the designated land use of 'Public and Semi-public Use' did not permit large-scale residential uses.

Additionally, based on a site visit, officials of DDA and the Sports Authority of India in October 2005 decided that the envisaged site of about 21 ha "appears to be inadequate in view of the increasing requirements and also keeping in view the possibility of organizing Asian Games - 2014".[156] Moreover, the DDA feared noise pollution from the nearby railway line and problems in the transfer of land from Uttar Pradesh Government. Therefore, the former proposed camping site was envisaged for developing the residential quarters of the Games Village and the site across the road was planned as the training venues (see Figure 31C). After this decision was made (again in secret), the DDA proceeded the change of land use. Two steps need to be mentioned here: As outlined above (see 14.1), for the construction of the Akshardham Temple Complex the Central Government had approved the change of land use of an area of 42.5 ha from 'Agricultural and Water Body' to 'Public and Semi-Public use' (PSP) in the Master Plan in 1999.

---

154 Concept Paper by DDA, 19th April 2005, 'Development Options for the Commonwealth Games Village' (Secret). Materials published as part of 'Relevant Documents (Vol-I)' by HLC (2011).
155 Ibid.
156 Internal documents DDA, File No. F3(73)2003MP. Provided by DDA as Annexure R-7 to its affidavit in the High Court WP (C) No. 7506 of 2007 and WP (C) No. 6729 of 2007. Materials accessed via LIFE.

While in those days the mega-sports event was still very much 'a pie in the sky' idea, a larger precinct area than just the proposed temple was sanctioned for future developments. In a second step and as a modification of the earlier notification in August 2006, the DDA changed the land use of an area of 11 ha to 'Residential' for the construction of the players' apartments and another 5.5 ha to 'Commercial/ Hotel' for the construction of a 5-star hotel complex (MoUD 2006, cf. FOLLMANN 2015: 219).

After issuing a public notice, the DDA received 10 objections and suggestions regarding the proposed change of land use. DDA did not convene a public hearing (as compulsory), but rather the comments were discussed in a Special Screening Board consisting of DDA members from Planning, Finance, and Engineering departments, the Chief Town Planners of the Town and Country Planning Organisation (TCPO) and the MCD in May 2006.[157] Several objections criticized the ad hoc manner of the proposed land-use change and the non-existence of an approved ZDP. One objection made criticized that the Commonwealth Games "get used to undermine legal process as has happened in earlier developments" pointing to the Asian Games in 1982.[158]

Furthermore, the objections indicated that the public notice was incomplete and flawed proposing change of land use from 'Recreational' to 'Residential' and 'Commercial' although the area was earmarked as 'Agriculture and Water Body' in the MPD-2001. Nevertheless, referring to the provisions of the Delhi Development Act, the Special Screening Board and the MoUD approved the change of land use. A high-ranked DDA planner involved in the process explained the technical decisions taken (DDA-01/2010-09-21):

> "NEERI suggested that development can be done in form of institutional [public and semi-public use], but once land has been proposed for institutional and is out of the river basin for all practical purposes, then the provisions of Master Plan and Delhi Development Act applies on that. And, under Delhi Development Act the Government of India changed the land use and made changes in the Master Plan to convert it into residential. [...] Once the area has been taken out from the river basin by construction of embankment, then it is as good as any other area for development."

As the quote by the DDA planner indicates, the vague land-use category 'Public and Semi-Public Use' opened up the River Zone for a wide range of uses and assured the state's own flexibility in changing the designated land use as needed at a later stage (cf. FOLLMANN 2015). The DDA never requested approval of the Games Village from the YSC.[159] Additionally, it needs to be pointed out that also the Central Groundwater Authority (CGWA) and the Yamuna Removal of Encroachments

---

157 Minutes of the meeting of the Special Screening Board, DDA, 26th May 2006. Provided by DDA as Annexure R-9 to its affidavit in the High Court WP (C) No. 7506 of 2007 and WP (C) No. 6729 of 2007. Materials accessed via LIFE.
158 Ibid.
159 An RTI response from YSC of July 2007 confirms this by declaring that "no specific permission was accorded to DDA for raising permanent residential multi-storey flats [...] in the flood plain of river Yamuna". Documents accessed by YJA via RTI.

Monitoring Committee were not informed by DDA at any point. Thus, based on DDA's logic, the site was purified as land and deemed for development.

14.2.2 Remaking environmental clearances

In order to seek environmental clearance for the Games Village, the DDA approached the MoEF in October 2006. The materials submitted by the DDA in the EIA process clearly stated that the site lies below the high flood level of the river "nearby" (DDA 2006b, EQMS 2006). Therefore the DDA proposed to fill the entire area with fly ash from the nearby power plants by about 1 meter to raise the level to the high flood level. While the EIA indicates that "the project site is located in the younger alluvial tract" of the river Yamuna having a high groundwater potential (ibid.), any risk of flooding is denied in the project report based on the argument that the site "is well protected by the high embankment marginal bund along the site to its west [the Akshardham Bund]" (ibid.). The EIA material also outlined that the project details included a complex of four hotels with 4,000 beds (5.5 ha), sporting venues for training purposes during the games declared as 'public and semi-public use' of 21 ha and a green area of 22 ha (DDA 2006b).

As per the rules of the MoEF, the project proposal was evaluated by an EAC. The EAC raised major concerns about the location of the project in the river floodplain, its interference with groundwater recharge, the use of fly ash for raising the level, and the adverse effects of the Akshardham Bund in terms of flood risks. Overall, the EAC concluded that "environmental concerns [...] had not been addressed adequately and that the site had been adopted without due investigation and assessments of environmental impacts".[160] Therefore, the EAC recommended that the DDA should look for an alternative site.[161]

During discussions between the DDA, the EAC and the MoEF, the DDA rejected any other site and pressed for an environmental clearance notification for the site due to upcoming time constraints. While in November 2006 the EAC was still requesting a change of the site, the EAC recommended that the MoEF should issue environmental clearance in early December 2006:

> "While the Committee does not doubt that time has already become a constraint, the Committee is not convinced that environmental impacts and their mitigation have been studied to a satisfactory level. Under the circumstances, the Committee will go by the 'Precautionary Principle' and emphasise the point that, **as far as possible, the proposed work should not be of a permanent nature**. Since the design of the structures is still to be made, it should be possible to take this point into consideration and adopt dismantled structures. Unless **detailed studies**

---

[160] Minutes of the 30th meeting of the Expert Appraisal Committee held on 2nd and 3rd November 2006. Materials accessed via RTI by YJA, also quoted in GILL (2014: 78).
[161] Ibid. The EAC supposed the site of the Safdarjang Airport but DDA dismissed the Safdarjang Airport site, reasoning this with the fact that possession of land was not with DDA and, thus, is likely to be not available. Furthermore, DDA argued the Safdarjang Airport is adjoining to a VVIP (very, very important person) residential zone making it a High Security Zone.

lead to the conclusion that the proposed structures **can be left behind permanently**, the proposal should **proceed with the assumption that the river bed may have to be restored to the river**. Since the **environmental significance and public open space amenity of the river flood plain** should be recognised, the Committee urges the concerned authorities that an extension of similar development in the area between the river Yamuna and its flood protection bunds **must not be proposed without due environmental planning and a prior environmental clearance**" [emphasis added].[162]

Based on these recommendations, the MoEF cleared the project under strict conditions on 14th December 2006. Following the recommendation by the EAC, the MoEF allowed only temporary structures to be built.[163] This restriction should have signified the end of DDA's plan to develop a high-class residential complex. But not willing to accept the clearance notification, the DDA aimed to provide detailed studies, which would allow them to build permanent structures. In doing so, the DDA procured another study from CWPRS to assess the impact of the already existing Akshardham Bund in terms of its impacts on floods. The MoEF in turn made a first amendment to its earlier clearance notification on 22nd January, stating:

> "DDA could go ahead with the **planning** of their construction works, permanent or temporary, subject to the conditions that the actual work on permanent structures will not start till such time that the mitigation/abatement measures against upstream flooding are identified [...] and their implementation begun in such a way **that the work is completed on or before the date when the buildings will be completed**" [emphasis added].[164]

In February 2007, CWPRS concluded that the Akshardham Bund does not cause "any undesirable flow condition in the vicinity of existing structures", since the maximum water level will increase during floods by maximal 10cm (CWPRS 2007: V). Based on the CWPRS-report, the MoEF for the second time amended its environmental clearance on 29th of March 2007, stating:

> "DDA could go ahead with the **planning** of their construction works, permanent or temporary, subject to the conditions that the actual work on permanent structures will not start till such time [the recommended mitigation measures] **are completed**" [emphasis added].[165]

The correspondence between the MoEF and the DDA clearly indicates that the DDA intended on going ahead with the construction of the Games Village, arguing that the MoEF had earlier indicated that DDA could start with the construction as

---

162 Minutes of the 32nd meeting of the Expert Appraisal Committee held on 1st and 2nd December 2006. Materials accessed via RTI by YJA (cf. FOLLMANN 2015: 219).
163 Approval letter from MoEF to DDA (14th December 2006). Letter received via RTI by YJA.
164 Letter from MoEF to DDA (22nd January 2007). Letter received via RTI by YJA.
165 Letter from MoEF to DDA (29th March 2007). Letter received via RTI application filed by YJA. The proposed mitigation measures included the raising and strengthening of the existing embankments and guide bunds along the river in Delhi to check flood discharge, covering the embankments with a layer of stone crates and applying geo-fabric filter to protect bunds and bridges between the Indraprastha Barrage and Nizamuddin Road Bridge (CWPRS 2007).

soon as the study is obtained from CWPRS and "their implementation [had] begun".[166] The MoEF, on the day DDA's letter reached the ministry, amended its environmental clearance notice for the third time:

> "DDA could go ahead with the **planning and construction works**, permanent or temporary, subject to the conditions" [emphasis added].[167]

In sum, within a period of about three and a half months' time, the MoEF eventually allowed the DDA to start with the construction of permanent structures after being pressured by other ministries of the Central Government and the organizing committee for the Commonwealth Games. The earlier approval for planning was amended into an approval for both planning and construction work.

This modification of the environmental clearance given by the MoEF was not done by the rules. The EAC was setup for screening, scoping and appraisal of the project and to "make categorical recommendations" to the MoEF either to grant clearance based on terms and conditions or to reject clearance based on reasons for the same (MoEF 2006: Section 5-6). As by the rules, the MoEF "shall normally accept the recommendations" made by the EAC (MoEF 2006: Section 7). If the MoEF disagrees with the recommendations of the EAC, the MoEF "shall request reconsideration" by the EAC before giving environmental clearance (MoEF 2006: Section 8). Most importantly, the MoEF shall only issue final environmental clearances to the applicant (ibid.). However, in this case the MoEF changed and adapted to the requests made by DDA three times. Furthermore, the MoEF did not contact the EAC for reconsiderations before issuing the changed notices to the DDA. Therefore, the MoEF went against its own regulation by issuing a series of amended clearances without renewed appraisal by the EAC. In the meantime, Delhi-based environmental activists started *Yamuna Jiye Abhiyaan*, the Yamuna Forever Campaign[168], and filed two separate PILs in the High Court of Delhi against the construction of the Games Village.

14.2.3 Formation of environmental protest: Yamuna Jiye Abhiyaan

Yamuna Jiye Abhiyaan (YJA), a consortium of a number of different environmental NGOs and individuals, was formed in February 2007. One of the activists explained the focus of the initiative on the land issues of the River Zone (NGO-04/2010-04-01):

> "Yamuna Jiye Abhiyaan is very focused on facts and less on emotions. Because it is very easy to speak in terms of a holy river – it's such an old river, Lord Krishna and all those things – which is fine and required. But it is better to have facts what is actually on the ground. That needs to be documented and worked together. And, of course that needs to be shared. [...] what is the primary importance at this point of time is to safeguard the remaining floodplain of the

---

166 Letter DDA to MoEF (2nd April 2007), Letter received via RTI application filed by YJA.
167 Letter from MoEF to DDA (2nd April 2007), accessed via RTI application filed by YJA.
168 '*Jiye*' literally means 'live'.

river in the city. Because there are technological fixes for the water, but there is no technological fix if once the floodplain is gone."

The environmental activist further highlighted that the environmentalists' inaction and silence in the Akshardham case was "the first charge" against them (ibid.). The Chief Minister of Delhi and other officials questioned them during meetings "where were you when Akshardham was coming up", and he had to admit that they "were sleeping" (ibid.).

The members of Yamuna Jiye Abhiyaan (YJA) presented their concerns (among others negative effects on groundwater recharge and risk of flooding) to the public agencies involved. They were given an audience with the empowered Group of Ministers in charge of the organization of the Commonwealth Games and concerned agencies including the DDA in November 2007. However, at this time the location of the Games Village had already been fixed (HLC 2011: 16). Additionally, the environmental NGOs, a number of voluntary (youth) organizations, and local farmers (among them members of Delhi Peasants) launched a 'Satyagraha' (a Gandhian peaceful protest camp) to demonstrate against the location of the Games Village in July 2007. The Satyagraha was widely covered in the daily newspapers. When realizing that the Satyagraha and the meetings with government officials would not stop the development of the athletes' village, individuals of the group together with the NGO INTACH approached the High Court of Delhi through PILs.

While the petitioning and lobbying for the river on first impressions corresponds with the arm-chair middle-class environmentalism which is often criticized to be anti-poor (BAVISKAR 2010, 2011a, KUMAR 2012, MAWDSLEY 2004, 2009, MAWDSLEY et al. 2009, NEGI 2011, TRUELOVE & MAWDSLEY 2011), a more detailed analysis shows multiple connections with non-middle class groups.

For example, the Yamuna Satyagraha started a mass planting of trees on the supposed construction site and included a sit-down strike (*dharna*) under a large sheesham tree (see Figure 33); modes of protest quite similar to other environmental movements in India (cf. GADGIL & GUHA 1994).

Figure 33: Protest camp 'Yamuna Satyagraha'
Source: Photo courtesy Yamuna Jiye Abhiyaan 2007

Socio-environmental action groups in India have traditionally used a wide repertoire of direct forms of protest drawing on Gandhian-style protest culture. Gandhian non-violent resistance, called *satyagraha* (literally 'truth force'), includes among other actions especially protest marches and processions (*pradhashan, yatra*), sit-down strikes (*dharna*), or even more militant forms of protest like roads blocks (*rasta rook*), as well as hunger strikes by charismatic leaders (*bhook hartal*). Direct protest in form of *satyagraha* aims at media coverage and related public pressure.

In the case of the Games Village, farmers, civil society organizations (including Jal Biradari, Ridge Bachao Andolan, Natural Heritage First, Swechha, CSE, Youth for Justice), and well-known individuals including Vandana Shiva, the lawyer Prashant Bhushan or the journalist Kuldep Nayar joined the protest. A key figure of the Satyagraha was Rajendra Singh, a well-known water conservationist from rural Rajasthan commonly known as *waterman of India*. His support was requested by members of Ridge Bachao Andolan, because none of them had led a mass public movement before. They were aware that to start a "real movement" they needed a spokesman who was able to communicate effectively with the farmers and other non-middle class groups (NGO-15/2011-12-08):

> "[…] we didn't have anyone who has seriously handled a public campaign on environmental issues. In Delhi, we didn't have any. Even today, I am not finding anyone else. Rajendra Singh has a name working on the ground. He has credibility. […] being more a rural kind of person, the farmers could empathize with what he said and could understand. The farmers generally don't join Delhi's middle class and the middle class generally doesn't join the farmers. […] Middle-class support didn't come, but through Rajendra Singh many NGOs joined. [There are] hardly any environmentalists in the city. Environmental awareness is very low. Activism is very low. Simple."

The protests failed in preventing the start of the earth works for the Games Village and the Yamuna Satyagraha camp had to shift to a site next to the construction zone. One of the farmers involved in the protest pointed out that the cooperation between the environmental NGOs and the farmers was difficult since the farmers were not interested (FARM-01/2011-03-09):

> "We [Delhi Peasants] worked together with the NGOs when we had this Yamuna Satyagraha. They came with us. Now we [NGOs and the farmer] regularly have meetings. […] But I am the only person, who is doing all these things. The rest of the people – even our president does not care for that."

Adopting these typical Gandhian forms of protest and combining them with own field research and petitioning the environmentalists of Yamuna Jiye Abhiyaan allowed for a series of channels to raise awareness. Engaging with several experts and scientists working on the river as well as carrying out own field research, Yamuna Jiye Abhiyaan developed a knowledge base on the river and its governance. Sharing this knowledge through their website and e-mail newsletter as well as engaging considerably with the print and television media, Yamuna Jiye Abhiyaan aimed to create public awareness for the protection of the floodplain. The environmental activists filed a series of applications under the RTI Act which was the crucial means to access the required information from state authorities to file the court case via PILs.

The case against the Games Village was not the first case filed by the activists for the protection of the land in the River Zone. The environmentalists Anand Arya and Manoj Misra had already sought immediate action by the High Court of Delhi and the Encroachments Monitoring Committee (set up under the High Court order discussed in section 13.2.3) for stopping the month-long 'Times Global Village Festival' in March 2007.[169]

The court case against the 'Times Global Village' involved environmental activists and bird watchers from Noida who were campaigning for the protection of the Okhla Bird Sanctuary. They had also filed a PIL against the 'Rashtriya Dalit Prerna Sthal and Green Garden' (National Dalit Inspiration Place and Green Garden) built under the direction of Mayawati, the Chief Minister of Uttar Pradesh directly adjoining to the Okhla Bird Sanctuary (see Map 2, page 387).

The Times Global Village festival was organized by the Dubai-based e4 Entertainment Company Pvt. Ltd. in collaboration with the Noida Toll Bridge Company Ltd., Delhi Tourism & Transportation Development Corporation Ltd. (DTTDC), and The Times of India Group. The month-long festival included restaurants, shops and joyrides. Stretching over about 32 ha (80 acres) of land in the River Zone on the right bank north of the DND Flyway, the organizers created "[m]any pucca roads along with access ramps" and undertook "huge earthworks" raising the level of the riverbed by 1–2 meters providing "underground sanitary, sewage water and electrical fittings" (report cited from DUTTA 2009: 106).

After a site visit, the Encroachments Monitoring Committee stated in its report that the festival has "caused serious environmental degradation and grave harm to the ground water recharge capacity of the river" (report of the YREMC cited from DUTTA 2009: 106) and the committee recommended the High Court to issue an order to "remove all temporary or semi-pucca structures from the site and dismantle all roads and access ramps within a month and restrain them from holding any event or festival in future" within the earlier declared 300 meter zone (report cited from DUTTA 2009: 106). Following the Encroachments Monitoring Committee's recommendation the structures were demolished after the event was over. The case of the Times Global Village indicated that the High Court might enforce its former orders also in respect to new development projects along the banks. The positive outcome of the case motivated some activists to file the PILs against the Games Village.

Other activists involved in the protest against the Games Village under Yamuna Jiye Abhiyaan had been active in earlier environmental campaigns in Delhi; especially in a movement to save local forest areas from large-scale development projects (Ridge Bachao Andolan). Their experience had told them that approaching the courts in the fight against projects sanctioned by the government is very difficult and often fails because the way of how the court takes up and responds to PILs is very inconsistent and heterogeneous (cf. BHUSHAN 2004, 2009, RAJAMANI 2007,

---

[169] High Court of Delhi, Writ Petition (Civil) No. 2344 of 2007.

RAZZAQUE 2004). They came to the realization in their earlier environmental campaigns that the outcomes of PILs were often highly unpredictable. Therefore, not all members of the campaign on the Yamuna were in favor of approaching the court against the Games Village. The different understanding of the role and impact of the courts remained a major area of contestation within the NGO network throughout the study period. The court case was nevertheless one major focus of the campaign from late 2007 to July 2009.

14.2.4 The Games Village in the courts

Vinod Jain from Tapas NGO filed the first PIL in September 2007, which requested legal protection of the river Yamuna based on the River Regulation Zone (RRZ) proposed by the MoEF in 2002.[170] This PIL was heard in the beginning by a single judge of the High Court of Delhi (Justice Gita Mittal), but was later transferred to a Division Bench (A.K. Sikri and Rekha Sharma) dealing with the second PIL filed by four petitioners.[171] The Division Bench, which was also engaged with other Yamuna-related cases, heard the case almost every Tuesday and Friday at the peak of the case. The respondents in the case were the GNCTD, the MoEF, the Ministry of Youth Affairs & Sports, the MoUD, the DDA, and the DMRC.

Referring to other related court cases with regard to the Yamuna, Justice Mittal acknowledged a series of critical aspects concerning the construction in the River Zone. Taking up the earlier discussed purification of the floodplain as water body by the High Court in the to the Okhla/Wazirpur Case (see 13.2.2), the judge stated that since it "is a 'water body'" and "not 'land'" the site "can be utilised only as a water body" (Justice Mittal, cited in DUTTA 2009: 28). Justice Mittal raised many questions also with regard to the Akshardham Bund, inquiring whether it is "artificially restricting the flood plain of the river" and has reclaimed land "which forms part of the river environment" (ibid.). Based on the attitude with which Justice Mittal took regarding the case, the environmentalists had more hope for positive judgements in their favor (NGO-04/2011-03-05, NGO-07/2011-03-02).

However, the case was transferred to a double bench in the High Court (Justice Sikri and Justice Sharma) and developed differently than previously expected by the activists. On 2nd February 2008, the judges reserved the judgement in the case, but the judgment was only pronounced on 3rd November 2008. The two judges gave two separate judgements. It needs to be mentioned here that Justice Sikri in 2005 had confirmed the legality of the acquisition of land for channelization (see chapter 10.5). While Justice Sikri gave a detailed insight in the whole court case in along order, Justice Sharma's judgements was comparably short. Since Justice Sikri agreed with the direction of Justice Sharma, her judgement became the direction of

---

170 WP (C) No. 6729 of 2007, Vinod Kumar Jain vs. Union of India & Others.
171 WP (C) No. 7506 of 2007, Rajendra Singh and Others vs. Government of NCT of Delhi & Others. The four petitioners in this case included Rajndra Singh, Manoj Misra, INTACH and Sanjay Kaul.

the bench. The discussion in the following paragraphs draws on both judgements. A. K. Sikri[172] summarized the case as the following:

> "[...] the entire matter revolves around the two distinguishing but contrasting interests: Development of the city coupled with National Prestige on the one hand and Ecology of the River Yamuna on the other. The respondents contend that the Commonwealth Games are a matter of national pride and any order prohibiting the construction would have an impact on the reputation of the country so far as international community is concerned. For the petitioners, that is secondary, and the prime consideration should be 'Save the Environment'. They argue that the construction would have irreversible impact and cause permanent damage to the ecologically fragile environment of the river Yamuna, its bed, banks, basin and flood plain. As per them, the concerns of reputation cannot be permitted to impact the constitutional rights of the citizens or to override the health of the citizens; the river and the environment."

The accessed materials of the case mirror this view.[173] The petitioners claimed that political pressures had caused the shifts in the environmental clearance issued by the MoEF and, thereby, the very purpose of the EIA was undermined. In contrast, the respondents (especially DDA and MoEF) claimed that all required environmental clearances based on detailed studies had been issued and any further delay in the construction of the Games Village would lead to serious problems for the event and a wasting of resources and time.

It needs to be mentioned that both PILs were not exclusively filed against the Games Village. The first PIL challenged the developments along the river in general, requesting legal protection for the floodplain. The second PIL contested, among other projects, the construction of the Games Village and the developments by Delhi Metro in the floodplain. The second petition also referred to the proposed RRZ and highlighted that NEERI (2005) had disqualified residential and industrial uses requiring permanent structures. In the light of the mega event the case entirely focused on the Games Village. The petitions set the 'whole government machinery' in motion defending the project of national prestige. In addition to all other arguments discussed, the court case narrowed down to the question whether the site is part of the floodplain or not.

14.2.5 A new boundary of the river? Multiple readings of the Akshardham Bund

The DDA stated in its affidavit that the project site is located "beyond the embankment" and, thus, is "not part of active river bed" and "no more part of the flood plain since the site has been reclaimed as per approval of the Yamuna Committee [YSC]".[174] Responding to the petitioners' observations, who argued that based on NEERI's (2005) recommendation, no residential structures should be built in the

---

172 High Court of Delhi, A. K. Sikri, 3rd November 2008, WP (C) No. 7506 of 2007.
173 The author accessed the detailed material of the case through the Legal Initiative for Forest and Environment (LIFE) based in New Delhi. The lawyers of LIFE advocated the case for the petitioners.
174 Affidavit DDA in High Court case WP (C) No. 7506 of 2007 and WP (C) No. 6729 of 2007. Materials accessed via LIFE.

River Zone, the DDA rebuked that these reports "are at best advisory in nature and have no binding force".[175]

Supporting the DDA's arguments, the MoEF's affidavit in the court stated that "neither the river nor the riverine species nor the flow of the river is adversely affected" by the project and the "only aspect, which needed consideration was the efflux by channelizing the river, which may flood the areas upstream".[176] Moreover, the MoEF argued that "the instant project cannot be said to be constructed on a floodplain".[177]

In contrast, the petitioners argued that "the area [...] continues to be part of the flood plain" since "a flood plain does not lose its identity merely because of bunds being constructed in it".[178] They highlighted the "inter-connectivity of the aquifers and its recharge continuously trough rains and natural river flood".[179]

In January 2008, the DDA requested NEERI for a scientific opinion whether the site was any longer part of the floodplain or not (NEERI 2008):

> "Is it correct that the land where Commonwealth Games Village is being constructed is not a part of the Yamuna Flood Plain, more so after the construction of embankment (Akshardham Bund)?"[180]

NEERI responded to the request by DDA with a brief report (8 pages) outlining that the site "is to be deemed to be no more a part of the flood plain zone", due to the "new boundaries" of the river after the construction of the Akshardham Bund (NEERI 2008). NEERI further indicated that its earlier report had not assessed the post-2002 situation including the Akshardham Bund (NEERI 2008).

Despite the new evaluation by NEERI, Justice Sikri aimed to establish an expert committee to interrogate the question again:

> "[...] whether the present construction is on the riverbed or not itself would be a disputed question. We are, therefore, of the opinion that so far as the issue raised by the parties herein as to the present site not being part of flood plain/ river bed after the creation of embankment in 2002, the same requires consideration of an expert committee."[181]

In contrast, Justice Sharma agreed with the petitioners that the site was undoubtedly on the river bed. She noted that the NEERI report of 2005 "describes the land in question as 'river bed'" and pointed out that also the Expert Appraisal Committee

---

175 Ibid.
176 Affidavit MoEF in High Court case WP (C) No. 7506 of 2007 and WP (C) No. 6729 of 2007. Materials accessed via LIFE.
177 Ibid.
178 Rejoinder affidavit petitioners to DDA affidavit in High Court case WP (C) No. 7506 of 2007. Materials accessed via LIFE.
179 Ibid.
180 Besides this question DDA sought further an opinion on whether DDA had compiled with the conditions for environmental clearance. Furthermore, NEERI should assess whether the project poses any further threat to the environment and would lead to a loss of groundwater recharge (cf. GILL 2014: 79, HLC 2011: 17).
181 High Court of Delhi, A. K. Sikri, 3rd November 2008, WP (C) No. 7506 of 2007.

in the environmental clearance process mentioned that "the river bed may have to be restored to the river".[182] Therefore, she came to the conclusion that:

> "[...] no doubt is left that the site in question is on the river bed. However, to my mind, even if it be taken that the site in question is not river bed, yet the urbanization of the site and colossal construction, under way may yet adversely affect the environment, the river and ecology."[183]

Eventually, the High Court of Delhi set up an expert committee, chaired by R.K. Pachauri[184], which was supposed to study whether the proposed constructions would adversely affect the ecology of river, its floodplain, its groundwater recharge capacity and whether the government agencies have failed to protect these.[185] The High Court allowed the DDA to continue with the construction of the Games Village under the condition of the supervision of the expert committee. However, the expert committee never met and being unsatisfied with the decision of the High Court, all parties went to Supreme Court of India through separate Special Leave Petitions. On a plea from DDA and other state agencies, the Supreme Court stayed the High Court's decision for setting up the expert committee in an interim order on 5th December 2008. When the state agencies had approached the apex court, also the environmentalists moved to the Supreme Court.

The Supreme Court of India dismissed the pleas made by the environmentalists on 30th of July 2009 – only about 14 months before the Commonwealth Games. The Supreme Court stated that since "there is no statutory definition for 'riverbed' and 'floodplain' from the statute" and the court had to rely on the dictionary meanings of the two terms. The dictionary definitions of 'floodplain' and 'riverbed' cited in the order drew on a very limited hydrological definition of floodplain without referring to the genetic history of river floodplain. For the court, the 'riverbed' was simply the area where the river flowed, a "hollow channel of a water course".[186] The 'floodplain' was understood as the land adjacent to rivers, which due to its low-lying nature gets flooded when the river "overflows".[187] Based on these simplified definitions and referring to the statements of the state agencies (including NEERI), the Supreme Court concluded:

> "The observation and conclusion of the High Court that the site is on a 'riverbed' cannot be sustained. The High Court disregarded and ignored material, scientific literature and the opinion of experts and scientific bodies which have categorically held that the site is neither located on a 'riverbed' nor on the 'flood plain'."[188]

---

182 High Court of Delhi, R. Sharma, 3rd November 2008, WP (C) No. 7506 of 2007.
183 Ibid.
184 R.K. Pachauri is the director general of The Energy and Resources Institute (TERI) based in New Delhi since 1982. He was the Chairman of the United Nations established Intergovernmental Panel on Climate Change from 2002 to 2015.
185 High Court of Delhi, R. Sharma, 3rd November 2008, WP (C) No. 7506 of 2007.
186 Supreme Court of India, 30 July, 2009, D.D.A. vs Rajendra Singh & Ors. Civil Appeal Nos. 4866-67 of 2009.
187 Ibid.
188 Ibid.

The Supreme Court further noted that its earlier ruling on the Akshardham Complex "is a binding precedent for all purposes".[189] Thus, DDA and its partner Emaar MGF could proceed with the construction of the Commonwealth Games Village. The role of NEERI in the case was highly criticized by the environmentalists:

> "NEERI changed the definition of the floodplain. [...] It has nothing to do with building embankments or having embankments. It is simple: The floodplain is that where you have a deposit of sand which has come over millions of years [and] which holds a lot of water and that's it. All you need to do is to look into the soil and that is the floodplain whether you build embankments or not" (NGO-03/ 2011-11-04).

During the case the environmental activists repeatedly argued that the Central Groundwater Authority (CGWA) had notified the floodplains of the Yamuna in Delhi in September 2000 to protect the aquifer. A former scientist of the Central Groundwater Board explains the CGWA's stand and the discussions within the government circles (SCIENTIST-02/2011-11-08):

> "Initially [...] Central Groundwater Authority did not give its clearance. They did not comment. [...] They were not asked, because their statement was clear from the beginning. [...] I remember talking to so many people and they said the only solution is that you make these buildings collapsible. You make all these structure temporary [...]. We senior government officers used to have so much of argument, no sir you think of a solution. They all agreed on collapsible structures, but in the end finance said, you invest some 400 or 500 crores, and then take it back? How will I permit this? That was a deadlock."

In addition to these scientific and economic questions, much debate after the court order concentrated on the way the Supreme Court had reasoned the 'green light' for the construction of the Games Village. Environmentalists argued the court should have named the real reason for permitting the construction and should have justified its verdict on grounds of the importance and national prestige of the project, not by accepting the stand of DDA that the site is not on the riverbed/floodplain (cf. BAVISKAR 2011c: 51). Even critical voices and opponents of the project might have accepted such on order one year before the games, if the Supreme Court had given an otherwise strong order for the protection of the river's floodplain.[190]

14.2.6 The river in between science, politics and time constraints

While the Supreme Court relied heavily on NEERI's statement that the site is no more on the floodplain, the judgement of A. K. Sikri in the High Court had earlier undoubtedly devalued NEERI's reputation based on the institute's changing opinion and contradictory recommendations:

> "[NEERI's] affidavit is the result of some of the loopholes in its earlier reports which were picked up by the petitioners and pointed out to the Court. From an institution of this repute, it

---

189 Ibid.
190 The author has discussed this point extensively with involved environmentalists and lawyers. This argument has further been developed out of a detailed discussion with Professor Ramaswamy R. Iyer.

was not expected that [a] report of this kind would be submitted. The Court gathers an impression that it is a tailor-made report given to suit the requirement of the respondents [DDA]. In the earlier report given by NEERI, [...] it had recommended only construction of temporary structures. It is not at all explained as to what were the changed circumstances which weighed with NEERI to opine that structures can be of permanent nature."[191]

In a similar way also judge Sharma noted:

"The reports of the NEERI [...] do not paint this body in bright colours. Rather, they show how it has changed colours and has not bothered to contradict itself. [...] It is not only NEERI, it is the Ministry of Environment and Forest also which is equally guilty of changing its position. [...] It is a sad story of men in haste fiddling with major issues and resultantly playing havoc."[192]

The detailed reading of the orders suggests that neither of the judges in the High Court were convinced that the environmental clearance was obtained based on independent scientific evidence and following the due process of law. Yet, the judges were also very much aware of the special importance and the pressing time limit given by the fixed deadline of the mega-sports event. They were unquestionably under high pressure from various government agencies:

"In this backdrop, though we cannot accept the plea of the petitioners to stop the construction, which may have far more serious consequences, more particularly when at this juncture no alternative is available. [...] the petitioners did not approach the Court well in time to enable it to direct the respondents to undertake further studies or think of alternate site, if it was ultimately to be found that the site in question was not very appropriate. There is hardly any time for such an exercise [...]."[193]

The Supreme Court later gave further merit to the state agencies' argument that the objections against the construction of the Games Village had been filed by the petitioners with "inordinate delay".[194] The court argued that the petitioners should have already objected to the change of land use in 1999 or could have opposed the developments in 2006 when the global tender process was issued by the DDA seeking private partners to develop the project.

After the High Court order was made public, media coverage also painted a very critical picture of NEERI in the national dailies across the country.[195] 'The Hitavada' (The People's Paper), the largest English daily newspaper of Central India from Nagpur, the city where NEERI is based, reported critically about the institute.[196] In response to the public backlash, a scientist from NEERI explained in a letter to the editor of newspaper NEERI's stand in the case and highlighted that NEERI's earlier reports "had not mentioned anything which would have facilitated" the environmental clearance given by the MoEF. The scientist further argued:

---

191 High Court of Delhi, A. K. Sikri, 3rd November 2008, WP (C) No. 7506 of 2007.
192 High Court of Delhi, R. Sharma, 3rd November 2008, WP (C) No. 7506 of 2007.
193 High Court of Delhi, A. K. Sikri, 3rd November 2008, WP (C) No. 7506 of 2007.
194 Supreme Court of India, 30 July, 2009, D.D.A. vs Rajendra Singh & Ors.
195 See among other: Hindustan Times, 3rd November 2008: 'Bench finds it difficult to reach a common ground'; Indian Express, 4th November 2008: 'HC: build on Yamuna bed at own risk'.
196 The Hitavada, 4th November 2008: 'HC refuses legal sanction to C'Wealth Games Village.

"Neither NEERI changed its stand nor had taken any cognizance of any construction [referring to the Akshardham Bund] in its final report submitted in 2005 [...] the conclusion drawn that NEERI's hand is tainted is not proper. [...] at no point of time, NEERI has revised its report to accommodate the construction of embankment on riverbed. NEERI's opinion only pertains to the prevailing condition which, however, was not created due to recommendation of NEERI at any point of time."[197]

The situation was generally described as being caught between science, politics and time constraints in discussions with involved environmentalists about the court orders. The environmentalists stated that they had clearly filed the case well in time before any major construction on the site had happened. Overall, it seemed that the opponents of the project were left with no option, since if they had filed the petition before the environmental clearance was granted, the court might have dismissed the plea on the ground "that the petition is premature" (Law-01/2013-02-15). From their perspective it has been the state machinery delaying the decision-making process both in the planning process and in the court. The state agencies' intension was to delay the decision that the construction becomes "a fait accompli situation, where the government will say we don't have any option and we have to conduct the Games" (Law-01/2013-02-15).

The discussion of the recommendations of the Expert Appraisal Committee and the development of four different environmental clearance notices issues by the MoEF further indicates that the MoEF has been under considerable pressure from DDA and other high level government agencies. The insights suggest that the exact formulation of the clearance notification and the mitigation measures were an outcome of a bargaining process between the project proponents and the EAC/MoEF, but also between the MoEF and the EAC. In addition to the outlined conflicts around the environmental clearance, the DUAC also denied its clearance to the project. When conflicts around the Commonwealth Games mounted, the Delhi Urban Arts Commission was put under high pressure to clear projects and was even accused of upholding projects for no reason (AGRAWAL 2010: 402). As a result, the entire Commission resigned in 2008.[198]

14.2.7 Agency of the river? Drainage problem versus floods

In September 2010 – only a few weeks before the Commonwealth Games – Delhi experienced the most serious flood since 1978. The flood waters reached the top of the embankments. Thousands of people living in the floodplain were seeking shelter along the embankments (see Figure 34). A dengue fever outbreak was challenging the city and, finally, the army was called in for disaster help. Due to seepage through the embankment, high level of groundwater, and heavy rainfall the Games Village

---

[197] Letter by Arindam Ghosh, Scientist, NEERI to Editor Hitavada, Nagpur, dated 10th November 2008. File No. RDPU/PR/22.54/2008. Materials accessed via YJA.
[198] The commission was led by Charles Correa, an eminent Indian architect. The Commission was re-constituted around its new chairman K.T. Ravindran, Professor and Head of the Department of Urban Design, School of Planning and Architecture.

was waterlogged and the basements of the residential towers were submerged. Additionally, construction workers were forced to stay in the apartments, because their nearby camps were flooded (see Figure 35).

Figure 34: Tents along the Noida Link Road during flood
Source: Follmann, September 2010

Figure 35: Flooded construction worker camps next to CWGs-Village
Source: Follmann, September 2010

Figure 36: Standing water behind the Akshardham Bund
Source: Follmann, September 2010

Figure 37: Standing water in front of the Akshardham Bund
Source: Follmann, September 2010

DDA officials nevertheless retained the standpoint that the site of the Games Village was not "flooded by the river" since the Akshardham Embankment "easily" withstood the flood waters and was not overtopped (DDA-01/2010-09-21).

> "No, there was no flood in the Games Village. [That was] water logging because of the poor drainage; because the drainage didn't work. […] this happens all over Delhi. […] But this does not proof that this was a bad site from drainage point of view or from flood point of view. […] This happens even across the bund in whole of Trans-Yamuna area. You go to any colony - even good colonies - you will find water is there, because it is a very poor drainage and even the pumping etc. whatever they do it never works. If subsoil water is so high, the soil never absorbs the water" (DDA-07/2011-11-23).

It needs to be mentioned that the DDA planner stood firm on the stance that the site was not on the riverbed or floodplain (DDA-07/2011-11-23). In his eyes, the land was soaked in water due to inadequate planning and infrastructure provision, and this was not an indicator of the area being flooded. Therefore, from his perspective the Akshardham Bund demarcated the boundary between the area affected by floods and the area affected by water logging (see Figure 36 and Figure 37). In the light of the floods and the problems in advance of the Commonwealth Games, more critical voices within DDA acknowledged that the site was not appropriate (DDA-06/2011-11-22):

> "It was almost flooded [...] If you go to Commonwealth Games Village there is seepage, because it is in our riverbed area. This is a problem. How it will be solved, I can't say. I personally feel that should not have been there, because it is part of river. We should not obstruct the flow. But this is personal – anyhow."

The activists of the environmental NGOs clearly viewed the 'drowning' of the Games Village as the agency of the river, which finally proved the location to be wrong (NGO-04/2011-03-05): "I think it was some kind of justice that in September 2010 the river actually proved us right."

14.2.8 Economic (ir)rationalities: the bailout package and corruption

Towards the end of the preparation for the Commonwealth Games, DDA was increasingly under pressure from time constraints and ongoing court cases. The project developer Emaar MGF argued that the court cases had adverse impacts on the project and, thus, affected its opportunities to raise the needed funds for the completion of the project from banks and potential buyers (HLC 2011: 46).

While the courts never ordered the suspension of the construction, in February 2009 DDA realized that Emaar MGF would be unable to complete the project in time. Emaar MGF further argued that they were unable to sell the apartments due to the court case against the project (HLC 2011: 57–58). A bailout package was approved by the government agencies for a "bulk sale" of 333 apartments by Emaar MGF to DDA for about 730 crore Indian rupees (ca. 110 million Euro) in May 2009 (HLC 2011: 58). Since the flats were purchased at a higher rate the bailout resulted in a loss for DDA of about 134 to 220 crore Indian rupees (ca. 20 to 33 million Euro) (HLC 2011: 80). The popularity of the Games Village remained unabated; even if the transfer of the apartments to the buyers was further delayed by other irregularities (e.g. the residential towers had more floor area ratio (FAR) than approved). Even members of parliament and later also medalists of the Indian team requested to be given priority in the allotment of the apartments.[199]

In sum, based on the irregularities documented by the Shunglu Committee in the reports on the overall planning and execution of the Commonwealth Games, and the development of Games Village in particular, the committee concluded that

---

[199] See for example: Times of India, 14th December 2011: 'CWG medallists want priority in Games Village flat allotment'.

DDA "failed to detect contract violations" by Emaar MGD, which resulted is high costs and "significant delays" (HLC 2011: 7). In using terms like "undue favours" or "unauthorized payments" (ibid.), the report avoids terming these practices as 'corruption'. Nevertheless, Suresh Kalmadi, Chairman of the Organising Committee of the Commonwealth Games, was identified and finally convicted as corrupt and imprisoned for nine months and later granted bail. Therefore, the underlying economic (ir)rationalities of the Games Village project indicate that the site might have offered specific financial gains, since it was 'available' at almost no cost to DDA, and higher profit margins could be realized, which in turn facilitated a higher margin in corruption.

14.2.9 Ground realities compared to planning schemes

When comparing the above described different schemes for 'pocket III' (see Figure 31, page 249) with the actual realties on the ground today (see Map 5, page 390) it becomes clear that the development of the two mega-projects encroached upon the proposed green belt. A huge adverting panel erected at the main road during the construction period had indicated that further infrastructure provision (utilities) including a sewage treatment plant, a water treatment plant, a police station, a fire station, etc. were to be constructed in the earmarked green belt area (see Figure 38, page 272).

Including these developments, the earlier sanctioned 21 ha for public and semi-public uses have increased to a total about 35 ha without any further procedure of change of land. The proposed green belt was reduced to the land along the railway tracks (ca. 9 ha), which was further degraded by the dumping of solid waste and construction debris as well as the remaining of the construction workers camp (see Figure 65, page 395).

The designated hotel complex was never realized. The land has since been re-designated as 'Agriculture and Water Body' in the ZDP approved in 2010, but it remains as a parking area.

The Akshardham Complex today covers an area of about at least 28 ha[200] including later extensions and parking area, compared to 23.5 ha originally allotted to the trust and 12 ha planned to be given to the trust in the mid-1990s. The area of the green belt towards national highway (NH-24) reduced to a narrow strip of about 5 ha of land.

Overall, both mega-projects, the Akshardham Temple and the Commonwealth Games Village, display several violations of planning norms and deviations of the earlier approved schemes. Existing monitoring processes, whether within the DDA or based on the conditions of the environmental clearances issued by the MoEF, have largely failed.

---

[200] The area increases to about 32 ha if construction activities within the green belt are included.

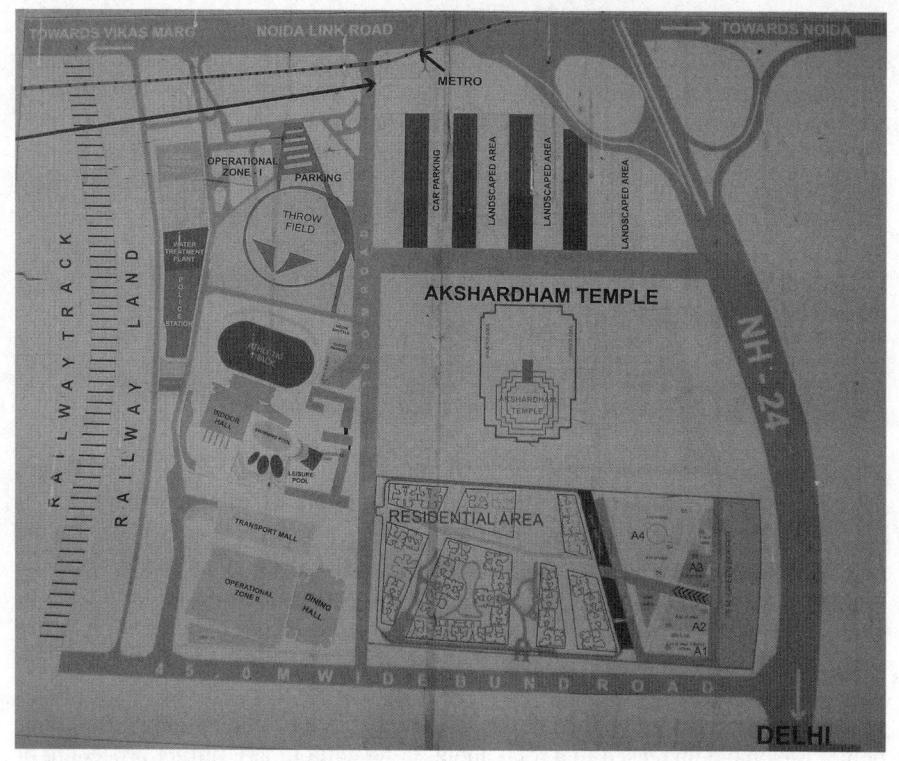

Figure 38: Signage erected during the construction period of the CWGs-Village
Source: Follmann, December 2009

14.3 'Development must take place' and 'leave the river to the experts'

In addition to the evident connections shown, the planning schemes of the DDA for pocket III were also part of the larger political debate about the Commonwealth Games Village and were still influenced by the legacy of channelization. The Chief Minister of the NCT of Delhi, Sheila Dikshit, was the key figure behind the Commonwealth Games. She continuously proclaimed she was in favor of large-scale riverfront development. When asked about the Games Village in the River Zone in an interview with the TV channel CNN IBN Live in October 2007, she stated: "Show me another city in the world, which has not developed its riverbanks. Development has to take place."[201] Other sources[202] have quoted her arguing that the city of Delhi "cannot afford of not doing anything with it [the riverfront]." The essence of Sheila Dikshit's claim "development has to take place" is certainly

---

201 CNN IBN Live, 19th October 2007: 'Commonwealth Games 2010: Delhi's death trap'. Online available: http://ibnlive.in.com/news/commonwealth-games-2010- "we cannot afford of not doing anything with it".delhis-death-trap/50715-3-1.html (2015-04-18).
202 Express India, 19th September 2007: 'Sheila game for Games Village on Yamuna bed'.

rooted in the wasted land story-line, which had gained importance, with regard to new infrastructure provisions for the 'world-class' city in the making, as outlined above (see chapter 12.3).

Furthermore, the legacy of river channelization is connected to technology-determined, expert-led decision making processes. The different views of the managing director of the Delhi Metro, E. SREEDHARAN (2009), and the former secretary of the MoWR and Indian water/river expert, RAMASWAMI R. IYER, are exemplary for the entire debate on channelization and riverfront development in Delhi in the recent decades.

In a newspaper article written by SREEDHARAN (2009) which was published in the Times of India in May 2009 – when the Games Village was still pending in the Supreme Court – he demanded channelization and riverfront development. He stressed that the river Thames was cleaned in London by the interception of sewage and channelization "which resulted in self-cleansing during high and low tides" and the development of "majestic and monumental buildings" behind the retaining walls (SREEDHARAN 2009). In referring to other Western cities, he asked: "Why cannot Delhi also learn lessons from the experience of these cities?" It is worth quoting the self-described 'technocrat' at length here as it highlights the underlying logic of why public agencies, in this case Delhi Metro have requested more and more land in the river zone over the years and blatantly violated environmental regulations by dumping massive loads of construction waste at various locations.

> "The Yamuna river has to be trained and confined to a width that is defined between abutments of the existing bridges by constructing appropriate guide bunds or retaining walls, and the large sprawling tracts of low-lying areas behind these walls utilized for high-end developments which can make the city rich, beautiful and prosperous. […] The resources needed for all these can be easily raised by exploiting the released riverbeds. In the development plan, a corridor of 300 metres should be reserved adjacent to the river bank for gardens, promenades and recreation centres. […] If development as suggested above can be undertaken, the river can be saved, Yamuna can be made clean and the river-fronts can be made the pride of the city" (ibid.)

He further denunciated the work of environmental activists and praised the 'clean' character of the urban mega-projects while also drawing on the *fear of encroachment* story-line:

> "A handful of self-styled environmentalists is stalling this idea. The result is rampant encroachments on the riverbed by jhuggis which catch fire at regular intervals every summer, often burning alive a few people. Sewage and untreated industrial waste are let into the river without treatment and nobody owns up responsibility for the same. The so-called environmentalists are vociferous against clean development schemes which are vital for the city, such as Commonwealth Games Village, Metro constructions, Akshardham Temple, etc." (ibid.).

RAMASWAMY IYER (2009b) in an open letter to SREEDHARAN wrote:

> "I share the national admiration for Dr. Sreedharan but find his recent article on the Yamuna disturbing. […] its [a] completely wrong approach to the river. Coming from a person of such eminence, the article cannot be ignored. It is necessary to deconstruct it."

RAMASWAMY IYER (2009b) then refers to SREEDHARAN's notion of 'a handful of self-styled environmentalists' obstructing the channelization and riverfront development arguing that his remarks are typical for "a downgradation of 'environmentalism', an exaltation of 'experts' over ordinary people, and a confinement of the term 'professionals' to engineers" [emphasis in original].

In contrast, IYER (2009b) claimed that "the Yamuna belongs to all of us, not merely to the experts" adding that "water has been left to the bureaucrats and technocrats for a century with consequences that are there for all to see." IYER (2009b) further aimed to deconstruct the notion of professionalism used by SREEDHARAN stating that "environmentalists, ecologists and social scientists are professionals too, like the engineers", and even if the environmentalist

"in the forefront of the popular movement [referring to Yamuna Jiye Abhiyaan] may or may not be 'professionals' in Dr. Sreedharan's sense, but there are solid professionals behind them" [emphasis in original].

The western examples cited by SREEDHARAN underline that at least some engineers and planners were of the opinion that monsoonal rivers could be controlled in the same way and their floodplains are "vacant land[s] to be played with" (NGO-04/2010-04-01). In the view of the environmentalists these statements further probed that these kind of engineers and technocrats "don't understand what rivers are" (NGO-04/2010-04-01).

In sum, while widespread opposition against channelization succeeded in preventing the DDA from channelizing the river, it could not stop the development of the Akshardham Temple complex, the Games Village and the Delhi Metro, since large-scale riverfront develop had remained "mainstream thinking" of politicians, engineers and planners and "this mind-set is not going to go away soon" (NGO-09/2011-11-12).

14.4 Interim conclusion

The political ecology of the Akshardham Bund highlights that when DDA's plans for channelization ran into wide opposition based on environmental concerns, the planning body approached NEERI to give its approval for its riverfront plans. NEERI did not sanction DDA's plan exactly, but a modified land-use concept for the area was developed (NEERI 1999). Pressurized to allot land to the trust by the Central Government and high-ranked politicians, the DDA realized that the development of Akshardham could drive forward the development of the riverfront through the 'back door'. The request of the BAPS trust to build the Akshardham Temple became the legitimization for the DDA to construct the Akshardham Bund and reclaim the land.

In summary, it can be argued that the development of a residential complex in the River Zone would not have been possible without the exceptional logic of the international mega-sports event, which pushed the project over the line. Thus, the 'legality' of the project was established not on the basis of existing environmental legislation, but rather based on the world-class vision putting enormous pressure on the courts and the MoEF (cf. FOLLMANN 2015: 220).

The state agencies were able to realize long-desired iconic flagship projects along the river based on the exceptionalist logic of urban mega-projects (cf. ALTSHULER & LUBEROFF 2003, FLYVBJERG et al. 2003, KENNEDY 2015, SWYNGEDOUW et al. 2002). While the state agencies first tried to defend the construction of the Games Village on the basis of the existing environmental framework pointing to obtained environmental clearance from the MoEF and the initial approval from the YSC, the projects were eventually cleared through an exceptionalist ruling locating it outside the 'active' floodplain of the river (cf. FOLLMANN 2015). The declaration as 'public and semi-public use' emerged as the flexible tool in the hands of the state to allow multiple uses of the land to be realized. Moreover, the terminology assured the state's own flexibility in changing the designated land use as needed at a later stage (ibid.).

The DDA, and later the Supreme Court, made use of the dichotomy of land and water by simplifying the complex ecology of the river floodplain to the existence of a man-made embankment. The hybrid land-use designation 'Agriculture and Water Body' was simplified, and therefore purified, into land which is suitable for large-scale development. This was done in stark contrast to the orders of the High Court in the Okhla/Wazirpur Case, which had earlier declared the River Zone as 'water body' rather than 'land' to evict the slums (see 13.2.2). In doing so, the Supreme Court avoided any further justification of its decision bowing to the urgency and national prestige of the project. Through the declaration and purification of the site as land, the arguments of the environmental activists became meaningless (cf. BAVISKAR 2011c: 51).

While the projects of Delhi Metro as well as the Akshardham Temple Complex and the Games Village are all closely related to the world-class city discourse, the exemptions undertaken by the state in developing these urban mega-projects as symbols of progress differed: the developments of Delhi Metro display an implementation process "well insulated from normal politics" (FAINSTEIN & FAINSTEIN 1983:248 cited in BON 2015: 224) and backed by exceptional power institutionalized in the parastatal agency of the DMRC (ensured by a specific act). The Akshardham Temple Complex and the CWG-Village needed to find their way through and around the existing institutional framework. Thus, while these two projects were cleared on the project level, the exceptionality of Delhi Metro had already been constituted on the policy level.

Furthermore, the political ecology of the mega-projects illustrates that the existing statutory provisions for land-use changes (under the Delhi Development Act) need to be critically reviewed. NEERI explicitly disqualified residential uses in the River Zone (NEERI 2005: 3.33). Nonetheless, DDA managed to get an approval of the change of land use from the MoUD. It is thus a paradox that the DDA and the MoUD could change the land use to residential in 2006 without prior environmental clearance. The non-integration of existing environmental and urban planning regulations in India is highlighted by this case. The DDA does not need environmental clearance from the MoEF in the plan-making, but rather a successive project specific environmental clearance process is the norm.

The floodplain conservation discourse, focusing on the protection of the land against development and promoted by the environmental NGOs, and Yamuna Jiye Abhiyaan in particular, started to gain wider attention with the development of the urban mega-projects in the River Zone. In some way the urban mega-projects emerged as a catalyst for protest. As mega-projects are generally closely followed by the public, the irregularities in the planning and implementation process were widely covered by the media. The legacy of the urban mega-projects with regard to the remaking of urban environmental policies in Delhi is reviewed in the final empirical chapters.

VII AFTER THE URBAN MEGA-PROJECTS: PURIFYING DELHI'S RIVERSCAPES FOR RESTORATION, CONSERVATION AND BEAUTIFICATION

15 'MANAGING DISSENT': SIGNS OF CHANGE

While the protest and the court case against the construction of the Commonwealth Games Village was unable to stop the project, this chapter outlines that it did nevertheless mark a critical point in the public discourse and policy-making process for Delhi's riverscapes. This chapter explains the changing policies for, and developments in, Delhi's riverscapes as a reaction to the protest against the Commonwealth Games Village.

With regard to the public discourse, for a long time the media was not interested in critically engaging with the state of the river beyond the aspect of pollution. However, as the large-scale construction in the river zone in the name of national prestige was sensitive to critique the media caught attention of the environmentalists' campaign (NGO-02/2011-12-09):

> "[...] the Commonwealth Games became a rhetorical place for political engagement, because it was the event of the games. It became a site for resistance. That is the reason why it became this high point of protest. [...] The language which you create for the protest has to find space in the rhetoric of the media. The Commonwealth Games event [...] invading the riverbed becomes a very strong point."

Prior to the environmental NGOs approaching the court in the matter of the Games Village in late 2007, the Prime Minister of India setup a High Powered Committee to develop a policy framework and action plan for the river Yamuna in Delhi in August 2007 (PRIME MINISTER'S OFFICE 2007). While the committee was actually requested to work out the basis for the constitution of a statutory body, referred to as the Yamuna River Development Authority (YRDA), the committee itself became known as the YRDA and a statutory body has not been constituted to date. In the following sections the High Powered Committee is equated with the YRDA always keeping in mind that it functions without an approved legal framework.

In September 2007, the Lieutenant Governor of Delhi issued a (temporary) no-development moratorium, banning any construction in the River Zone with the exception of the ongoing construction of the Commonwealth Games Village and the projects of Delhi Metro. Furthermore, a legal planning document for the River Zone was sanctioned with approval of DDA's reworked ZDP by the MoUD in 2010.

15.1 A new authority and a moratorium for the river?

The High Powered Committee, known as the YRDA, is chaired by the Lieutenant Governor and the Chief Minister of Delhi is designated as the vice-chairperson. Members of the committee are the Secretaries of the Ministries for Urban Development (MoUD), Environment (MoEF) and Water (MoWR); additionally, the Chief Secretary of the Government of NCT of Delhi, the Chief Secretary of the Delhi government's Department of Urban Development (DoUD), the Chief Executive Officer of the DJB, and the Vice-Chairman of the DDA.

The terms of reference issued by the Prime Minister's Office outline the mandate of the committee:

- To "commission studies on different aspects of the development of the river, viz. hydrology, ecology, environmental pollution, sustainable use of the river front, etc. to feed into the policy framework";
- To "develop a policy framework and prepare an integrated plan addressing issues of both quantity in terms of river flow and quality";
- To "develop an operational plan for implementation of the river action programme";
- To "effect intersectoral coordination for planning and implementation until such time a statutory arrangement is in place"; and
- To "suggest the design for a statutory framework" (PRIME MINISTER'S OFFICE 2007).

The task of the YRDA was challenging. The body was to function as part of the YRDA while proposing its statutory framework. Additionally, the YRDA had to function under time pressure and was repeatedly confronted with the ongoing court case and the developments around the Games Village. The protest against the Games Village through the Satyagraha was ongoing and supported by well-known environmentalist like Vandana Shiva at that time. Activists shouted slogans in the Delhi Assembly and blocked the Games Village construction site. Some activists were sitting on the site fasting, while others were willing to violate laws to be taken into custody to make their voices heard. When trying to enter the National Groundwater Congress convened by the Prime Minister to submit a memorandum, a group of activists were arrested by the police (NGO-15/2011-12-08). Overall, the Games Village site developed into a political issue and leaders of the opposition party (Bharatiya Janata Party) started supporting the campaign against the Games Village. The media covered the protest extensively, requesting responses from the concerned agencies and politicians.

The environmental NGOs also sent letters to the Prime Minister before requesting to relocate the Games Village.[203] Furthermore, members of the environmental NGOs under YJA made several presentations to, and had meetings with, inter alia

---

203 Letter YJA to Man Mohan Singh, Prime Minister of India, 21 March 2007. All letters send by YJA are online accessible via Peace Institute Charitable Trust: http://www.peaceinst.org /publication/book-let.html (2015-03-23).

the President of India and Minister of Sports and Youth Affairs. On the same day when the Prime Minister constituted the YRDA, members of YJA had also a meeting with the Vice-Chairman of DDA and other officials (NGO-03/2010-03-26; NGO-04/2010-04-01).

The first interim report of the YRDA from November 2007 confirms a direct link between the setting up of the YRDA and the protest of the environmental NGOs (YRDA 2007):

> "[…] in view of the increasing oppositon to the construction of the Commonwealth Games Village on the flood plains of river Yamuna in East Delhi from some envionmental groups and well known environmentalists, it was decided that the issue relating to the sustainable use of river front could be considered by the Authority after having studied in detail how the river stretch in Delhi could be revived as an eco-system."

The environmental activists viewed the setting up of the YRDA and the no-development moratorium, as an important step in the protection of the River Zone from development. In their eyes the "temporary moratorium was a victory" (NGO-15/2011-12-08). However, they were aware that the YRDA and the moratorium were set up in order to show action and keep the opposition against the Games Village quiet (NGO-04/2011-03-05):

> "They were trying to manage dissent till the games happened. The LG [Lieutenant Governor] utilized the moratorium to keep us kind of happy about it. We were very much aware of it and we are even today aware of it. We were being managed. But these are the games that politicians keep playing."

The Lieutenant Governor (LG), Tejendra Khanna, acknowledged the efforts by the environmentalists in personal communication with the author and even stated that he was "on board basically even mentally" that something needed to be done with regard to the river (LG/2011-12-05). Referring to his personal background coming from Patna, Bihar, he outlined that since he grew up "practically next to a river" he has developed an understanding of living with rivers (LG/2011-12-05).

The LG further explained that the constitution of the YRDA was the outcome of a detailed discussion between him, the Prime Minister, the Chief Minister and other ministers on the deterioration of the Yamuna in Delhi (LG/2011-12-05). He emphasized that the YRDA was not solely a response to the requests by the environmental NGOs against the Games Village, but was rather a reaction also to the various claims received by him as well as the Prime Minister "to clean the river and put it into urban uses" (LG/2011-12-05).

Interestingly, the name of the proposed statutory authority became a matter of discussion reflecting the different views on the river. Initially, "Yamuna River Regulatory Authority" and similar formulations were discussed in the meeting chaired by the Prime Minister (LG/2011-12-05) but the LG, as outlined by him, emphasized the term "development" instead of "regulation" as "regulatory may mean that you put some regulatory restrictions on uses and other things" (LG/2011-12-05). In the eyes of the LG a regulatory authority for the river would have been "very limited" since the idea was to think about how to revive the river and to rejuvenate the river and "make it again a real kind of ecosystem" (LG/2011-12-05). Yet, it was specifically the termed 'development' which produced dissonances in the ears of the environmental activists. Confronting the Lieutenant Governor with the critique of the environmental NGOs, he emphasized that maybe a different nomenclature could have been chosen, but for him "to develop means to really bring the river back to life" (LG/2011-12-05).

At the first meeting of the YRDA the environmentalists were invited to present their concerns. Furthermore, four main points were discussed:

1) Water recharge potential of the floodplain in Delhi and its optimization;
2) Minimum perennial and environmental flow;
3) The capacity for impounding river waters in Delhi or upstream; and
4) The reduction of pollution in the river (TAG-YRDA/1/2007-11-22).[204]

Thereafter, a technical advisory group (TAG) for the constitution of YRDA was set up in order to work out the technical details and prepare a report for the Prime Minister's Office. The technical advisory group consisted of members from the National Institute of Hydrology, CWC, CWPRS, CGWB, NEERI, Inland Waterways Authority of India, MoEF, CPCB, IFCD, DDA, DJB and the well-known and widely respected environmental NGO, INTACH. The TAG was requested to prepare a detailed report based on the existing data as well as conduct additional technical studies. The TAG was further asked to make recommendation regarding a more efficient water abstraction for the Yamuna (upstream storages) as well as an estimation of flows in regard to a rejuvenation of the river and the risk of flooding in the Delhi stretch.

The TAG met fourteen times between November 2007 and April 2010 (scc Appendix III). The first meeting on 22nd November 2007 started with a brain storming on the various aspects requested by the YRDA to be explored by the TAG. The minutes of the meeting indicate that the commissioner planning of DDA and the official of NEERI jointly remarked that it would be important to first of all define what is to be understood under river Yamuna in Delhi including "its riverfront, riverbed, river channel and river floodplain" (TAG-YRDA/1/2007-11-22). They further raised the question whether the Yamuna in Delhi "could be called a 'river' at all?" as it has no flow (ibid.). Therefore, the minutes of the meeting indicate that the DDA (and NEERI) were expecting that the TAG would confirm their development plans for the riverfront, while the draft ZDP-2006 had been denied approval earlier in May 2007 (see chapter 11.4). It remains unclear to what extent the three morphologies defined in the draft ZDP-2008 were already discussed within DDA at this time. Nevertheless, the request by DDA to engage in defining the boundaries of the riverbed/floodplain indicates that they were reviewed by the DDA and NEERI by then.

Other members of the committee requested to assess the economic value of the floodplain in terms of water recharge capacity drawing on the monetary valuation of the ecosystem services story-line also outlined above (see also 11.4).

---

204 See list of TAG-YRDA minutes in Appendix III.

The deliberations in the following months narrowed down to the matter of flow and pollution, while the question of riverfront development played a marginal role in the discussions (own evaluation based on the available minutes). In this context, the DJB pushed two issues to the fore: the extraction of groundwater from the river aquifer (which was promoted also by a certain wing of the environmentalists based on the above cited World Bank study (BABU et al. 2003) on ecological valuation) as well as the construction of the interceptor drains along the three major drains (Najafgarh, Supplementary, and Shahdara). Furthermore, DJB officials highlighted the need for upstream storage capacity to realize year-round minimum/environmental flow in the river requesting the construction of upstream dams (Renuka, Kishau and Lakhwar-Vyasi) as well as proposing a new barrage upstream of Delhi at the village of Palla.

The connection between water pollution and riverfront development was also discussed in the meetings, stating in early 2008 "that any development of the riverfront would not be possible with the present condition of pollution in the river".[205] The water recharge capacity of the floodplain remained a major issue of deliberations of the TAG. A study of the water-yielding potential of the Yamuna was conducted by the CGWB, which outlined that the recharge capacity of the floodplain is about 100 million cubic meters.[206] These figures remained disputed and members of the environmental NGO Natural Heritage First (NHF) in cooperation with other experts indicated a much higher recharge capacity (SONI et al. 2009). Nevertheless, the water recharge potential and its economic benefits found some support in the debate. After the initial invitation of individuals from environmental NGOs, INTACH and NHF in particularly, the NGO-representatives were not invited anymore. Nevertheless, the team of NHF, Prof. Vikram Soni and Diwan Singh, worked as advisors to the DJB in the Palla area on the estimation of the groundwater potential of the floodplain.

A second major concept gaining importance in the course of the meetings was the idea of transforming the River Zone into a biodiversity zone. In the 8th meeting of the TAG in August 2009 – only about a month after the Supreme Court had given the green light for the Games Village – the main points of the report were defined (TAG-YRDA/8/2009-08-26):

> "No concretization of the Yamuna flood plain should be permitted. [...] A Bio diversity park on the entire 22 km. stretch of the Yamuna river flowing through Delhi should be established."

Additionally, the temporary moratorium was reassured by the LG to remain in effect even after the Supreme Court had allowed the construction of the Games Village. The LG further gave a clear statement against channelization (TAG-YRDA/9/2009-11-10):

---

205 Report Subgroup TAG-YRDA, 2008 (no date), attachment to letter Principal Secretary, Department of Urban Development, GNCTD, 18th February 2008. Materials accessed through RTI filed by YJA in 2010.
206 Ibid.

"Non-Channelization of the river needs to be highlighted. The considered opinion of YRDA experts is not to go for such channelization of the River Yamuna. Accordingly, DDA Zonal Plan for Zone O (for the river bed) has been finalized. Keeping the Yamuna bed as a natural Bio-diversity area."

This is remarkable, since materials brought into the discussion by the Delhi Government still argued for riverfront development and channelization in 2008:

"Besides removal of habitation within the bed of river to prevent faecal coliform, DDA has to take up the river front development work which includes channelization of River [sic] to maintain proper flow and development of banks, creation of water body and beautification."[207]

The above quote indicates that the questions of flow, pollution and land use were all related to channelization; and channelization was still viewed by some agencies as the appropriate development strategy able to 'cure' all the problems of the river. Thus, the rejection of channelization by the YRDA must certainly be interpreted as a major shift in the overall discursive framing of Delhi's riverscapes.

Closely related to this shift, the Landscape Unit of the DDA was requested by the technical advisory group to prepare a "landscaping plan" for certain areas of the River Zone (TAG-YRDA/9/2009-11-10). Later this initial request combined with the biodiversity idea evolved into the biodiversity zone concept (for details see chapter 15.2). Furthermore, the TAG discussed amendments to Interstate Water Dispute Act 1956, incorporating a special status to NCT of Delhi. It is remarkable that these interstate aspects were discussed in the meetings without involving the neighboring states in any regard.

The report for the Prime Minister's Office was finalized and submitted in September 2010. It by in large listed projects already proposed earlier by the DJB and DDA. The main "policy prescription" of the report included:

- The construction of upstream storages to ensure year-round flow;
- The construction of the interceptor project along Najafgarh, Supplementary, and Shahdara Drain by DJB;
- A new public awareness campaign along the lines of Yamuna Action Plan; and
- The development of "a biodiversity zone for conserving the natural heritage" of the River Zone by DDA (DoUD 2010: 60–76).

With regard to the statutory framework of the YRDA the report suggests three options for implementation (DoUD 2010: 92–93):

- 'Option I' proposed a "Yamuna River Board" under the provision of the River Boards Act and a second-tier authority established by the GNCTD, which should have no powers over land-use planning. Land-use planning would therefore remain under DDA and the provisions of the Delhi Development Act.
- 'Option II' outlined that the Parliament should enact a "Yamuna River Basin (Development and Management) Authority" (requiring a new act), which should also have powers over land-use planning. This option followed the

207 Ibid.

model of the National Ganga River Basin Authority (NGRBA), which was constituted by the MoEF in February 2009 under the Environment Protection Act.
– 'Option III' proposed the creation of the "National Yamuna Basin (Development and Management) Authority" with powers to regulate minimum water flow, pollution control, master planning and land-use planning.

Each option proposed a second-tier authority on the state level, which would implement the recommendations and guidelines of the national authority. All interview partners emphasized that statutory backing for the YRDA is needed to give powers to the authority and make it accountable. For example, a DDA planner explained that while "some committees have legal back-ups [...] most of the committees don't have legal back-up and they become a kind of non-governmental organizations" (DDA-07/2011-11-23). He added that the YRDA needs to be "translated into a legal framework" to give it a clear "mandate" (DDA-07/2011-11-23). In a similar way, a CPCB official highlighted that the YRDA should have "legal teeth" (CPCB/2011-11-16). The environmental NGOs largely agreed to this view, while rejecting the specific recommendation made in the report.

As the report and its recommendations were not made public, no public debate on the recommendation of the report occurred. It further needs to be mentioned that the environmental NGOs, which attended the first meeting of the TAG, were not involved in the preparation of the report. After they had accessed the report via RTI in early 2012, they heavily criticized the proposed policy prescriptions. Moreover, the activists were astonished of the inaction by the Prime Minister and the LG. It is noteworthy that the YRDA and the TAG did not meet anymore after the report was finalized and sent to the Prime Minister's Office.

Discussing the matters with the LG in December 2011 – more than one year after the report was forwarded by the YRDA – he indicated that he was still waiting for any response from the Prime Minister's Office (LG/2011-12-05). He seemed surprised that the report had not been published by the Prime Minister's Office and declined any responsibility for inaction. In his eyes, the YRDA – under his leadership – had done a good job, and any future action needed to be initiated by the Prime Minister, since he had requested the report in order to set up a statutory body (LG/2011-12-05).

Several interview partners highlighted that they view the YRDA as yet another government committee, which often come and go without having any major impact on policy-making. The impact of these committees is also highly depending on the individual capacity and commitment of a leading figure, in this case the LG (NGO-04/2011-03-05):

> "[The YRDA] is a good first step, but all these steps are very individual oriented. As long as you have a right thinking or an open-minded chief executer - say LG at this point of time - these things happen. Otherwise a lot of these committees and a lot of these authorities are just paper tigers. They never meet, they never do anything – you may cry [...] that they are useless, they are doing nothing, but it does not make a difference to them."

In sum, since the Central Government – and the Prime Minister in particular – remained reluctant in initiating any further action, the inter-sectoral committee, which had convened several meetings and started a dialogue, remained only a short interlude of inter-sectoral knowledge exchange on the river Yamuna. Moreover, the policy recommendation of the final report suggests that each government authority remained focused on its sectoral tasks, seemingly unwilling to start intensified collaboration. The vague recommendations for a statutory framework further highlight that also within the TAG a common decision was not reached. Since the recommendation of the report focused to a great extent on interstate concerns without even involving the neighboring states, the initiative of the YRDA was doomed to fail.

Furthermore, running parallel to the efforts of the YRDA, the DDA proceeded with the finalization of the ZDP. As the discussion of the approved ZDP outlines below, the proceedings under YRDA had by and large no influence on the development guidelines for the River Zone.

15.2 A plan for the river

In March 2010, about 12 years after the first draft was circulated, the MoUD approved the ZDP for the River Zone prepared by the DDA.[208] With the public pressure remaining high and the Commonwealth Games coming up in fall 2010, the ZDP post-facto created the legal basis for the construction of the Commonwealth Games Village and the projects by Delhi Metro. At the same time, the ZDP aimed to signalize to the general public that the 'rest' of the floodplain will be preserved through eco-friendly development by the DDA. The final approval of the ZDP is thus to be interpreted as a direct outcome of the protest against the large-scale development projects in the previous years.

In contrast to the earlier discussed drafts of the ZDP and in addition to the earlier studies for channelization, the word 'channelization' was removed from the approved ZDP based on the MPD-2021 recommendations. Table 10 gives an overview of the existing and proposed land uses in the River Zone (see land-use plan in Appendix IV). The approved residential uses in the River Zone include the Games Village and the Madanpur Khadar Resettlement Colony. The unauthorized colonies and urban villages are not earmarked in the land-use plan and shown either as 'River and Water Body' (e.g. Kalindi Kunj area in south Delhi and Wazirabad/Jagatpur in north Delhi) or as 'Recreational' (e.g. Usmanpur Village, Old Garhi Mendu Village, Yamuna Bazaar). As indicated above, the DDA intends on changing the land use from 'River & Water Body' to 'Residential' for the regularized unauthorized colonies (see 13.3). The Yamuna Bazaar area, which is shown as a District Park in MPD-2021, is still designated as 'Recreational' and only public and semi-public facilities have been approved, while it remains a dense residential area.

---

208 Approval MoUD vide letter No. K-12011/23/2009-DDIB to DDA dated the 8th March, 2010.

Land use	existing ha	existing %	proposed ha	proposed %	remarks / restrictions
Residential*	62.2 980.0	0.6 10.1	62.2 (980)	0.6 (10.1)	No additional areas other than existing/ earmarked (unauthorized colonies will be dealt with as per policy decision GNCTD)
Commercial	39.5	0.4	39.5	0.4	No additional areas other than existing/ earmarked. Includes existing IT Park (6.0 ha), Bottling Plant (28.0 ha), at Madanpur Khadar, Commercial/Hotel (5.5 ha) site at CWGs Village
Industrial (fly ash brick plant)	34.0	0.4	34.0	0.4	No additional areas other than existing/ earmarked
Recreational (green)	528.4	5.2	2045.0	21.1	Proposed recreational uses will be considered as green use zone in which green stretches, bio-diversity park, forest, botanical park/ herbal park, science park, theme park, etc. will be permitted without any pucca/ permanent construction
Transportation	345.7	3.6	582.9	6.0	-
Utilities	166.0	1.7	172.7	1.8	-
Government	1.8	0.02	1.8	0.02	No additional areas other than existing/ earmarked
Public and Semi-Public (PSP)	179.8	1.9	181.7	1.9	No additional areas other than existing/ earmarked
River & Water Body (including agriculture)	7362.6	75.9	6591.1	67.8	The area of "River & Water Body" may decrease by about 980 ha after the regularization and subsequent change of land use of the unauthorized colonies
TOTAL	9.700	100	9.700	100	

Table 10: Existing and proposed land use for River Zone (Zone O)
Source: own compilation based on DDA (2010b: 3, 24)
Remarks: *existing unauthorized colonies: ca. 980 ha

The loosely termed, and flexibly used, category of 'Public and Semi-Public" has been limited to the already existing uses including the Akshardham Temple Complex and the training facilities of the Games Village. With regard to commercial uses, only the Metro IT Park, the bottling plant at Madanpur Khadar (which is supposed to be shifted) and a petrol/CNG station have been approved. Besides the three fly-ash brick plants no further industrial (manufacturing) activity will be allowed in the River Zone.

The approved ZDP further stipulates that after the coal-based power plants are shifted or converted to gas-based power plants, the area of the fly-ash brick plants "will be developed for green/recreational areas" (DDA 2010b: 24). The land-use plan indicates that the redundant power plant (Raj Ghat Power Station) and the fly-ash ponds are deemed to become 'Recreational' (see e.g. today's site of the DTC Millennium Bus Depot) or remain as 'River & Water Body'. Clear policy recommendations (with timelines) are missing in the ZDP regarding these developments. Overall, the category 'River & Water Body' (white) emerges as a residual category and covers the agricultural uses. The ZDP does not specify any regulation for agriculture in the River Zone; yet suggests "organic and biological farming" to curb the pollution from pesticides (DDA 2010b: 18).

The ZDP stipulates the importance for increased accessibility to the river and proposes the development of a pedestrian promenade (in line with the existing ghats), the setting up of benches, and ornamental street lighting in order "to create a new opportunity for citizens of the community to exercise and engage in active recreation" (DDA 2010b: 25). Development priority is given to the river segment between Wazirabad and Nizamuddin Railway Bridge. Yet, certain phrasings indicate that the DDA did not entirely abandon the idea of developing certain pockets in the River Zone. For example, the ZDP stipulates that "[c]ertain highland pockets can be identified in the river basin for any kind of proposed development" and that "[s]pecial zoning controls" should enable a "regulated three-dimensional comprehensive development" (DDA 2010b: 26). These phrases are rooted in the earlier drafts for the ZDP and reflect the channelization discourse. The controversial floodplain zoning prepared by DDA and NEERI also remains the basis for the ZDP despite its incorrectness.

In sum, while channelization has been abandoned, the approved ZDP by and large remains a 'development' plan rather than a 'restoration' or even 'conservation' plan. Riverfront development remains on the agenda of the DDA. By holding on to the three morphologies (riverbed, river floodplain, riverfront) the DDA apparently restricted its development plans to areas protected from the river by embankments (riverfront). The approved ZDP outlines the following "strategies for riverfront development" (DDA 2010b: 16):

- "[E]stablish riverfront walkway, trails, parks;
- Create visually pleasing order to the river's edge;
- Attract people and investment to the riverfront;
- Develop an arts/entertainment/cultural district;
- Expand leisure and recreational use of the river and riverfront;
- Emphasize pedestrian streets that connects to the riverfront;
- Provide outdoor activities for the people; and
- Provide opportunities for boat launching and storage".

The rationality of this kind of riverfront development was further specified (DDA 2010b: 16):

> "Reclaiming the riverfront ensures not only its environmental integrity but also stimulates the economy. The aim is to achieve maximum economic benefit with minimum public investment to ensure a self-sustaining set up. Thus, the riverfront development will be based on financial and environmental catalysts. The revenue generated by the various activities will allow the riverfront to further develop parks and trails, commercial sites and infrastructure improvement

within the area for an economically and environmentally healthier riverfront. This healthier vision is critical, as it is the foundation for new, long-term economic growth in the city and region."

The biodiversity concept, already discussed by the technical advisory group of the YRDA in 2009, did not find its way into the approved ZDP. This is remarkable, since apart from proclaiming the 'no-development zone', the LG had requested the DDA to await the recommendations of the YRDA before finalizing the ZDP. However, the DDA finalized the ZDP and the MoUD approved it even before the YRDA report was submitted to the Prime Minister's Office. This indicates that the deliberations of the YRDA and the approval of the ZDP were two disconnected but overlapping policy-making processes. While the ZDP has legal backing based on the Delhi Development Act, the recommendations of the YRDA remain without any 'legal teeth'.

15.3 Interim conclusion

The new language of the ZDP-2010, with an emphasis on the aesthetic and recreational value of riverfront development, reflects the citywide clean/green discourse. In this context, *bringing back biodiversity* emerges as the major new story-line used by the DDA and other state agencies to promote a 'new' vision for Delhi's riverscapes. Recreation and biodiversity have become the new reference points for riverfront development replacing the idea of channelization from the policy documents. The rejection of any further large-scale construction work in the River Zone emerged, apparently as a direct outcome of the public outcry against the Games Village, particularly the protest from the environmental NGOs. For the first time, the YRDA meetings created an inter-sectoral platform for deliberations on the River Zone beyond the courts. As the YRDA meetings came to an end and further actions were not taken, the opportunity for ongoing coordinated action with regard to land-use and pollution policies was not seized by the involved agencies; they rather have resumed the implementation of their sectoral approaches.

The deliberations in the YRDA meetings additionally indicate that the pervasive water shortage discourse, which was earlier discussed with regard to the groundwater recharge capacity of the floodplain (see section 11.5), also influenced the YRDA policy prescriptions drawing on another well-known story-line in the Indian context. In line with other authors (BAGHEL 2014: 109, NÜSSER & BAGHEL 2010: 238) this discursive framing is labelled here as the *wasted water* story-line, which goes along with the "'denaturing' of rivers into moving water bodies" and "the recurring complaint of 'water running waste to the sea'" as it is not used by humans (NÜSSER & BAGHEL 2010: 238).[209]

In the context of the reduced flow in the Yamuna, the wasted water story-line is recurrently used by the state agencies in order to request the construction of large-scale upstream dams (e.g. Renuka Dam). The riparian states have used the non-existence of upstream storages as an excuse for not realizing a minimum flow in the river, as the seasonal allocation of surface Yamuna water MoU from 1994 were agreed upon by the riparian states only as interim seasonal allocation in the light of the pending construction of upstream storages. Besides the implantation of the interceptor project, the construction of upstream dams was therefore a major interest of the DJB in the course of the deliberations of the Technical Advisory Group of the YRDA (DJB/2011-11-21). The DJB and the Delhi Government aimed to use the YRDA in order to push forward their request for more water for Delhi. The final report states (DoUD 2010: 61):

> "Since a substantial part of the monsoon flows are not utilizable, all the basin states have to suffer the shortage of Yamuna water during the non monsoon season in absence of upstream storage."

The Government of Delhi, through the DJB, tried to legitimize large-scale dams by superficially linking them to the ecological needs of the river, in addition to demanding drinking water for Delhi.

This story-line is also found in the approved ZDP-2010 which requests the "construction of storage reservoir upstream to collect the monsoon excess flow" (DDA 2010b: 14). The story-line was also earlier deployed to establish reasons for channelization proposing the dams as flood protection measures. Thus, while new story-lines (e.g. bringing back biodiversity) arose, the deeply rooted logic of the pro-channelization era following the discursive logics of wasted land and wasted water remained influential in setting direction for policy-making also after the Commonwealth Games.

---

209 Supreme Court of India (2000), Majority Opinion, B.N. Kirpal and A.S. Anand, Narmada Bachao Andolan vs Union of India and others, Supreme Court of India, Docket no. 10 SCC 664. "The quantity [of water] going waste to the sea without doing irrigation or generating power should be kept to the un-avoidable minimum"[209] (quoted also in BAGHEL 2014: 64)

16 REDEFINING THE RIVER'S BOUNDARIES

The DDA aimed to develop Delhi's riverscapes into a biodiversity zone based on the recommendations of the YRDA and the approved ZDP. However, the implementation of the biodiversity concept was challenged by court cases before the newly established National Green Tribunal (NGT). The courts again emerged as an important platform in the policy-making process in their response to PILs filed by environmental activists against the dumping of solid waste in the River Zone and the covering of storm water drains across the city. The NGT rejected the biodiversity concept of the DDA and initiated a court-driven policy-making process for Delhi's riverscapes. The following sections review the policy-making process for the biodiversity zone. The analysis highlights the changing interactions between environmental NGOs, the judiciary, and the involved state agencies.

16.1 The model: the Yamuna Biodiversity Park

The biodiversity zone concept of the DDA evolved out of the Yamuna Biodiversity Park, a joined project developed by the DDA in collaboration with the Centre for the Management of Degraded Ecosystems (CEMDE) from the University of Delhi. CEMDE was established in 2001 and is part of the School of Environmental Studies (SES), University of Delhi. CEMDE is a Centre of Excellence under the MoEF involved in strengthening awareness, research and training in regard to the environmental management of degraded ecosystems. CEMDE operates the Yamuna Biodiversity Park at Wazirabad in collaboration with the DDA. DDA and CEMDE further established the Aravalli Biodiversity Park on the Ridge in south Delhi.

The Yamuna Biodiversity Park was the brainchild of Professor C.R. Babu in the late 1990s, who was also involved in the valuation study of the eco-system services of the floodplain discussed above (BABU et al. 2003). As outlined by the interview partners, the DDA and CEMDE choose a site upstream of the Wazirabad Barrage to develop the Biodiversity Park due to the high level of pollution in the river south of the barrage (CEMDE-01/2011-03-01, CEMDE-02/2011-03-09). The DDA managed to acquire private, degraded farmland (salinized soil) for the project near Jagatpur Khadar Village. Aside from the initial funding for the park, the DDA bears the operational costs and therefore the financial aspects are fully controlled by DDA.

Work on the Yamuna Biodiversity Park started with phase I in 2002 and included the creation of wetlands, grasslands and forests areas. Phase I of the park covers 63.5 ha and is separated from the river by the Jagatpur Bund. With the development of phase II (121.5 ha), stretching across the embankment into the open floodplain, the park has been extended towards the river since 2009. The park is gated and not open for the general public. CEMDE offers guided tours and about 10,000 people visited the park annually (CEMDE-01/2011-03-01). Besides the 'renaturated' wetlands, grasslands and forests areas, the features of the park consist of a nature interpretation center, an herbal garden, a field nursery, a butterfly conservatory and an amphitheater (see Figure 66–Figure 69, page 395).

The collaboration between CEMDE and the DDA was not conflict-free as involved scientists and planners admitted during interviews (CEMDE-02/2011-03-09, DDA-08/2011-12-07). The concepts of the natural scientists for the restoration of wetlands and forests etc. were difficult to be matched with the landscape designs drawn by the DDA Landscape Unit. Moreover, the scientist of CEMDE work on a temporary project base running the park for the DDA. The involved scientists of CEMDE aim for the creation of a "biodiversity unit" within the DDA to manage the biodiversity parks in Delhi. They fear that the biodiversity park otherwise "will be destroyed", since the "DDA does not have a scientific man-power and the university cannot run thousands of hectares" (CEMDE-02/2011-03-09). Yet, the biodiversity unit has not been established at the time of this writing.

To govern the Biodiversity Parks in Delhi the DDA has setup the "Delhi Biodiversity Foundation", which is dominated by DDA-members, scientists from CEMDE and one delegate each from the Delhi Government and the MoUD. With the modification of the MPD-2021 in 2014 the DDA has outlined the development norms for biodiversity parks permitting a maximum built up area of 0.5% in the parks (absolute maximum 10,000 square meters) (DDA 2015: 9.9). The guidelines further permit orchards, butterfly parks, water harvesting structures, open air-theatres, food courts, scientific laboratories, an interpretation center, administrative offices, camping sites and toilet blocks, among others (DDA 2015: 9-9).

16.2 Scaling up: from the Biodiversity Park to the Biodiversity Zone

The DDA issued a public notice in April 2010 setting out to transform the whole River Zone into a biodiversity zone based on the initial project of the Biodiversity Park in north Delhi (DDA 2010a):

> "The river front development provides a unique opportunity for developing a **strong city image** along with providing an opportunity to scientifically design systems for **recharging ground water** from the **only large natural resource of the city**. An environmentally conscious approach for **integration of the river into the urban fabric** development has been devised. There has been a [sic] appropriate consideration of the natural potential of the land for developing into a biodiversity zone for **conserving the natural heritage** of the river basin as well as the local and transient requirements of facilities at the city level, like large level city greens of varying nature along with some recreational facilities. [...] the total area of 9700 hectare on the banks of the river Yamuna [...] **is proposed to be developed as a biodiversity park**" [emphasis added].

The structure plan (see Appendix VII, page 383) proposed a three-part zoning concept including core biodiversity areas ("protective biodiversity zones"), buffer zones ("interactive biodiversity zones") and the transitional zone ("public recreational zone") (DDA 2010a). So-called "green linkages" and a riverfront walk were planned to connect these different "bio-diverse pockets" (ibid.).

The information from interviews and the accessed materials prepared by the DDA Landscape Unit highlight the limitations inherent to the way the biodiversity concept developed by the DDA. First of all, the Landscape Unit had very limited resources for an inventory of the existing land uses in the River Zone. In the study period, the DDA Landscape Unit consisted of a regular staff of four to five landscape architects. They were supported by junior architects and landscape planners hired by the DDA on temporary basis for the study. The temporary staff consisted of recently graduated architects and planners from the SPA Delhi (DDA-04/2011-11-18).

The exercise of the Landscape Unit to a great extent depended on the materials provided by the Planning Unit (especially the ZDP) and available satellite images (especially from Google Earth). Accessed materials, however, highlight that the assessment was flawed by inaccuracies and for many stretches the actual ground realities were unknown. The Landscape Unit was not provided with sufficient funding to conduct a detailed field survey. Furthermore, other government agencies withheld available and required high resolution spatial data due to security issues (DDA-04/2011-11-18).

By matching the ZDP regulations with the information from Google Earth images, the landscape architects of the DDA prepared the zoning concept and landscape design studies. A landscape architect during the discussion highlighted the problems faced by them (DDA-03/2011-03-03):

> "This is the O-Zone-Plan of the Planning Department [...] Here we have little patches of blues, reds and whatever. And you have a whole white area. Ask any of the planning people, what white means. They have no answer. It is watershed, but you can't just leave such a large chunk of land as watershed. You need to define it for some kind of land use. It is neither green, it is neither yellow. It is nothing. It is white. What does white mean? [...] As far as I am concerned, white means a question mark, because it is free for the public to come and [...] pick it up."

As the above quote indicates, the fear of encroachment story-line echoed in many of the statements made by the landscape planners during the interviews. The planners were unsatisfied with the provided financial resources, the provisions by the Planning Unit and the strict time constraints imposed by the YRDA. When analyzing the work of the Landscape Unit, these briefly mentioned limitations faced by the DDA staff need to be taken into account.

The landscape architects prepared a booklet of maps and colorful design studies which articulated the desires of beautified recreational spaces along a clean river (see Appendix VIII, page 383). They produced a powerful urban imagery which was also highlighted by the media. For example, the Times of India celebrated DDA's plans headlining "Holiday by the Yamuna in 2015" and the positive article stated:

> "DDA's much-needed Yamuna Riverfront Development plan is finally off the ground [...] If administrative lethargy doesn't set in, the banks of Yamuna will be full of theme parks, promenades, amphitheatres, wellness centres and biodiversity zones for ecological study, camping

and bird-watching, giving Delhiites a refreshing break without really having to go far and away."[210]

The Times of India narrowed down the multiple challenges associated with the implementation of the concept to the question of 'administrative lethargy'.[211] Furthermore, the Times of India article clearly expresses the middle-class desires, which the recreational zones of the proposed concept aimed to satisfy. However, the challenges faced by the scheme were much greater. With regard to inter-sectoral coordination the DDA was depending on the DJB to 'clean' the river before both biodiversity parks and recreational areas could be developed. However, the DDA initiated the development of a series of large-scale recreational projects along the river before any new pollution abatement measures were installed by the DJB. Solely focused on the availability of land the DDA selected different sites along the river to be developed in the first phase of the biodiversity zone concept (DDA-03/2011-03-03).

The internal DDA materials accessed suggest that the planning for the biodiversity zone was not based on physical characteristics and actual requirements of rejuvenation of the river and its banks, but rather determined by the land availability and the possession of land with DDA. A land ownership map prepared by the Land Management Department of DDA in 2010/2011 distinguished five types of land (legend of the map) in the River Zone:

1) 'DDA Land – Possession with DDA';
2) 'DDA Land – Possession not with DDA';
3) 'Land under Possession with other Govt. Agencies';
4) 'Habitation'; and
5) 'U.P. Conflict'.[212]

Without going into the details of the land ownership status it needs to be mentioned that DDA concentrated on developing only the land under category 1) 'DDA Land – Possession with DDA', which comprises about 1,450 ha (compared to 9,700 ha in total for Zone O).[213] These included three zones on the western bank including the Kudsia Ghat, an area north of the DND-Flyway along the Ring Road, and the area between the samādhis and the river south of the Old Railway Bridge. On the east bank, the DDA proposed to transform the segment between the DND-Flyway and the Nizamuddin Road Bridge (NH 24) into a biodiversity park. The proposed features of the parks included butterfly parks, herbal gardens, camping sites, water ponds, theme parks, play grounds, orchards, jetties, etc. The four zones in total covered an area of more than 1,500 ha and the execution of the work was envisaged in

---

210 Times of India, 4th October 2011: 'Holiday by the Yamuna in 2015', by Neha Pushkarna.
211 Ibid.
212 Land Ownership Map for Zone O, prepared by DDA Land Management Department of DDA 2010/2011. Material accessed via interview partners.
213 Ibid. See also National Green Tribunal, 3rd May 2013, Manoj Misra vs. Union of India and others. Original Application 6 of 2012. Quoted from the original application.

2011 to be completed by December 2016.[214] By the end of this study, development work had only started for the so-called 'Golden Jubilee Park' south of the Old Railway Bridge.

16.3 Actions on the ground: the Golden Jubilee Park

The idea of the Golden Jubilee Park (earlier referred to as 'Pushta Park') dated back to the demolition drive period which occurred under Jagmohan (see section 13.2.5). The park was renamed in the context of DDA's 50 year anniversary in 2007. Stretching south from the Old Railway Bridge, the Golden Jubilee Park today covers parts of the areas which were densely populated until the demolitions of 2005 (see Map 6, page 391). The park project is located in the Bela Estate (Nazul lands) stretching from Yamuna Bazaar to Raj Ghat covering more than 400 ha of land.

As per information of DDA Land Management Department about 320 hectares (802 acres) are with DDA, out of which more than 165 hectares are 'vacant' land in the eyes of DDA, but remain to a large extent in possession of "unauthorized cultivators".[215] The rest of the land (about 80 ha) is with other government agencies including MCD, PWD, Indian Railways and L&DO. Other sources reveal that about 125 ha were given to the Delhi Peasants under the previously described lease agreement between the DIT and the farmers' society (see chapter 13.1.1).[216]

The development of the first phase of the park started in 2005/2006 aiming to establish an "amphitheatre, arrival plazas, information centre, exhibition spaces, food courts, children's play area, maintained greens, pedestrian walkways, cycling tracks etc." (DDA 2006a: 50). As per the terminologies used by the DDA in those days, these recreational uses were deemed as the "Active Recreational Zone" and were complemented by a "Passive Recreational Zone" including "a number of water bodies with pedestrian trails and cycle tracks meandering through the site" (ibid. 50–51). The development of the park at this site was outlined in the draft ZDP-2006 reserving 83 ha land for recreational uses west of Vijay Ghat (DDA 2006d: 19). Newer planning materials outline that the development concept for the Golden Jubilee Park includes an entrance complex, a food plaza, ghats with jetties, an amphitheater, a mythological park, and an access road with side parking. Phase I of the park covers about 83 ha and is supposed to be enlarged by another 14 ha after the removal of the batching plant operated by CPWD for the construction of the Geeta Colony Bridge and the By-Pass Road. Restricted entrance and the selling of entrance tickets is planned.[217]

In light of the planning history for this central stretch described earlier (see chapter 13.2.5), the Golden Jubilee Park is inevitably connected to the earlier ideas

---

214 DDA Landscape Unit, Presentation at the 6th Biodiversity Foundation Meeting. 8th September, 2011. Internal materials accessed by the author via interview partners. Information confirmed during interview (DDA-04/2011-11-18, DDA-08/2011-12-07).
215 Internal materials DDA, accessed via interview partners.
216 Internal materials DDA, accessed via interview partners.
217 Internal materials DDA, accessed via interview partners.

of Jagmohan to develop a riverfront promenade and tourist destination exactly at this location. The development of the park was only later reframed as part of the biodiversity zone concept based on the decisions of the YRDA and the directions issued by the LG. After the demolition of the slums, the area was under police control to prevent "encroachment" (personal communication with police officer on site, February 2010). During the first field work phase (2009/2010) the author faced difficulties in accessing the area due to police patrolling. During the later fieldworks access to the site was not restricted but remained difficult due to the road construction and traffic situation.

As the Golden Jubilee Park was planned on land which was, and to some extent still is, used for agricultural purposes, the park development threatens the livelihood of the farmers from Bela Estate. During the slum demolition drive these farmers were also partly displaced. Some of them came back after the evictions and repossessed parts of the barren land towards the river (FARM-02/2011-12-04, FARM-03/2011-12-04). The interviewed farmers outlined that they are living in a constant fear of eviction and demolition since DDA destroyed their fields and huts when the establishment of the park started (FARM-04/2011-12-04). DDA planners confirmed that the Land Management Unit of DDA is clearing the area from encroachments. The author documented demolished and abandoned huts in the area.

The farmers of Bela Estate belong to the caste of the *Mallaah* (FARM-02/2011-12-04), which is the traditional boatmen caste of the people living along the banks of the Yamuna and the Ganga in Delhi, Uttar Pradesh and Bihar. They are listed as either a 'Scheduled Caste' or an 'Other Backward Class'. They outlined that their families have been living on the land for many generations and they were granted the lease of the land under the Delhi Peasants. During interviews, individual farmers indicated that they were willing to shift, if land is provided to them. They made several appeals to the LG in this matter, but a meeting was not granted to them (FARM-02/2011-12-04).

When the DDA had developed a lake, lawns, and walkways, and planted a series of trees by late 2009, access was blocked to the park because of massive construction by CPWD for the Ring Road By-Pass Road from Salimgarh Fort to Raj Ghat (see Figure 39, Figure 40 and Map 6, page 391). While the original design of the park and the alignment of the new road were different, DDA had to adjust the park design to the new road. Furthermore, due to the blocked access, the DDA failed to maintain the park. The DDA planners involved complained that CPWD "just went ahead with what they were doing" without coordinating their work with the DDA (DDA-04/2011-11-18) and, in turn, "we [DDA] didn't have access to our site. So the whole focus was shifted" (DDA-03/2011-03-03).

Figure 39: Construction of Ring Road By-Pass Road near Old Railway Bridge
Source: Follmann, December 2009

Figure 40: Ring Road By-Pass Road at Geeta Colony Bridge intersection
Source: Follmann, March 2013

While the development of the by-pass road was part of the Commonwealth Games infrastructure development and was given a fast-tracked approval, the development of the park was brought to a standstill and the farmers of Bela Estate repossessed parts of the land again (see Figure 64, page 395). Furthermore, the park had been completely inundated during monsoonal floods in September 2010 which covered the lawns and flower beds with a layer of silt about 20–30 cm thick. Only after the completion of the road in 2010 did development of the park resumed in early 2011 (see Figure 70 and Figure 71, page 396). In 2011, the DDA had aimed to complete construction works for the park by the end of 2012 but failed to do so (DDA-08/2011-12-07).

Since concretization in the River Zone was banned under the moratorium of the Lieutenant Governor, the main structures were designed to be made of bamboo. The development of the bamboo structures was supported by the National Bamboo Mission under the Ministry of Agriculture and the cost of the structures were estimated with 3.5 to 4 crore Indian rupees (more than 500,000 Euro).[218] By cut and fill earth works the DDA aimed to make the basic infrastructure safe for 25-year flood events. The DDA Engineering Unit provided the engineering feasibility for the whole project and the landscape design was made by the DDA Landscape Unit based on the land availability taken care of by the DDA Land Management Unit (DDA-08/2011-12-07).

CEMDE played no role in the development of the Golden Jubilee Park. CEMDE's inputs, as stated by DDA, were limited to "the pure biodiversity zones" (DDA-08/2011-12-07). The scientists of CEMDE confirmed that they were not consulted for the development of the Golden Jubilee Park and unmistakably opposed the designs made by the DDA (CEMDE-02/2011-03-09, CEMDE-03/2011-12-10).

The Golden Jubilee Park was also widely opposed by the environmental NGOs. YJA especially submitted several appeals to the Lieutenant Governor to stop the

---

218 Minutes of the 6th Biodiversity Foundation Meeting. 8th September, 2011.

development of the park arguing that the site belongs to the river, as shown by the floods of 2010.[219] Furthermore, the environmentalists critiqued that the park "does not provide any habitats" but rather "consumes resources – water, man power, fertilizers" and it would be a wiser approach to leave the site to nature providing "much more ecosystem services" (NGO-11/2011-11-18).

> "They [DDA] are using this word [biodiversity] to create public parks, because they don't know anything other than public parks. [...] They are creating those lawns, and some landscape area, and a few trees here and there. But it is nonsense. How can you maintain those lawns without all the poison that you have to put?" (NGO-04/2011-11-07).

However, due to its green labelling and promotion, it emerged to be very difficult for environmentalists as well as the farmers to criticize developments under the biodiversity zone concept. Outlining the powerful discursive framing of the biodiversity concept, an involved activist succinctly summarized the discursive power (NGO-02/2011-12-09):

> "If you take the whole riverbed as a park [...] who will say this is a bad idea? Who will be on your side? You have to create the dialogue. For creating the dialogue you need to have a more intelligent, more sustained intervention."

Spaces of dialogue were not offered by the state. The state agencies hardly coordinated their actions among themselves after the last meeting of the YRDA in April 2010. Dialogue between the different state agencies was again reduced to a minimum; a dialogue with the environmental NGOs and the farmers on an action plan for Delhi's riverscapes was non-existent.

The main forum with regard to the biodiversity zone was, and still is, the Delhi Biodiversity Foundation functioning under the aegis of the DDA. Public comments on the proposed developments, as outlined above were requested by DDA in April 2010. While the author was unable to review the original comments made by the public, the summary of the comments provided by DDA outlined that several suggestions were made with regard to the legal basis of the River Zone and the planned development. These included requests to designate the whole area as a city forest (disagreed by DDA in tabular summary) or declare the River Zone as a protected area under the Wildlife Protection Act (agreed by DDA) as well as requests to fence the sites in order to avoid encroachment (agreed by DDA).[220] While DDA agreed in internal materials to declare the River Zone under the Wildlife Protection Act, no action in this matter was taken till date.

These public comments remained the only interaction of the DDA with the public with regard to the biodiversity zone concept. The environmental NGO YJA was largely supportive of the Yamuna Biodiversity Park at Wazirabad, referring to it as "a fine example of what Delhi's river front [...] can actually become" and requested

---

219 Letter YJA to Tejendra Khanna, Lieutenant Governor of Delhi, 8th October 2011: 'DDA's plans for Zone O are flawed - Pictorial evidence'.
220 DDA Landscape Unit, Presentation 6th Biodiversity Foundation Meeting: 'Suggestions as received through comments on uploaded information on website'. Internal materials accessed by the author via interview partners.

the Lieutenant Governor to develop more of these parks along the river.[221] When the drawing of the Landscape Unit were circulated in the press and development work started for the Golden Jubilee Park, a letter by the environmental NGO to the DDA, however, states that "by no stretch of imagination [the Golden Jubilee Park] can be called as a 'Biodiversity Park'".[222] The NGO requested DDA to convene "a public presentation and debate" of the proposed developments but any larger public debate on the concept was denied by DDA.[223]

The environmental NGOs therefore initiated the space for dialogue themselves. In November 2011, artists and environmentalists from Delhi and Hamburg[224] jointly organized the "PROJECT Y: A Yamuna-Elbe Public Art and Outreach Project" on the site of the Golden Jubilee Park. After some debate DDA had agreed that the project could be held on the site. DDA displayed the model of the Golden Jubilee Park (see Figure 72, page 396), but was not willing to participate in the debate about the site or the river in the course of the project.[225]

The future workshop convened by the author in March 2013 aimed to provide another forum for the discussion on the biodiversity zone and the Golden Jubilee Park. The idea of the biodiversity zone was critically discussed face-to-face by environmental activists and DDA planners of the Landscape Unit. A broad consensus was reached that the Yamuna Biodiversity Park is a good example, which should be replicated. The environmental activists though requested the DDA planners to alter the specific design of the Golden Jubilee Park accordingly to develop a 'real' biodiversity park rather that a recreational theme park (YFW-2013-03-08).

These appeals, as well as detailed discussions with the different interview partners on the Golden Jubilee Park and the biodiversity zone concept, support the argument that the environmentalists certainly feared that the biodiversity zone was only meant to be diversionary tactics by the DDA. The fear was that by co-opting the biodiversity discourse to DDA actually aimed to develop the riverfront in the fashion always promoted by the authority in the past.

Therefore, the bringing back biodiversity story-line, which the environmental NGOs originally drew on and even pushed, was later difficult to be deconstructed

---

221 Letter YJA to Tejendra Khanna, Lieutenant Governor of Delhi, 7th July 2009: 'Your decision to declare the 22 km of flood plain as a Biodiversity Zone' (YJA/CORRES/7/09).
222 Letter YJA to G.S. Patnaik, Vice Chairman DDA, 14th December 2011: 'Please make public the DDA plans for the BD parks in river Yamuna flood plains in the city' (YJA/CORRES/12/11).
223 Ibid.
224 "PROJECT Y: A Yamuna-Elbe Public Art and Outreach Project" has been a public art and outreach programme initiated by the Ministry of Culture, Hamburg, and carried out in the framework of "Germany and India 2011-2012: Infinite Opportunities". The public art project, which was curated by Ravi Agarwal (Delhi) and Till Krause with Nina Kalenbach (Hamburg), was held from 10th to 20th November 2011 in Delhi and combined with a series of talks on, and walks along, the river. The project had earlier started with a symposium on the Yamuna and Elbe organized by Max Müller Bhavan in Delhi in 2010.
225 Personal communication with the curators Ravi Agarwal and Till Krause during and after the project in Delhi.

and reframed by them when co-opted by the DDA. The structuration and institutionalization of the conservation discourse had already started based on this storyline. Yet, policy decisions (e.g. whether the floodplain would be designated as a protected area) were by and large avoided in the process pushing through project-based agendas. The farmers of the Delhi Peasants, which were facing evictions, also tried to organize protest against the biodiversity zone concept and were supported by members of YJA and Rajendra Singh. They also rejected the development of large parks along the river, arguing that they have been the green keepers of the floodplain over the last decades.

16.4 Evicting the 'green keepers'? An attempt to develop a counter-story line

Connected to the biodiversity zone concept and the development of the Golden Jubilee Park in particular, the DDA issued a series of new eviction notices to the Delhi Peasants since 2007. While earlier eviction notices had covered certain areas proposed for development, the new eviction notices were issued for the whole River Zone.

In addition to these eviction notices, farming in Delhi's riverscapes is superimposed by a discourse on the environmental risks associated with the consumption of the vegetables being grown on toxic soil and irrigated by contaminated bank filtrate. Health risks have emerged as a strong argument banning farming from the floodplain as long as water and soil quality remain degraded (SEHGAL et al. 2012, SEHGAL et al. 2014). The discourse, as COOK et al. (2014: 5) outline, is coined by the "concern that farming along the Yamuna River posed a threat to the environment due to chemical fertilizer and pesticide runoff into the Yamuna". Therefore researchers (cf. SEHGAL et al. 2012, SEHGAL et al. 2014) and journalists have argued for a prevention of toxic vegetables entering the city's markets. Additionally, the pollution caused by the farmers using fertilizers and pesticides (even if it might be insignificant compared to the total pollution load of the river from non-agricultural uses) increases the pressure on the local agencies to resume possession of land and terminate vegetable farming along the river.

While the farmers' society has long been fighting cases in the courts, it is difficult to assess the cohesion of the farmers' society and their willingness to remain in possession of the land. The findings of this study rather indicate that many of the society members are not primarily involved in farming 'their' leasehold lands anymore. Rather migrant workers cultivate the lands while the society members have other occupations earning most of their income from non-agricultural sources (farming is often of secondary occupation). This might also explain why both NGO-activists and active society members indicated in interviews that the majority of the farmers do not actively challenge the forcible reclamation of their lands (FARM-01/2011-12-06, FARM-02/2011-12-14, NGO-04/2011-12-06).

As field visits and interviews further indicated, individual farmers have sublet 'their' lands to nursery operators at different locations. An NGO-activist working closely with the farmers outlined (NGO-04/2011-12-06):

"[…] this cooperative society is so much involved in litigation matters […] that to really start a kind of a self-regulation where they actually go all organic is not there. [We are telling them] you create this blueprint [organic farming] and stick to it. That will be your strength. That will bring the civil society by in large to your assistance. Today, it is just some lease holder fighting DDA for some personal interest. That is the general impression."

Overall, it must be stated that the appearance of (migrant) farmers in the very heart of a city aiming to become 'world-class' is viewed by the interviewed planners and politicians as backward and not modern. From this point of view, their activities are prone to being overhauled by a remaking of Delhi's riverscapes under the biodiversity scheme. The most recent developments indicate that agricultural uses in the River Zone might finally be terminated not on grounds of their violation of the lease agreement, but rather because of their inability to change farming practices and the health risks posed by the agricultural products from the floodplain.

Individual members of the Delhi Peasants have made attempts to make their voices heard in the process by writing to their earlier patrons in the Congress Party. In the course of the slum demolitions the farmers had written to the court-appointed committee highlighting that they have been the predominant 'green keeper' of Delhi's riverscapes:

"[…] the farmers of the society have saved this Yamuna bed from the encroachers. It is due to these farmers that this belt of Yamuna Khadar is still safe and alive. The DDA wants to take passion [sic] of this land from the farmers, sometime in the name of common wealth games, some time [sic] in the name of cleanliness of Yamuna and now in the name of Bio-diversity Park. […] There is no wisdom to invest money for making Bio-diversity park in flood plains."[226]

During interviews the farmers explained that the government agencies have not once contacted them for a dialogue. The only interaction between the Delhi Peasants and the state has remained through the courts. Challenging the biodiversity story-line the secretary of the Delhi Peasants emphasized the role of the farmers as the green keeper (FARM-01/2011-03-09):

"It is due to these farmers that this land is saved till today. Otherwise this whole land would be under the jhuggi mafias and all over you would see the jhuggies and it would be impossible for the DDA to get the land cleared. […] The farmers not only have saved the land, but they have also kept it green. […] the farmers are also creating a biodiversity park. We are doing seasonal farming and we are maintaining it green, but the main matter is that they want to get money out of that land."

Also the environmental activists emphasized the role of the farmers creating some kind of "agro-biodiversity" (NGO-04/2013-02-11) and protecting the floodplain from encroachment (NGO-15/2011-12-08):

"The city needs the farmers to secure the floodplain from encroachments. They are the only ones who have managed to secure it from encroachments and continued to have a natural use

---

226 Letter Delhi Peasants Co-operative Multipurpose Society Ltd. to Usha Mehra, Chairman of the Yamuna Removal of Encroachment Monitoring Committee, dated 23rd February 2011. Letter accessed through Delhi Peasants Co-operative Multipurpose Society Ltd.

of the floodplain, to allow flood water to spread over it. The city certainly needs the farmers [otherwise] the floodplain security is lost."

There remains a big challenge to be overcome to unite the interests of the farmers and environmental activists. As outlined in several talks, the main complain of the activists is that the farmers do not change to organic agriculture banning pesticides and fertilizers (NGO-04/2013-02-11):

> "Our only complain – our only point of difference – with these people [Delhi Peasants], which we have been telling them again and again in every meeting, is that if you continue not doing organic farming, then we are not with you."

While farmers may try to make a shift to organic farming, they feel that there exists hardly any market for organic vegetables in India and especially not on the local markets (FARM-01/2011-03-09). The NGO activists, however, also admitted that they "have not found time to really engage in much more serious manner with the farmers" with regard to farming practices (NGO-04/2013-02-11). Several interview partners outlined that the role of the farmers is therefore ambiguous. One the one hand, it is widely agreed that the farmers have played a crucial role in protecting the River Zone from unauthorized encroachment. On the other hand, DDA planners and other officials highlighted the role the farmers play in selling land illegally to slum dwellers or subleasing it to nursery operators; a fact that has not been denied but rather regretted by the interviewed farmers (FARM-01/2011-03-09).

The Delhi Peasants have so far not been able to make their voices heard to develop a clear counter story-line, since (besides their internal problems) their capability of actively engaging in the larger policy discourse remains very limited. Appeals by them to the DDA were generally rejected and the only platform for interaction remains the court. Furthermore, by being 'blamed' by the media and to some extent also scientific studies for polluting the river and selling toxic vegetables, as well as supporting migrant laborers coming to Delhi and settling along the river, the farmers credibility and legitimacy claiming to be the true green keepers of the floodplain has been undermined. Nevertheless, interviewed farmers want to be involved in today's planning for Delhi's riverscapes.

During the future workshop, participants agreed that the eviction of the farmers might have unforeseeable consequences, since DDA does not have enough man power and resources to manage and protect the whole floodplain as a biodiversity zone (YFW-2013-03-08). An eviction of the farmers might in turn result in new encroachments and development of more slums. With regard to the quality and health risks of the vegetables, a heated debate underlined that scientific facts can be questioned and can be differently interpreted. There was a consensus that the farmers cannot 'bear the blame' of pollution, but farming right up to the edge of the water was critiqued during the discussions as it negatively affects the riparian ecosystems (YFW-2013-03-08).

Therefore the 'green keepers' might gain more support from the environmental NGOs if they agree to preserve these habitats and shift to organic farming. The conservation discourse on Delhi's riverscapes not only endangers the practices and livelihoods of the farmers and their migrant laborers, but the 'fronts' between environmentalists and local uses might not be as hardened as in the case of other conflicts on urban nature conservation in India (ZÉRAH & LANDY 2013, ZÉRAH 2007). This is accounted for by the fact that the conservation discourse has not been institutionalized with regard to Delhi's riverscapes, which are thus not (yet) designated, and therefore purified, as spaces of conservation. The farmers – as intermediate caretakers – benefit from this hybrid status of the land, which DDA has been unable to control with its purified land-use categories. Furthermore, one might argue that the farmers are tolerated based on the inter-discursive complexity of the riverfront development and the conservation agenda as both discursive framings value the role of the farmers protecting the land to some extent but also co-opt the farmers' practices in order to strengthen their arguments. The green keeper story-line has largely been without much impact on the overall discourse of Delhi's riverscapes as powerful actors (e.g. DDA) have rejected and deconstructed the story-line. The environmental NGOs also rarely drew on the story-line in order to support the farmers.

16.5 Remaking the plan: the influence of the NGOs and the courts

Public participation and several attempts of the environmental NGOs were largely blocked by the state agencies to facilitate a dialogue on the future of the biodiversity concept. Furthermore, the state agencies remained reluctant in dealing with ongoing encroachments on the River Zone. As a result, environmental activists, with Manoj Misra of YJA leading the way, again set their hope in the courts to direct the state agencies to ensure the protection of the floodplain.

16.5.1 The evolvement of the Debris Case at the NGT

After the establishment of the National Green Tribunal, Manoj Misra (convenor of YJA) filed one of the first applications to the tribunal in February 2012. The application focused on a local water body in the River Zone located close to the Yamuna Pushta Embankment near Geeta Colony in East Delhi. Based on the Delhi Eicher City Map, a Google Earth satellite image and pictures from the field the applicant highlighted that due to dumping of the solid waste the "long living natural water body" had been degraded "into a dumping zone" (see Figure 41, page 302).[227]

The applicant, after field visits in November 2011, had informed the LG and involved government agencies including the IFCD and the MCD by writing e-mails and the matter was also reported in the daily newspaper. Therefore, the agencies were made aware and action was requested. Yet, further field visits by the applicant in December 2011 revealed that no action was taken:

---

[227] National Green Tribunal, Manoj Misra vs. Union of India and others. Original Application 6 of 2012. Quoted from the original application.

"To the utter shock of the applicant [...] a huge lot of fresh debris had been dumped at the site and the encroachment and dumping into the natural water body had increased manifold. [...] it was also found that a few small 'Jhuggies' (hutments) have also been erected at the site."[228]

The environmental activist again informed the LG, requesting a site visit by the YRDA. Since no action was taken by the authorities, the applicant moved to the NGT in early 2012 (the case is herein referred to as 'the Debris Case'). While the application challenged the degradation of a particular local water body, the practice of dumping of solid waste and debris was documented by the author in several stretches along the river (see Figure 42 to Figure 44).

Figure 41: Debris along the Yamuna Pushta Embankment near Geeta Colony
Source: Follmann, November 2011

Figure 42: Debris dumping along guide bund near Yamuna Bank Metro
Source: Follmann, February 2013

Figure 43: Solid waste recycling near ISBT Bridge, Shastri Park, East Delhi
Source: Follmann, February 2011

Figure 44: Slums on debris material in front of Shastri Park Metro Depot
Source: Follmann, November 2011

228 Ibid.

Government agencies had long discussed the issue of solid waste dumping. The problem was also clearly articulated in the course of the relocation of the slums in the mid-2000s.[229] Therefore, the concerned the landowning agencies must have been well-aware of the issue, which was already highlighted by the National Capital Region Planning Board in the late 1980s (NCRPB 1988: 107). In this context, GOPAL et al. (2002: 24) noted that the disposal of solid wastes and debris "is the major method of encroachment upon riparian land" in India and has "occurred practically everywhere in the country along all rivers and lakes".

The ZDP also refered to the need for "preventing dumping of urban waste"; e.g. by installing metal railings along bridges (DDA 2010b: 15). Nevertheless, large areas in the River Zone continued to be filled up with debris materials unloaded by trucks. The piled up materials further encouraged the establishment of new slums in the River Zone as the higher elevation provides some security during floods (see Figure 44). The problem of solid waste dumping along the river therefore emerges as a promising reason for the environmentalists to initiate action by the NGT (LAW-02/2013-03-06).

The NGT heard the case for the first time on 12th March 2012 and issued an order to the responding state agencies "to take steps to stop further encroachment and dumping of solid waste and debris in the river bed within seven days".[230] The Debris Case was entertained by the court in the following months in various hearings and orders.[231] The state agencies remained reluctant in responding to the petition and later accused each other of being responsible. The DDA in particular skirted responsibility, since solid waste management is the task of the municipal authorities.

Nevertheless, after some debate about the costs and possible sites for storing the removed materials, the land-owning agencies including DDA, DMRC, CPWD, IFCD and Uttar Pradesh Irrigation Department started removal of debris from the floodplain in early 2013. Within about 6 months' time the landowning agencies removed almost 50,000 tons of construction debris and solid waste from the River Zone (cf. MISRA 2015: 18).

---

229 See minutes of the eleventh meeting (30 June 2006) YREMC. The NRCD, commenting on the draft ZDP of the DDA in 1998 also already pointed to solid waste disposal directly into the river as well as the discharge of "fly ash slurry into the river".
230 National Green Tribunal, 12th March 2012, Original Application 6 of 2012, Manoj Misra vs Union of India and Others. The respondents included from the very beginning the 1) Union of India, through the Secretary MoEF; 2) the Government of the NCT of Delhi, through the Chief Secretary, 3) the DDA, MoUD through its Vice Chairman; 4) the Delhi Pollution Control Committee through its Member Secretary, 5) the YRDA (originally intended to involve the Lieutenant Governor as the chairman, but issued through Principal Secretary, Urban Development, Government of NCT of Delhi) and 6) the Irrigation Department of Uttar Pradesh, Government of Uttar Pradesh through its Principal Secretary. After the second hearing (27th May 2012) the respondents 7) State of Uttar Pradesh, Government of Uttar Pradesh and 8) Municipal Corporation of Delhi (East Zone) were added as respondents.
231 The Debris Case was heard at the NGT in 2012 on 12th March, 25th May, 17th July, 30th August, 17th September, 10th October, and 19th December.

Furthermore, the state agencies installed signboards prohibiting the dumping of waste (see Figure 45) and erected barriers (e.g. steel barricades, walls made out debris material) to hinder trucks from entering the floodplain.

Figure 45: Public notice boards posted along the embankments
Source: Follmann, September 2014

In March 2013, in addition to the removal of the debris, the court demanded an action plan from the state agencies on "how and in what manner the banks of river Yamuna can be developed rather than making it mere garbage dumping site".[232] In this course, the NGT requested the DDA to prepare "a complete proposal for development, beautification and upliftment of Yamuna river banks in terms of horticulture and eco-friendly site free from pollution and other allied fields".[233] Prior to this, the case was limited to the removal of debris and the prevention of new dumping along the river. Therefore, the court considerably enlarged the scope of the case.

In their first affidavit, the DDA had already outlined that the DDA "would be glad to protect and restore the biodiversity of Yamuna" including the particular site along the Geeta Colony Road "provided it comes under the possession" of the DDA. Since the land ownership of the dumping site was contested, the DDA argued

---

232 National Green Tribunal, 21st March 2013, Original Application 6 of 2012, Manoj Misra vs Union of India and Others.
233 Ibid.

that DDA was not able to protect the land due to the land ownership conflict between the NCT of Delhi and the state of Uttar Pradesh.[234] Trying to make DDA responsible, the agencies of the NCT of Delhi involved in the case had pointed to DDA's responsibility to plan for the river.[235] Therefore, the widening of the case was likely as the DDA and the agencies of the NCT of Delhi also viewed the case as an opportunity to resolve the land conflicts with Uttar Pradesh. The state agencies agreed that the DDA Landscape Unit should present its riverfront development plan in front of the NGT.[236]

In July 2013, the court stated that "still dumping continues [...] particularly at Geeta Colony site" and observed "an utter confusion in regard to the coordination and cooperation between various public authorities".[237] The court requested a meeting of all concerned agencies on the next day to "submit a final beautification programme".[238] After this meeting the MoEF requested more time to finalize the "Yamuna preservation and beautification plan". The NGT granted more time and soon thereafter, the MoEF formed an Expert Committee comprising of Prof. C.R. Babu (Chairman; CEMDE, Delhi University), Prof. Brij Gopal (Member; retired professor from Jawaharlal Nehru University) and Prof A.K. Gosain (Member; IIT Delhi), to analyze DDA's riverfront development schemes in the light of the environmental concerns raised in the Debris Case and related to the Maily Yamuna Case dealt with by the Supreme Court since 1994.

Thus, while the DDA had prepared schemes for riverfront development based on the biodiversity concept outlined above (see chapter 16.2), the court – based on the inaction of the DDA in the case – requested the other state agencies under the leadership of the MoEF's National River Conservation Directorate (NRCD) to critically assess DDA's and other existing development plans for Delhi's riverscapes. The ministry in turn delegated the task to three eminent experts – all of them long involved in the planning and policy-discourse on the river Yamuna in Delhi. In doing so, the court clearly challenged DDA's planning and development sovereignty for the River Zone. In turn, the MoEF's National River Conservation Directorate, which earlier had remained absent in the policy discourse on the Yamuna in

---

234 Ibid. The land dispute is subject also of a pending Supreme Court case (SLP (C) No. 25782/2009).
235 Affidavit on behalf of respondents NCT of Delhi, DPCC, and YRDA. 17th July 2012. Materials accessed via YJA. The agencies of the NCT of Delhi further used DDA's plans for the riverfront in their affidavit, to suggest that they are not responsible but rather DDA is to be responsible.
236 Minutes of the meeting held under the chairmanship of Secretary MoEF on 4th April 2013. Copy of the minutes provided to the author by YJA. The meeting was attended by MoEF officials including members of the NRCD, and officials from the DDA, CPCB, EDMC, SDMC, Uttar Pradesh Pollution Control Board, DMRC, Uttar Pradesh Irrigation Department, and Delhi Parks and Garden Society.
237 National Green Tribunal, 17th July 2013, Original Application 6 of 2012, Manoj Misra vs Union of India and Others.
238 Ibid. In this order the NGT put pressure on the involved officials by considering to "impose exemplary costs which shall be recoverable from salary of the responsible Officer/s concerned".

Delhi, was given an important role in the process. The language of the court indicates that it did not draw on the biodiversity story-line developed and promoted by the DDA, but rather the term 'beautification' emerged as the central term in the court's orders.

After a first round of meetings of various agencies, field visits and the collection of data, the Expert Committee outlined the major constraints for developing an action plan for the Yamuna in Delhi:

> "[An] Action Plan with locations along the river can be formulated only after [the] boundaries of 'O' Zone are clearly demarcated and maps for the flood zoning with elevation contours are finalized and analysed for flood levels at different discharge values."[239]

The Expert Committee further highlighted that "even on basic issues for understanding river morphology, hydrology and hydraulics such as, contour maps [...], flood zone maps, depth of the channel, meandering of the river, wetlands along the river system, etc." data was not available from DDA or any other state agency.[240] Most importantly, the Expert Committee indicated that "there was no clarity on what constitutes the flood-plains and flood-ways and their boundaries."[241]

With regard to governance issues, the Expert Committee pointed out that "no coordination between various departments" and the "lack of coordination and non-availability of [...] data to the planners" has been resulting in encroachments.[242] Responding to the pleas of the Expert Committee, the court granted more time until March 2014, for gathering and analysis of data, and the preparation of an action plan. Later the court extended the time given to the Expert Committee until April 2014.

---

239 Request made by the Expert Committee to the NGT, no date, in National Green Tribunal, Original Application 6 of 2012, Manoj Misra vs Union of India and Others. Materials accessed via YJA.
240 Ibid.
241 Ibid.
242 Ibid.

16.5.2 Rezoning the river – DDA's move to redefine the boundaries of the river

While the NGT was delivering orders in the Debris Case and had several times requested the DDA to take action, the DDA issued a public notification for the re-delineation and rezoning of the River Zone on 28th September 2013 without informing the NGT (see Figure 46).

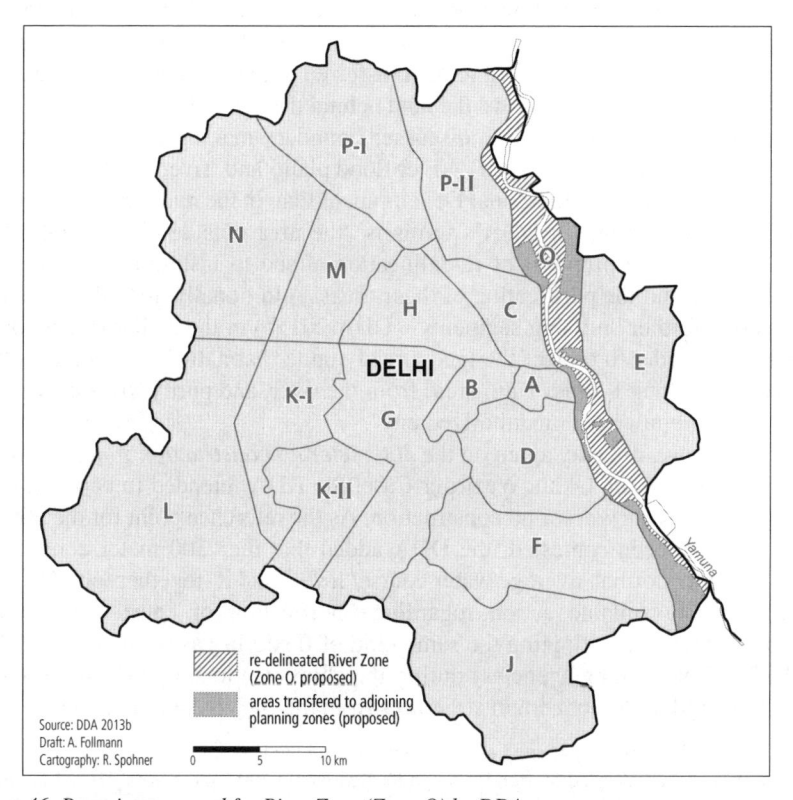

Figure 46: Rezoning proposal for River Zone (Zone O) by DDA

The DDA proposed a reduction of the planning zone by about 3,100 ha; almost one third of the original 9,700 ha (DDA 2013b). This proposal was reasoned by the DDA on the basis of the mid-term review of the Master Plan 2021, outlining that the residents of the unauthorized colonies falling in the River Zone were "not getting any permission for reconstruction/repair of their buildings from the MCD" due to the no-development moratorium of the Lieutenant Governor on the River Zone (DDA 2013b). Further, the DDA claimed that the moratorium obstructed the development of basic infrastructure like schools or community halls in these settlements. Accordingly, the DDA officially backed the "re-delineation", as it was termed by them, on the process of regularization of the unauthorized colonies processed by the Delhi Government (for details 13.3).

Taking the "already constructed embankments [...] as the boundary of the re-defined Zone O [River Zone]" the DDA also intended to remove the urban mega-projects developed along the river including the two metro depots, the Akshardham Temple Complex, the Commonwealth Games Village and the DTC Bus Depots from the River Zone. Additionally, the area separated from the river by the Ring Road By-Pass Road (including the existing samādhis, the power plant) as well as the land of the IP Thermal Power Plant and IP Gas Power Plant south of the ITO Barrage was proposed to be excluded from the River Zone. Thus, a major part of the former Zone O was supposed to be transferred to the adjoining planning zones.

The DDA intended to remove the land behind the embankments from the River Zone, based on the logic of earlier discussed boundary-making and the invention of the three morphologies 'river-bed', 'river flood plain' and 'river front' by the DDA in the drafting process of the Zonal Development Plan in the mid-2000s (see section 11.7). The 'river front' in DDA's terms is "the area outside the embankments" (DDA 2013b). The proposal of re-delineation aimed to institutionalize the three morphologies and the purification of these areas. Additionally, the DDA proposed to construct further "missing segments" (DDA 2013b) in the embankments on the right bank in order to protect the land already under habitation (e.g. the unauthorized colonies along Kalindi Kunj area) from the river, and purify these areas as the realm of the 'regularized unauthorized city'.

Furthermore, drawing again on the *300 meter no construction zone*, defined by the High Court in the Okhla/Wazirpur Case, the DDA intended to earmark a 300 meter belt from the river for no construction. As the reference point for the 300 meter has always been contested, the DDA added that the "300 meter zone can be identified from normal/ average water course/ level" and it (together with GNCTD) would "initiate suitable action regarding clarification of court order" (DDA 2013b).[243] Moreover, adopting the same kind of flexibility as applied by the DDA (and other land-owning agencies) during the slum evictions in the mid-2000s the DDA proposed to except certain structures from the 'no construction zone':

> "DTC, DMRC have constructed embankments/ raised the level of land for locating depots so that they are not affected by floods though located within 300 meters from River Yamuna. In view of this, the similar areas/ locations, based on ground realities could be examined for excluding from 'no construction zone'" (DDA 2013b).

In doing so, the DDA again tried to simplify the complex nature of a monsoonal river in order to gain clear boundaries between the city and the river. While the DDA had earlier managed to reclaim and purify the land under the regulations for the Zone O, the DDA now tried to once and for all purify the area behind the embankments as land, by taking them out from the River Zone itself. This would have also given DDA the opportunity to post-facto regularized the DTC Millennium Bus Depot, which is yet to be relocated (see Figure 47).

---

243 DDA refers in the Public Notice to the High Court order of 29th March 2006, which was a later order in the Okhla/Wazirpur Case, WP (C) No. 2112/2002 and W.P.(C). NO. 689/2004.

Figure 47: DTC Millennium Bus Depot
Source: Follmann, February 2011

Figure 48: 'Land bank' protected by Ring Road By-Pass Road
Source: Follmann, February 2011

An approval of the de-lineation of the zone would have allowed the DDA to further develop a massive 'land bank' of more than 300 ha protected from the river by the embankment of the Ring Road By-Pass Road (earlier Mughal Bund) in front of the Red Fort. This area, today still used by the Rajghat Power Plant and its surrounding fly-ash ponds, is certainly one of the future prime (re)development locations in the city (see Figure 48).

The de-lineation of the O-Zone was vehemently opposed by the environmental NGOs. Yamuna Jiye Abhiyaan criticized the proposal in a letter to the Lieutenant Governor to be: "most audacious, incredulous and unacceptable as it is not only against the interest of the river Yamuna and the city but also is illegal, irregular and based on a very flimsy excuse".[244] An activist of the South Asia Network on Dams, Rivers and People (SANDRP), Himanshu Thakkar, warned in an article that "such large-scale post facto regularization of illegal encroachments in an ecologically sensitive area will send a clear invitation for more such encroachments" adding that "DDA needs to justify every hectare of diversion with convincing reasons which it has clearly not done".[245] The Times of India quoting both environmental NGOs headlined "DDA courts disaster, plans to cut Yamuna zone by 40%".[246]

The proposal for the re-delineation of the Zone O is to be reviewed with regard to the Debris Case at the NGT. The court had clearly jeopardized DDA's earlier unchallenged planning and development powers by requesting the MoEF's National River Conservation Directorate to evaluate DDA's riverfront development scheme. The DDA aimed to redefine the boundaries of the organization's undisputed planning and development hegemony by re-delineating the boundaries of the

---

244 Letter YJA to Lieutenant Governor of Delhi, 29th September 2013: YJA - Sir, prevent the DDA from slicing half of the river Yamuna in the city - Please direct DDA to withdraw its PUBLIC NOTICE on Zone O (River Zone).
245 Thakkar, Himanshu, November 2013: DDA chokes the Yamuna. Civil Society Online. http://www.civilsocietyonline.com /pages/Details.aspx?434 (last access 25 March 2015).
246 Times of India, 30th September 2013: 'DDA courts disaster, plans to cut Yamuna zone by 40%', author: Jayashree Nandi.

River Zone. The re-delineation proposal was immediately discussed in the hearing of the Debris Case and the NGT prohibited the DDA from taking further action for the re-delineation of the River Zone in October 2013.[247] Once again, the courts intervened in the policy-making process for Delhi's riverscapes.

16.5.3 The action plan for 'Restoration and Conservation'

In April 2014, the three members of the Expert Committee under the MoEF submitted their report titled 'Restoration and Conservation of River Yamuna' (EXPERT COMMITTEE MoEF 2014). The information given in the acknowledgements of the report indicate that the Expert Committee, in contrast to the DDA Landscape Unit earlier, was able to prepare high resolution maps for the River Zone.[248] The Expert Committee was, however, also put under strict time constraints to prepare the report and maps for the River Zone.

Reviewing DDA's riverfront development proposals, the Expert Committee set out that the scheme "is untenable and should be stopped" (EXPERT COMMITTEE MoEF 2014: 5). With regard to the Golden Jubilee Park, the committee requested that "the current programme of development must be stopped forthwith and the area should be restored to the river and its natural floodplain" (EXPERT COMMITTEE MoEF 2014: 44). Moreover, the Expert Committee highlighted the incoherencies in DDA's argumentation: The DDA had always referred to the 'river front' to be located outside the embankments, "but the area of the proposed YRFD [Yamuna Riverfront Development] is within the active floodplain" (EXPERT COMMITTEE MoEF 2014: 5). Therefore, DDA's scheme was contradictory to the organization's own purification logic. However, to some extent the Expert Committee agreed with the approach taken by the DDA, declaring (EXPERT COMMITTEE MoEF 2014: 34):

> "The Zone O defined by DDA does not have well-defined boundaries, as the new bunds/embankments/roads have been developed within the old bunds/embankments/roads for human settlements and infrastructure development. Consequently, Zone O is open ended zone with boundaries changing as the encroachments and infrastructural development continue to take place in the ecologically critical functional floodplain ecosystems."

Accordingly, the committee articulated that in order "to manage the riparian ecosystems sustainably and to maintain their ecological integrity, it is essential to delineate the present river boundaries", and recommended redefining of the boundaries of the River Zone based on the existing embankments (EXPERT COMMITTEE MoEF 2014: 30). The Expert Committee also proposed a purification of Delhi's riverscapes into the separated realms for the city and the river. In line with the DDA, the Expert Committee suggested the existing embankments and roads as the defined boundaries of the River Zone. Furthermore, "in the absence of regulatory measures

---

247 National Green Tribunal, 28th October 2013, Original Application 6 of 2012, Manoj Misra vs Union of India and Others.
248 The data sets were developed by Geospatial Delhi Limited, a Govt. of NCT of Delhi Company, funded by the DDA (EXPERT COMMITTEE MoEF 2014: V).

and legal protection" the Expert Committee recommended that the MoEF should declare the entire open floodplain in-between the existing embankments as a "Conservation Zone" under the provisions of the Environmental Protection Act, 1986.

With regard to the interstate disputes, the Expert Committee highlighted that "the river does not recognise the administrative boundaries", and "it is therefore necessary to consider together" the subzones of Delhi and Uttar Pradesh in the planning process (EXPERT COMMITTEE MoEF 2014: 36). The measures proposed by the Expert Committee also set forth the purification agenda including the relocation of "the encroachments, the villages, the cattle farms and nurseries located on the active floodplain i.e. within the embankments" (EXPERT COMMITTEE MoEF 2014: 30).

The DDA vice-chairman Balvinder Kumar responded to the report in the media clearly drawing on the fear of encroachment story-line, but accepting the provisions of the Expert Committee:

> "We will go by the expert panel's recommendations. We started building recreational spots to save the ecologically-fragile floodplains from encroachment. But we respect the panel's views and will implement the alternative ways it has suggested to protect the floodplains".[249]

While in the terminology used by the DDA 'encroachment' generally refers only to unplanned structures, the Expert Committee – in line with the use of this terminology by the environmental NGOs – broadened the understanding of encroachment to include also 'planned' developments in the River Zone under this terminology (cf. EXPERT COMMITTEE MoEF 2014).

16.5.4 The orders of the NGT: 'Maily se Nirmal Yamuna'

In early 2014, the Debris Case was combined with a second application filed in October 2013 against the covering and concretization of storm water drains in Delhi (herein referred to as 'the Drains Case').[250] In the Drains Case the environmentalists challenged the ongoing concretization and covering of two storm water drains in South Delhi and the plan to cover the Shahdara Link Drain in East Delhi to enable commercial developments. They stressed that today's city drains, which without exception carry wastewater, need to be included in the river restoration project as they are originally the seasonal tributaries of the river originating from the Delhi Ridge.[251] Thus, the environmental activists aimed to broaden the spatial extent of the restoration efforts.[252] Their petition was also a response to the critique that river restoration (and the associated environmental activism) had earlier narrowly focused on the river, and thereby largely neglected the socio-ecological problems

---

249 DDA-vice chairman quoted in Singh, D. (April 26, 2014): Will junk Yamuna projects to save floodplains: DDA. Hindustan Times. http://www.hindustantimes.com/india-news/newdelhi/will-junk-yamuna-projects-dda/article1-1212637.aspx (15 April 2015).
250 National Green Tribunal, Manoj Misra & Madhu Bhaduri vs. Union of India and Others. Original Application 300 of 2013.
251 Ibid.
252 Personal communications with the two petitioners.

associated with high levels of water pollution and inadequate sanitation in the residential areas (KARPOUZOGLOU 2011, MEHTA & KARPOUZOGLOU 2015, ZIMMER 2012b, 2015a).

On 13th January 2015, the National Green Tribunal delivered its judgement in the joined cases pronouncing it as "Maily se Nirmal Yamuna rejuvenation project, 2017" [from polluted to pristine Yamuna Revitalisation Plan, 2017].[253] Over almost 100 pages the order outlines a road map for the rejuvenation of the river within two and a half years:

> "We are not oblivious of the herculean task which will be required [...], but we are of the firm view that any further deferment in taking stern and serious steps for preventing and controlling pollution of river Yamuna, is bound to expose Delhi and its residents to grave environmental disasters."[254]

Challenging the state agencies inaction and failure in cleaning the Yamuna, the court declared that "a lack of institutional will to implement various policies, schemes and decisions" for the protection and restoration of the river, and an "attitude of planning, waiting and watching" had led to the current state of the river in Delhi.[255] The court therefore outlined an action plan for the restoration of the river and set up a Principal Committee to oversee the progress of the action plan's implementation (cf. MISRA 2015).

The action plan outlined by the court closely follows the recommendations of the Expert Committee and the expert's recommendation with regard to the pollution control derived from the Maily Yamuna Case in the Supreme Court. The court requested the DJB to make all 23 existing sewage treatment plants fully functional within two months and construct 32 new sewage treatment plans within two and a half years so that the river and all the about 200 drains in the city become free from pollution. As the GNCTD had already provided special funds (20,000 crore Indian rupees) for the upgrade of the sewage system in Delhi, the court did not "foresee any difficulty in provisioning adequate funds for timely completion".[256] The court further requested that Haryana should ensure the minimum environmental flow in the river.

Furthermore, the court stopped DDA from proceeding with its development of recreational areas under the biodiversity zone concept requesting a reworked scheme "preservation, restoration and beautification".[257] The court set out that the re-delineation of the River Zone should be done in accordance with the experts'

---

253 National Green Tribunal Judgement, 13th January 2015, Original Application 6 of 2012, and M.A. Nos. 967/2013 & 275/2014 Manoj Misra vs Union of India and Others; and Original Application No. 300 of 2013 and M. A. Nos. 877/2013, 49/2014, 88/2014 & 570/2014 Manoj Misra & Madhu Bhaduri vs. Union of India and Others.
254 Ibid. page 81.
255 Ibid. page 22.
256 Ibid. page 59.
257 Ibid. page 54.

recommendations taking the existing embankments and roads as the new boundaries of the river. Therefore, the court sanctioned DDA's intention to remove certain areas from the regulations of the River Zone.

> "The river zone so designated should be preserved and protected for the conservation and restoration of the river and no development activity should be permitted within the river zone that encroaches upon the active floodplain, obstructs the flow or pollutes the river."[258]

In line with this clear institutionalization of the conservation discourse and the purification of the active floodplain as the realm of the river, all residential areas within the River Zone unprotected by embankments "should be relocated at the earliest".[259] If this order is implemented the urban villages (e.g. Old Usmanpur Village and Garhi Mandu Village on the eastern bank between the Wazirabad Barrage and ISBT Bridge) would be demolished 'in the interest of the river' despite their long existence predating almost all other settlements located on the Yamuna *khadar*.

The court further banned all "recycling units, farm houses, cattle farms and nurseries" from the River Zone as well as disallowed the construction and widening of any new roads or embankments.[260] In doing so, the court order draws on the model bill prepared for floodplain zoning by the Central Water Commission requesting the DDA again "to demarcate the floodplain for the flood of once in 25 years and to prohibit any kind of development activity in the area in question".[261] With clear reference to the scarcity of land discourse earlier deployed by the state agencies, the court stated:

> "No authority or person before us has even taken up the plea that why development/ construction activity cannot be carried on in other parts of NCR, Delhi. As of now, sufficient land is available, may it is [sic] expensive, but that cannot be a ground for destroying the ecology, environment and biodiversity of River Yamuna [...] The result of indiscriminate, unregulated and uncontrolled development activity are widely visible [...]. It would not only be unwise, but may prove fatal, if such approach is continued any further."[262]

The court viewed the agricultural practices on the floodplain as "one of the glaring examples of indirect impacts of environmental pollution" and as the vegetables "are bound to be contaminated" the court directed that "agricultural activity needs to be stopped immediately [and forbidden] till the time Yamuna is restored to its original status and carries only wholesome water".[263] The court argued:

> "Even if these persons [farmers] have an interest in the land, they cannot carry on an activity which is environmentally improper and is completely injurious to human health, just to make some money."[264]

---

258 Ibid. page 55.
259 Ibid. page 47.
260 Ibid. page 46.
261 Ibid. page 74.
262 Ibid. page 77–78.
263 Ibid. page 44.
264 Ibid. page 52.

The concerns of public health and the wholesomeness of the river are recurrent framings in the order. The court outlined that the "aquatic lifeline for millions of people" should be "restored to its crystalline and pristine form".[265] Thus, the court requested the deepening and enlargement of existing wetlands and water bodies, and the construction of culverts in the approach roads and guide bunds of the bridges to ensure their connection.[266]

Besides the 'restoration' and 'conservation' terminologies used by the court, the court also heavily emphasized the need for 'beautification'. Generally, as was already identifiable to some extent in the report of the Expert Committee (cf. EXPERT COMMITTEE MoEF 2014), the court mentions "restoration, preservation and beautification" in the same breath. For example, the court also adapted the idea of the green belt, which could fulfil many functions at once:

> "[...] a greenbelt/greenway should be developed on both sides of the embankment, for controlling erosion, reducing sediment load of the main channel, reduce pollution, and beautification."[267]

With regard to governance issues, the court set up the Principal Committee based on the "unanimity" of all involved actors to have "one organisation to look after the entire project".[268] All concerned national and state agencies, the neighboring states, and the experts of the committees are members of the committee, which will report to the tribunal on a quarterly basis.[269]

At the time of writing, the implementation of the recent orders of the NGT by the state agencies remained pending. The state agencies have already failed to comply with the strict deadlines of the court to issue their reports of how they will implement the orders. Specifically, the demarcation of the 25 year flood levels has not been provided by the Irrigation and Flood Control Department.[270] The most recent developments further indicate that the ambitious task set out by the court that the agencies of the NCT of Delhi, the DDA and the agencies of Uttar Pradesh come up with 'one' demarcation of the River Zone, remains challenging due to the interstate conflicts. The NGT requested that the Government of Uttar Pradesh provides the data for the demarcation of the floodplain to the DDA, which in turn shall define the "physical demarcation of the flood plain area".[271]

---

265 Ibid. page 8 and 69.
266 Ibid. page 70.
267 Ibid. page 70.
268 Ibid. page 75.
269 The Principal Committee includes high-ranked officials from the MoEF, MoWR, DDA, DJB, IFCD, and the three municipal corporations. Furthermore, the other riparian states (Haryana, Uttar Pradesh, Himachal Pradesh and Uttarakhand) have a representative in the committee. The Proncipal Committee further includes the three expert members Prof. Brij Gopal, Prof. C.R. Babu, Prof. A.K. Gosain, and Prof. A.A. Kazmi (IIT,Roorkie), who had together with Prof. A.K. Gosain also been involved in another expert committee in the Maily Yamuna Case in the Supreme Court.
270 National Green Tribunal, 27th March 2015, Original Application 6 of 2012 and Original Application No. 300 of 2013.
271 Ibid.

Serious doubts remain that the state of Uttar Pradesh will follow the order and allow the DDA to demarcate the limits of the floodplain within the territory of Uttar Pradesh. The state of Uttar Pradesh is expected to deny compliance of the order based on the fact that the DDA has no legitimate powers to interfere in the land-use planning of Uttar Pradesh. These most recent developments thus once again underline the complex governance set-up for Delhi's riverscapes and highlight the interference of certain court orders with existing governance structures.

16.6 Interim conclusions

The DDA aimed to appease growing and consistent opposition by developing the biodiversity zone concept. In doing so, the DDA changed its language: channelization was converted into rejuvenation, and riverfront development was now about developing biodiversity. The language and images used in the biodiversity zone concept thus aimed to "delegitimize dissent" (HOLIFIELD & SCHUELKE 2015: 301) by apparently linking the 'needs of nature' to the desires of the city, and the urban middle classes in particular.

The term biodiversity itself is a social construction of nature (ESCOBAR 1998, MARTIN et al. 2013), but "offers many illustrations of the inseparability of the non-human and the human" (ZIMMERER 2009: 61). Biodiversity is "an influential multi-faced concept crossing sciences and humanities" (ZIMMERER 2009: 62). Therefore, it seemed to be suitable to be transformed into a powerful story-line for the remaking of Delhi's riverscapes. In SMITH's words (1984: 57), the biodiversity zone concept proposes urban nature "neatly packaged" in biodiversity reserved zones (eventually fenced off to be protected from the city) fading out that these riverscapes are hybrid, produced natures connected to multiple city-river relationships.

The discussed Golden Jubilee Park mirrors the middle-class centered vision of a world-class riverfront and matches with the often described middle-class bourgeois environmentalism (BAVISKAR 2011a, BOSE 2013, ELLIS 2012, MAWDSLEY 2009, TRUELOVE & MAWDSLEY 2011). Due to the history of the site and its location in the heart of the city, the Golden Jubilee Park represents more than a park. Adopting the words by GANDY (2013: 1307), the Golden Jubilee Park rather needs to be viewed as "a designed fragment of nature that inscribes social and political power into the urban landscape." The social and political power of the Golden Jubilee Park is rooted in the hegemonic world-class discourse, which has been newly promoted as a tourism and recreational attraction under the name of creating biodiversity. The concept of biodiversity emerges here as a social construction of nature that is highly ambiguous when applied to an environmentally-degraded urban 'wasteland' (see GANDY 2013: 1311 for a similar remark on urban wastelands more general).

The biodiversity concept connected the ideas of DDA's landscape architects, civil engineers and planners to the interests of ecological scientists involved in the restoration of degraded ecosystems and rivers. The concept, however, embodied the knowledges and desires of these two groups exclusively. While it was a response to growing critique, the critical voices (especially the environmental NGOs) were not involved in the policy-making process. Only after an initial institutional framework (the Biodiversity Foundation) was created, and the structural plan outlining the zonal concept was prepared by the DDA Landscape Unit, the 'public' was requested to comment. Thus, the public at large, including the environmental NGOs, could make their comments only at an advanced stage and after first developments had already been carried out (e.g. Golden Jubilee Park). There was no 'real' participation even by the concerned (middle-class) citizens in whose interest the DDA aimed to redevelop Delhi's riverscapes as a major recreational area. The local population, including the farmers, were facing eviction and they were never consulted to contribute to the plan-making. For the planners and scientists they did not exist (as legitimate uses), but were rather still defined as the 'encroachers'.

In such a mind-set, the counter-discourse defining the farmers as the green keeper of the floodplain did not find merit. When scientific data supported the health risks of the vegetables from the floodplain, the ties between the environmental NGOs and the farmers were further weakened. A broad consensus was reached in the deliberations under the NGT that agricultural uses needed to be immediately stopped to ensure public health. The livelihood of the farmers remained of no concern for the court.

With regard to actions of purification, the Expert Committee clearly aimed to define the boundaries and purify the floodplain to protect the river. In contrast, the DDA aimed to redefine the boundaries in order to be able to develop the land behind the embankments without major interference. Thus, while the intention of the Expert Committee was different from the intention of the DDA, both applied the logic of purification in a similar way. As already outlined for the Akshardham Bund (see 14.2.5), the logic that embankments separate the 'active' from the 'passive' floodplain prevailed.

VIII GOVERNING DELHI'S RIVERSCAPES: A SYNTHESIS

17 REFLECTION AND DISCUSSION

The empirical insights in the discursive and material remaking of Delhi's riverscapes have outlined major discrepancies and contradictions between the socio-ecological importance of Delhi's riverscapes and their physical state as well as between the desired urban imaginaries of the planners and environmentalists to reconnect the city to the river and the existing city-river relationships. This chapter summarizes and synthesizes the empirical insights from Delhi's riverscapes. It revisits the main discourses and story-lines, the processes of hybridization and purification, and the associated transformation in the urban environmental governance of Delhi's riverscapes.

17.1 Revisiting the main planning discourses and story-lines

In the course of the empirical chapters, six story-lines have been introduced (fear of encroachment, wasted land, slums as polluters, farmers as green keepers, ecosystem services, bringing back biodiversity). These story-lines are related to the three main discourses which appeared in the discursive framing of Delhi's riverscapes: the recreation/beautification discourse, the channelization discourse, and the conservation discourse (see Figure 49).

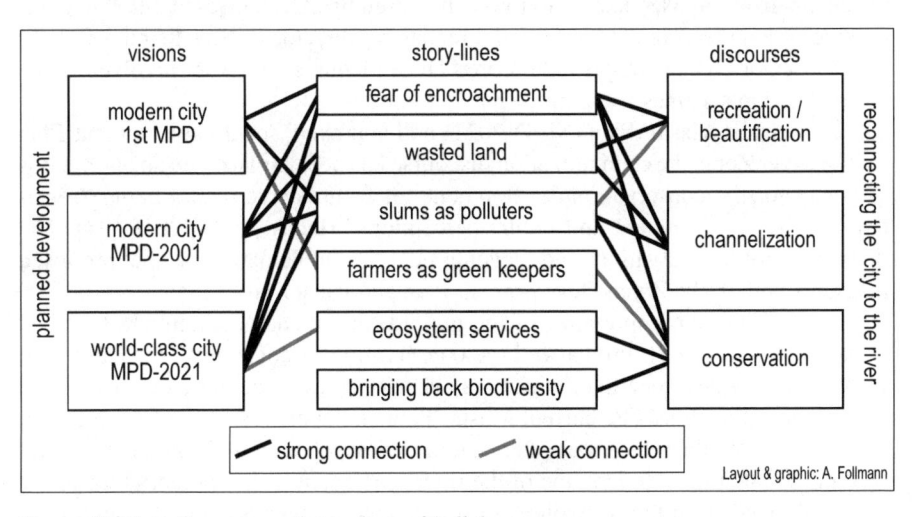

Figure 49: Main discourses and story-lines of Delhi's riverscapes

In all three discourses the DDA planners outlined the overall goal to reconnect the city to the river evoking the importance of river rejuvenation and a 'planned' development of the riverfront. Furthermore, all three Master Plans are coined by the planners' goals to ensure 'planned' development and control 'haphazard growth' of the city.

17.1.1 The main discourses and story-lines

As outlined in the empirical chapters, the planners' visions for Delhi's riverscapes shifted over time. The Town Planning Organisation already proposed beautiful parks to reconnect the historic monuments to the river in 1956. The DDA planners largely left out Delhi's riverscapes in the first Master Plan (1st MPD) in 1962. They proposed some recreational uses in the form of parks along the riverfront in line with the Interim General Plan of 1956. The recreation/beautification discourse is therefore rooted in the planning schemes of the 1950s and 1960s and remains an important discursive framing of Delhi's riverscapes.

In the second Master Plan (MPD-2001) channelization and riverfront development were outlined as "a project of special significance for the city" (DDA 1990). Yet, no projects were defined by DDA as the opponents of the channelization idea (in those days especially the neighboring states and the Central Water Commission) had requested DDA to obtain further studies of the river hydrology and ecology before river channelization could occur. The channelization scheme was propagated by the DDA to generate a 'post-independence gift' for Delhi's population of more than 7,000 hectares of green space for recreational activities (cf. NEERI 2005). Through this project, the channelization and recreation/beautification discourse are closely linked. The wasted land story-line has remained the dominant framing in the channelization discourse up to today. It is used by a discourse-coalition encompassing of engineering circles (e.g. the managing director E. Sreedharan of Delhi Metro), senior DDA planners, and retired DDA planners who were involved in the channelization schemes.

In the latest Master Plan (MPD-2021) and approved Zonal Development Plan for the River Zone, the channelization discourse has been replaced by an ambiguous "strategy for the conservation/development" of Delhi's riverscapes being "sensitive" to the "environment and public perceptions" (DDA [2007] 2010: 94). The dichotomy of conservation and development is emblematic for the proposed measures both in the Zonal Development Plan and latest Master Plan.

By outlining the importance of the groundwater recharge potential, the latest Master Plan partly institutionalized the conservation discourse by drawing on the ecosystem services story-line, which was developed as a counter story-line to the wasted land story-line. The current Master Plan and approved Zonal Development Plan also evoked the recreational and beautification potential of Delhi's riverscapes by outlining that the rejuvenation of the river must include the "potential for reclamation derived from the foregoing considerations, designation and delineation of appropriate land uses and aesthetics" of the riverfront (DDA [2007] 2010: 94).

Through the use of this vague terminology, the DDA allows room for interpretation. The reclamation of 'wasted land' is still considered even if important arguments against channelization have found their way into the Master Plan.

It needs to be highlighted that the 2007 approved MPD-2021 was finalized simultaneously to the change of land use for the Commonwealth Games Village and the large-scale land reclamation for Delhi Metro's lines, depots and the IT Park. The language and formulations of the MPD-2021 therefore need to be understood as flanking support and legitimization of the large-scale reclamation of land, which was further linked to the notion of the 'world-class' city. While not explicitly drawing on the wasted land story-line, the latest Master Plan still embodies the logic of this influential story-line. The MPD-2021 predates widespread opposition by local environmental NGOs against the environmental clearance for the Commonwealth Games Village. Lobbying for the river by the environmental NGOs resulted in the setting up the Yamuna River Development Authority and the no-development moratorium issued by the Lieutenant Governor in 2007. In order to 'manage dissent' the politicians and planners adopted the language of the conservation discourse by drawing especially on the bringing back biodiversity story-line.

The conservation discourse surrounding Delhi's riverscapes has been outlined using the ecosystem services and the bringing back biodiversity story-line. The ecosystem services story-line was developed against the channelization and riverfront development. The World-Bank funded study by BABU et al. (2003) aimed to deconstruct the wasted land story-line by valuating the services provided by the river to the city highlighting that the land of the floodplain is closely interlinked with the ecology of the river. The story-line is closely connected to the idea of bringing back biodiversity to Delhi's riverscapes. As the larger idea of the biodiversity zone concept emerged out of the joined projects between the DDA and CEMDE, the story-line was used by both planners and scientists. The story-line also made sense to the environmentalists as it promoted the protection of the river.

Overall, this resulted in a structuration of the conservation discourse (rudimentarily already institutionalized by the MPD-2021 as discussed above) when the Yamuna River Development Authority, under the leadership of the Lieutenant Governor, requested the DDA to develop the biodiversity zone concept for the entire River Zone. The biodiversity zone concept drew on the bringing back biodiversity story-line. This story line, in turn, is closely connected to the development of the Yamuna Biodiversity Park which was seen as the positive role model for the development of the entire River Zone.

The DDA drew on the bringing back biodiversity story-line in order to gain widespread acceptance for the development of the riverfront. The story-line was accepted and adapted by the central actors in the planning process for Delhi's riverscapes. Following HAJER's (2006: 70–71) conceptualization of discourse structuration and discourse institutionalization, the conservation discourse based on the bringing back biodiversity story-line (closely linked the ecosystem services story-line) became the hegemonic discursive framing of Delhi's riverscapes. A strong discourse-coalition of planners (DDA), politicians (in particular the Lieutenant

Governor) and scientists (CEMDE) seemed to make another "communicative miracle" (HAJER 1995: 46) possible which based on mutual understanding would eventually lead to an urban environmental policy-making in line with the larger conservation discourse.

The adoption of the bringing back biodiversity story-line by the DDA and the collaboration with CEMDE suggested a common understanding which would allow the actors to develop a holistic rejuvenation plan for Delhi's riverscapes. However, the discourse-coalition was not stable for a long time. Rather, difference in opinion between DDA and CEMDE became evident, particularly because the DDA Landscape Unit developed the Structure Plan for the Biodiversity Zone without a proper consultation of CEMDE. Furthermore, the development of the Golden Jubilee Park, which predated the biodiversity concept and draws directly on Jagmohan's plan for a riverfront promenade as well as the larger recreation/beautification discourse of the world-class city, was massively opposed by interviewed scientists from CEMDE and the entire group of environmental NGOs.

The latter mentioned groups were not involved in the process of planning for the biodiversity zone at all and only requested once to make comments to the draft concept in 2010, even when their protest had resulted in the overall structuration of the conservation discourse. In the course of the Debris Case the in the National Green Tribunal the Expert Committee deconstructed DDA's co-optation of the bringing back biodiversity story-line and stated that the biodiversity zone concept developed by DDA "is untenable and should be stopped" (EXPERT COMMITTEE MoEF 2014: 5). The discourse-coalition of the planners and the scientist had finally dissolved. The final order in the case indicates that also the court drew simultaneously on the recreation/beautification and the conservation discourse by suggesting "preservation, restoration and beautification".[272] Furthermore, the expert committee and the court agreed to the re-delineation of the River Zone based on the existing embankments following the logic of purification of spaces – either being part of the city or the river.

Before turning to hybridization and purification the following paragraphs outline the deeper logic of the story-lines fear of encroachment and slums as polluters. While ideas for Delhi's riverscapes shifted, these two story-lines remained important discursive framings throughout the whole debate on the remaking of Delhi's riverscapes. The following paragraphs recapitulate the underlying discursive framings of these story-lines.

---

272 National Green Tribunal Judgement, 13th January 2015, Original Application 6 of 2012, and M.A. Nos. 967/2013 & 275/2014 Manoj Misra vs Union of India and Others; and Original Application No. 300 of 2013 and M. A. Nos. 877/2013, 49/2014, 88/2014 & 570/2014 Manoj Misra & Madhu Bhaduri vs. Union of India and Others.

17.1.2 The persistent story-lines

The 'integration' of the river as recreational space into the urban fabric has been envisioned and discussed in all major planning documents for the city from the 'Interim General Plan for Greater Delhi' in 1956 until today. Nonetheless, the early proponents for developing Delhi's riverscapes into green and recreational spaces had already foreseen the difficulties to implement these ideas in the light of "haphazard" development along the river (TOWN PLANNING ORGANISATION 1956: 57). Therefore, the fear of encroachment story-line has deep historical roots and continues to influence the planners' overall approach to Delhi's riverscapes today. The encroachment discourse goes back to the colonial period and the term 'encroachment' is contested as it denotes certain practices or structures as being harmful and undesired.

The dominant discourse in the judgements of the courts since the late 1990s has equated encroachment with illegality (BHAN 2009: 139, RAMANATHAN 2006: 3193). This shift even included trying to suspend existing relocation policies to fast-track the reclamation of land from encroachment in Delhi's riverscapes and beyond. As outlined in detail in chapter 13, the poor have been deemed as "encroachers" (DUPONT & RAMANATHAN 2008, RAMANATHAN 2006) on the public land of Delhi's riverscapes, destroying the river's ecology by their disposal of untreated sewage, open defecation and other polluting practices such as washing of cloths (*dhobis*) or waste recycling. The fear of encroachment story-line and the slums as polluters story-line are closely combined here in a "salvage" and "scapegoating" discourse (COELHO & RAMAN 2010, 2013). Both aim to create legitimacy for slum demolition and the reclamation of land along urban water bodies.

Within these discursive logics, riverfront beautification and river rejuvenation appeared as legitimizations for the demolition of slums in the public interest and the interest of the river, which in turn offered opportunities for the land-owning agencies – especially the DDA – to reclaim land for 'planned development' (see similar for Chennai COELHO & RAMAN 2010 and for Ahmedabad DESAI 2012b). In recent years, the discursive framing of the 'world-class' city has come to bolster these story-lines further, as the 'slum-free' vision and the riverfront areas are a top priority in the remaking of Delhi (cf. BAVISKAR 2011c, GHERTNER 2011b, SHARAN 2015).

The slums as polluters story-line is emblematic for the discursive framing of the consequences of 'haphazard and unplanned growth' and is a narrative throughout the planning documents for the River Zone. The latest developments indicate that the logic of the slums as polluters has been broadened into a general discrimination and marginalization of traditional, 'unplanned' activities along the river. This study has identified this by describing the struggles of the farmers on the Yamuna floodplains against evictions. In this context, the farmers aimed to develop a counter story-line portraying themselves as the green keepers of Delhi's riverscapes.

This counter story-line, however, was deconstructed by the DDA and other state agencies on the grounds that the farmers are either themselves responsible for encroachment or have supported the development of slums in the past. The environmentalists only hesitantly agreed with the farmers as green keepers story-line as they have been critical of the pollution caused by the farmers' using of agrochemicals, as well as of the health hazards connected to the vegetables irrigated with Yamuna's polluted water. The story-line therefore remained largely defined by the Delhi Peasants who have, however, largely failed to make their voices heard as the farmers remained largely unorganized and could not convince the environmental NGOs to fully support them. Their historical political ties with the Congress Party were of very limited help, due to the larger shifts in the party's politics aiming to increasingly address middle-class interests in Delhi.

17.1.3 'Planned' versus 'unplanned' encroachment

In the eyes of the land-owning agencies, and the DDA in particular, the slum demolitions made way for 'planned development' of Delhi's riverscapes finally cleared from 'encroachments'. As in other cases dealing with the reclamation of wetlands for urban development in India (cf. COELHO & RAMAN 2010, 2013, D'SOUZA & NAGENDRA 2011, DEMBOWSKI 2001, RANGANATHAN 2015, ROY 2003), the government's nomenclature 'encroachment' is narrowly linked to slums and unauthorized colonies – 'unplanned developments' – while the state-led 'encroachment' into these eco-sensitive areas are overlooked in the policy debates. A retired DDA planner involved in the planning for the River Zone put this logic in a simple formula (DDA-10/2013-02-08):

> "Akshardham has come - very good temple, very good temple. Commonwealth [Games] Village has come - that is also a good one. Any unauthorized construction should not be there – no shanty, no *jhuggi*. That should not be there. […] Nothing should come unplanned. Planned can come!"

The managing director of Delhi Metro N. Sreedharan spoke of the large-scale projects as 'clean' developments in a similar way (SREEDHARAN 2009). Moreover, by highlighting rainwater harvesting or on-site wastewater treatment facilities of the metro infrastructure, the state agencies deployed the rhetoric of guaranteeing environmentally sustainable developments in order to legitimize the projects in the environmentally sensitive zone and distinguish them from the 'polluting' encroachments of unplanned development. This allowed for differential treatment of the non-conforming, polluting land uses (e.g. JJ cluster and unauthorized construction) in comparison to the planned, desired developments (cf. DE MELLO-THÉRY et al. 2013: 230) – which were, as the environmental NGOs argued, just as non-conforming with regard to the 'health' of the river. The detailed analysis of the land-use changes in the river zone highlighted the spatial effects of this controversial and discriminatory treatment.

The legality of any development or existing structure along the river was therefore not determined by whether it actually conformed with the existing urban planning or (urban) environmental regulations, but rather depended solely on whether it could be deemed to fit in the envisioned 'planned' development along the river based on DDA's ideas of riverfront development. Such a reading of the developments is in line with GHERTNER (2011e: 280), who has argued that "if a development project looks 'world-class', then it is most often declared planned; if a settlement looks polluting, it is sanctioned as unplanned and illegal." While the process of sanctioning these 'world-class' projects as 'legal' needed the final approval of the courts (as shown for the Commonwealth Games Village) and has sometimes been a long process, the fundamental logic of GHERTNER's argument has been observable along the river in the pure form. Justifying the developments the DDA (and the other government agencies) created an environmental myth of 'formal', 'plan-ned' developments being not harmful to the river and not resulting in any increased flood risk for the city. 'Planned' development was overall perceived to be *better* than 'unplanned', informal encroachments which would otherwise prevail sooner or later.

As described in more detail elsewhere (FOLLMANN 2015), the flexibility ensured by the court orders followed the logic of *unmapping* outlined by ROY (2003, 2004, 2009b). The tool of unmapping has been used in the River Zone in two consecutive ways. First, by not officially mapping the existing land uses (e.g. informal settlements, agricultural uses) as well as not defining the eco-sensitive River Zone clearly, the DDA has intentionally disregarded these existing uses and the sensitive ecology in the long process of plan-making, since planning was dominated by the channelization discourse. Or to put it differently, DDA was reluctant to draw any plan for the River Zone if it was not concerned with channelization. Second, DDA changed the land uses in the River Zone later on an ad hoc basis without actually having finalized the Zonal Development Plan. The DDA unmapped and remapped the banks of the river without having the legal plans approved. Therefore, the logic of the calculated informality has been inevitably affecting the 'planning' for Delhi's riverscapes.

In contrast to the narrow understanding of the planners referring only to slums and other unplanned structures as encroachment, opponents – especially interviewed environmentalists – denoted the large-scale developments as planned 'encroachments' on the city's lifeline. By denoting the state as 'the worst encroacher', the environmental activists adopted the language of the planners and the courts in their campaign for the protection of Delhi's riverscapes. The activists for example equated 'illegal encroachment' with 'planned encroachment' in their petition against the Commonwealth Games Village arguing that "the river flood plain in the city has seen a gradual land use change through encroachments on it either by illegal encroachers or most unfortunately by State created structures".[273]

---

[273] Petition in WP (C) No. 7506 of 2007, Rajendra Singh and Others vs. Government of NCT of Delhi & Others.

Even the High Court[274] in 2005 setting up the 300 meter benchmark expanded the narrow understanding of encroachment by originally stating that "[n]o encroachment either in the form of jhuggi jhopri clusters or in any other manner by any person or organization shall be permitted." However, in the demolition actions taken by the landowning agencies the narrow 'encroachment' discourse prevailed, based on the division of the 'planned' and 'legal' versus the 'unplanned' and 'illegal' development. In doing so, the planned/unplanned dichotomy allowed a flexible, differential treatment of slums and mega-projects.

'Planned' encroachment in the form of riverfront development through mega-projects was even justified by the DDA as terminating and preventing 'unplanned' encroachments. Such an understanding of planned development underlines SHATKIN's (2011: 82) argument that mega-projects in urban Asia have been realized based on "the perception that the state had lost control of the city". Thus, mega-projects appear as a strategy of the state to respond to the omnipresent urban informality by bringing back large-scale 'planned' development into the Indian city (FOLLMANN 2015: 220). In this context, COELHO & RAMAN (2010: 19) indicate that large-scale infrastructure projects in Indian cities "compress multiple rationalities of neoliberal governance". In the case of Delhi's riverscapes they have been both the tools for, and the results of, the purification of the hybrid spaces.

While the hegemonic clean/green/slum-free city discourse coupled with the 'world-class' imaginations stipulated the beautification of the riverfront areas based on the logic of bourgeois environmentalism (BAVISKAR 2003, 2011a), the projects of Delhi Metro and the construction of the DTC bus depots, but also the construction of new electrical substations are clearly contrary to the vision of an aesthetic riverfront. They have no connection to the river. In contrast, they are cutting the river off from the city. Therefore, these projects have been realized along the river drawing solely on the logic of the wasted land story-line nestled within the larger shortage of land discourse.

17.1.4 DDA's reluctant planning for the River Zone

In the analysis of the policy-making process for Delhi's riverscapes, the DDA has emerged as being driven largely by the idea of channelization and riverfront development, and thereby has overlooked developing more appropriate policies to govern Delhi's riverscapes. As DDA planners interviewed in the course of this study outlined, planning for Delhi's riverscapes is only meaningful when it directs immediate development actions, since the omnipresent practices of 'encroachment' on barren land can only be controlled and prevented when the land is 'properly' developed *before* it gets encroached (DDA-01/2010-09-21, DDA-06/DDA-2011-11-22, DDA-07/2011-11-23).

---

274 High Court of Delhi, 08 December 2005, WP (C) 2112/2002 Wazirpur Bartan Nirmata Sangh vs Union of India and Others, and W.P.(C). NO. 689/2004 Okhla Factory Owners' Association vs. Government of NCT of Delhi.

The DDA's planning efforts concentrated thus on the reclamation of land along the river for immediate 'planned' development based on the fear of encroachment story-line. In this context, the channelization discourse had long envisioned to make available the 'wasted land' in-between the two fast-growing parts of the megacity. However, the 'great scheme' of channelization and riverfront development which was supposed to cure the problem of the slums and reconnect the city to the river never materialized. The DDA nevertheless reclaimed land from the river through large-scale development projects based on exceptionalist logic for infrastructure developments, especially connected to the Asian Games in the 1980s as well as for the construction of the Delhi Metro and the Akshardham Temple Complex/ Commonwealth Games Village in the 2000s. The development of sports complexes and high-profile infrastructure (e.g. Delhi Secretariat) was in line with the larger riverfront development schemes.

Both mega-sport events emerge as extraordinary moments in the socio-ecological transformation of Delhi's riverscapes. While the Asian Games resulted in the development of high profile infrastructure and the growth of the slums along the banks in the late 1970s and early 1980s, the Commonwealth Games brought about the demolitions of the slums in the course of city-wide urban transformations and the development of urban mega-projects. The 'planned' development for Delhi's riverscapes proceeded therefore often in an ad hoc, informal manner. The developments by the DDA for the Asian Games clearly violated the Master Plan. For the Commonwealth Games the DDA changed the Master Plan, but the construction violated existing environmental regulations (including EIA). Many commentators have viewed the DDA thus as the worst violator of the Master Plan itself (cf. BAVISKAR 2011b, BHAN 2013, CSE 2004).

Furthermore, when the DDA failed to proceed with the 'great scheme', the authority's interest in planning for the river zone diminished. Firstly, because the immediate opportunity to development the land was not given. Secondly, the DDA was never interested in planning for the conservation or protection of Delhi's riverscapes for the sake of the ecology of the river, but rather aimed for large-scale urbanization. The DDA has always been preoccupied with developing land for 'planned' urban extensions (e.g. East Delhi, Narela, Rohini, and Dwarka). The DDA's legacy is therefore somehow contrary to the conservation of land or rivers. In this sense, the DDA always considered the Zonal Development Plan to outlay 'development' – not 'river protection' or 'environmental conservation'. After the Central Government had declared 3,500 ha of land in the River Zone as 'development area' in 1989, the large-scale 'planned' riverfront development seemed to be only a matter of time. However, the channelization discourse was finally rejected and even DDA planners indicated that channelization was a 'dead proposal' already in the MPD-2001 in 1990.

It is important to highlight the role of the media which promoted DDA's channelization and riverfront development in the beginning. In particular, an artist's impression of the riverfront development was published by the DDA in the late 1980s and the media took over the idea (ARCH-01/2011-12-09):

"They [DDA] were slowly planting it [riverfront development] into media. There is a slow contamination of the public mind by this idea. It is a slow process by which they release little by little the toxins. And finally, when the environment is right, they bring out the project"

In line with this argument, it is remarkable that the DDA did not discard the idea of riverfront development and channelization but rather slowed down the planning process for the River Zone. When riverfront development faced further opposition the 'great scheme' was redesigned by the DDA into smaller 'pockets' of land to be reclaimed from the river. Furthermore, when the draft proposals for the Zonal Development Plan aiming to institutionalize these reworked channelization schemes were rejected on ecological grounds, the DDA tried to slowly adjust its plans without changing the main direction. By and large the DDA has been reluctant to plan for the River Zone.

In doing so, the DDA neglected its duty to prepare a legally binding Zoning Development Plan in order to retain the flexibility to proceed with the channelization and riverfront development plans in the future. Thus, the River Zone remained officially 'unplanned' until 2010 and even the land-use designation in the 2010 Zonal Development Plan appeared to some extent to be randomly chosen (e.g. the change of color for the green patches on the east bank discussed in chapter 11.7) and reflect DDA's development interest (e.g. flawed floodplain zoning, see chapter 11.6). The boundary of the River Zone remains contested (see 16.5.2) and the DDA has been reluctant to integrate sectoral environmental regulation (e.g. to conserve the forests, to regulate groundwater) into the Master Plan and Zonal Development Plan. In doing so, the DDA kept land-use and environmental regulation intentionally vague for the River Zone to ensure its own flexibility in the future development process. The DDA used this informality as "a practice of planning" (ROY 2012: 698) through flexible instruments (e.g. the declaration of 'Public and Semi-Public Use') and exceptions (e.g. Commonwealth Games Village) realizing urban megaprojects according to the earlier schemes for riverfront development. Nevertheless, the DDA was put under pressure by other government agencies to accommodate 'world-class' infrastructure in the River Zone; most importantly Delhi Metro (metro depots and lines) and DTC (bus depots).

17.2 The hybridization and purification of Delhi's riverscapes

In dealing solely with the territorial category of land, DDA's tools of master planning and land-use zoning completely failed to pay attention to the ecological processes and the fluidity of Delhi's riverscapes as well as the multiple city-river relations. The DDA has treated the river with a site-based rather than system-based approach in order to formalize the separation between the city and the river in terms of land-use categories (WHATMORE & BOUCHER 1993: 169). In doing so, DDA has purified the hybrid riverscapes into the domains of land and water – plus a remaining, hybrid intermediary space of the so-called 'active' floodplain. DDA's underlying rationality of the invention of the three river morphologies ('river-bed' = area under river water, 'river flood plain' = area between the river water course and the embankments, and 'river front' = area outside the embankments), outlined in chapter 11.7, was to claim planning autonomy for the land protected by embankments.

The conflict between the DDA and the Yamuna Standing Committee/Central Water Commission with regard to floodplain zoning is crucial for the overall understanding of the purification logic. While the physical construction of embankments reclaims the land from the river and produces socio-natural hybrids in the form of reclaimed spaces, the DDA aimed to obtain full control over the thereby reclaimed land through discursive purification. This would have allowed developing the land based on DDA's urban land-use categories. Purification work is therefore closely connected to conflicts between urban and environmental governance institutions on different scales. Thus, separating the domains of the river and the city is inevitably connected to governing power over these spaces.

The analysis has shown that the wasted land story-line was the baseline rationality for the construction of embankments and cut and fill intended to purify land for urban development. The deliberations on the channelization schemes from the 1980s to 1990s, DDA's weakening of the recommendations made by NEERI in its reports on the environmental restoration of the River Zone, and most especially the categorical demarcation of the realm of river by embankments in the case of the Akshardham Bund, have highlighted the 'co-production' of science and policy. In this context, the separation of the 'active' floodplain from the 'passive' floodplain emerges as a social construction. Floodplains are the result of complex interaction of fluvial processes. These processes are only partially obstructed by the construction of embankments and the creation of land as a socio-natural hybrid which is generally not flooded by the river anymore. However, groundwater flows and seepage through the embankments highlight that 'passive' floodplains are not *passive* in terms of their interaction with the 'active' river and its remaining 'active' floodplain.

While the notion of the 'active' floodplain makes sense in epistemological terms to separate different categories of land in order to make urban environmental policies, the ontological separation of the 'active' floodplain from the land reclaimed by embankments – as it has been argued by the DDA and NEERI in the case of the Akshardham Bund – is not only wrong, but actually curtains the underlying forces which drive the metabolizing processes of urban environmental change. Much debate in the case of the Commonwealth Games Village concentrated on whether the site is on the floodplain or not. NEERI's scientific opinion simplified the hydrological system of the river to the existence of the man-made embankment and all other ecological aspects (e.g. groundwater recharge potential) were becoming meaningless (cf. BAVISKAR 2011c: 51).

Technical studies by different scientific bodies (e.g. CWPRS, NEERI) legitimized partial channelization and riverfront development and established the perceived ecological feasibility for the reclamation of land for urban development. Therefore, natural science acted as a powerful advisor in this process of boundary-making between nature and society, the river and the city. As especially evident in the case of the Commonwealth Games Village, other claims aiming to denaturalize and problematize these scientific facts and create a more holistic understanding of the city-river relationship failed. This failure highlights the interlinkage of truth and power (FOUCAULT 1984).

While in the case of the Commonwealth Games Village the purification was applied in order to reclaim land for 'planned' development, in the case of the slum demolitions the floodplain areas were purified as the realm of the river by the court. Moreover, in the latter case, the court engaged in purification work in order to bypass (or change) existing policies of slum relocation by arguing that the space under 'encroachment' is a *water body* and not *land*. Therefore, as also argued by FORSYTH (2003: 89), drawing boundaries by deploying purification work is able to establishment rationalities for policy frameworks. These two examples of different purification logic highlight the flexibility as well as the highly politicized nature of drawing boundaries between the realms of humans and nature, the river and the city, based on scientific 'evidence'.

The latest approach by the DDA to rezone the river by drawing new boundaries also highlights that boundary-making is a political exercise, which is shaped by truth claims and power struggles. While traditional uses of the riverscapes are denoted as being illegal and polluting – as shown for farming in the course of this study – other more recent 'encroachments' on the river are deemed being not harmful to the river. Again simplification based on the existence of embankments has been deployed in this process. In this context, the findings of this study further underline MURDOCH's (2006: 109) argumentation that it is paradox how environmentalism spatializes nature by separating it into different zones.

The Structure Plan for the Biodiversity Zone developed by DDA follows this paradox logic. While in the past the spatial classifications were applied to *preserve* nature from urbanization (cf. MURDOCH 2006, WHATMORE & BOUCHER 1993), in the current process of re-zoning the River Zone the topographies and territories along the river in Delhi are re-classified based on the idea of bringing back biodiversity. While this emerges as an approach to link human-nature relationships in the realm of the city, the outlined zoning classification of "protective", "interactive" and "public recreation" as well as the proposed fencing of these areas highlights that the realm of nature and humans along the river is intended to be purified not only to save (the remaining) nature but to bring back nature. However, as evident from the monsoonal floods of 2010 and 2013, when the Golden Jubilee Park was flooded and covered with a layer of silt, the application of these fixed spatial classifications and its materialization as lawns, flower beds, butterfly parks and bamboo huts fundamentally failed to be resilient to the fluidity of the riverscapes.

In contrast to the above summarized practices of purification, the land-use classification of 'Agriculture and Water Body' in the Master Plan expresses the hybridity of Delhi's riverscapes very well. On the scale of the Zonal Development Plan, this category is, however, simplified, and thereby purified, into "river & water body/pondage" (DDA 2010b).

Overall the analysis underlines that urban planners have remained uncomfortable in planning for the River Zone as its hybridity of water and land posed multiple difficulties to control land uses (fear of encroachment story-line) and to create fixed boundaries for 'planned' urban development. The purification of the hybrid riverscapes into the firm spatial categories of land (riverfront) and a well-behaved river (water channel) and the production of "neatly packaged" nature (SMITH 1984: 57) in the form of riverfront parks thereby deliberately ignored the multiple city-river relationships.

The river, as the non-human actor in the Latourian sense, raised a claim to be incorporated in the plan-making and policy-making for Delhi's riverscapes by silting up the floodplain, shifting its bed, and overflowing the space assigned to it by planners and engineers. The river's actions metabolized with the human (structural) intervention and created new hybrids. New barrages, embankments and bridges (with guide bunds) reclaimed land and exercised a certain degree of control of these 'natural' processes. The land temporarily abandoned by the river is a result of the metabolizing interaction between human interventions and the changing behavior of the river.

It has been evident that Delhi's riverscapes as metabolized hybrids have been discursively reproduced and new meanings, perceptions and ideas have been attached to Delhi's riverscapes over time. The purification of the riverscapes has been deployed in a highly flexible manner by the state agencies. On the one hand, the riverscapes were 'cleansed' as the realm of the river by e.g. prohibiting bathing and washing in the river in the colonial times, or most evidently, evicting the slum population from the banks in order to control pollution. In doing so, powerful human actors (discourse-coalitions often coined by scientists, planners and politicians) have aimed to control the river, both materially and discursively in order to exercise power not only about nature (the river) but also about the human-nature relationship and the people living along the river's banks. COELHO & RAMAN (2013: 151) speak in this context of a new "teleology of ecological enlightenment in urban governance" which is marked by a shift in the perception and dealing with urban lakes, wetlands and rivers. For decades, they argue, urban water bodies were seen as "land-in-the-making, to be filled and reclaimed" for urban developments (COELHO & RAMAN 2013: 146). However, nowadays, these hybrid spaces are seen as "lakes-in-the-making, to be cleared, dredged, desilted and beautified" (COELHO & RAMAN 2013: 146).

Urban water bodies are thus not only discursively purified as (natural) water bodies, but rather their socionature is (re)hybridized through material actions like dredging, de-silting, and beautification. However, as shown in this study – and by others for different water bodies across India (cf. COELHO & RAMAN 2013, D'SOUZA & NAGENDRA 2011) – the purification of the river takes place not primarily for the sake of the ecological functions or to bring back nature, but as an aesthetic component of the 'world-class' beautification agenda and the 'green agenda' (cf. GHERTNER 2011b, 2011e). Overall, this raises multiple doubts of an urban environmental awakening (especially with regard to socio-ecological justice) in urban environmental governance in India.

17.3 Urban environmental governance: prevailing insights

Delhi's riverscapes are governed by a multiplicity of agencies with multiple linkages between the different levels of government from the national to the local level. The different tasks (e.g. land-use planning and pollution control) are spread among a large number of state agencies, but the responsibilities and liability of the concerned agencies is often very limited. Sectoral approaches to the governance of Delhi's riverscapes prevail. Urban environmental governance with regard to Delhi's riverscapes is therefore characterized by fragmentation and lacks coordinated action to upgrade the physical conditions. The fragmentation of the governance set-up emerges as both cause and effect of sectoral approaches in governing the urban environment in Delhi.

The DDA holds the power over land-use planning for the River Zone based on the provisions of the Delhi Development Act and has eminent domain over land. As outlined, the DDA has been reluctant to plan for the River Zone after the rejection of the channelization schemes. Therefore, the most important state actor never intended to protect the environmentally sensitive floodplain but rather envisioned its material transformation to fit in the urban development trajectory. In this context, the study has outlined how the DDA was requested by other state agencies operating in the sphere of environmental governance (e.g. CWC and the YSC) to commission studies to government-sponsored technical engineering research institutes (NEERI, CWPRS) in order to legitimize the authority's river channelization and development schemes. While originally the scientific 'facts' more or less explicitly rejected channelization and riverfront development, these 'facts' eventually were interpreted and instrumentalized by the DDA as legitimizing the reclamation of land from the river for urban development at least in selected pockets.

The limited powers of the YSC and the CWC to prevent the DDA from reclaiming land were outlined in detail throughout the study. While the YSC has the oversight over the hydrological interventions along the river, and the CWC issued a draft for floodplain zoning already in the 1970s, state-level interests prevailed and the riparian states largely disregarded the decisions of the interstate and central body. While all major projects need approval from YSC, the interstate body emerges as a rather weak authority, since the YSC has often been side-lined and projects have been developed without approval from the YSC.

The YSC has no direct powers to enforce its decisions and emerges powerless if projects are simply not placed before the committee. Furthermore, it needs to be pointed out that approvals by CWC and YSC are solely based on technical details regarding water flow and flood management. These clearances do not include any other environmental components and therefore do not equate to ecological clearances and must not be confused with EIAs. EIAs are equally inadequate tools for the protection of the riverscape. This was demonstrated by the circumstances of the environmental clearance for the Commonwealth Games Village (cf. GILL 2014). Thus, while the existing institutions are in place to deploy oversight and regulate the environment, powerful actors such as the DDA (or Delhi Metro) are able to bypass certain existing environmental rules and regulation.

Generally environmental science and urban planning hold key roles in the (re)making of urban environmental policies with regard to Delhi's riverscapes. However, new counter-discourses are gaining importance in the wake of urban environmental activism since at least the early 1990s. In this process, the Indian judiciary has played an increasingly important role responding (in often contradictory ways) to PILs transforming itself into an arena of urban environmental policy-making. The courts have set up several committees and expert groups, which have largely failed to stipulate long-term changes in the urban environmental governance of Delhi's riverscapes.

The courts are nevertheless an important platform for knowledge exchange even between state agencies which are pressed to cooperate to a certain extent. Yet, the PILs filed in the courts against state-led, large-scale development projects failed to stop the projects. These insights highlight that, as BHUSHAN (2004: 1772) argued, in "cases where the cause of the environment [has been] pitted against 'development projects'" of great or even national importance the courts have most often ruled in the interest of development even if "the legal soundness of the case was also evident from the fact that some of the judges gave dissenting judgments or that the court went against the advice of its own expert committees." The Commonwealth Games Village case is emblematic for BHUSHAN's argumentation and underlines also the reasons why the effectiveness of environmental protest through the courts against large-scale projects has been viewed critically by some interviewed environmentalists and environmental lawyers.

The Indian courts have remained an important way to challenge environmental degradation and to facilitate changes is (urban) environmental governance (cf. WILLIAMS & MAWDSLEY 2006b: 666). The newly established NGT has emerged as a progressive new judicial space considering PILs on the river. The planning of the Biodiversity Zone highlighted a general shift in dealing with the city-river relationship triggered by environmental PILs and enforced by the NGT. While rejuvenation of the river had already become the central term in the MPD-2021, much action had concentrated on the water pollution issue. The developments between 2009 and 2014, and especially the deliberation in the NGT since 2012, indicate that for the first time issues of water pollution and land-use change were discussed in their close interconnection.

It is important to note that in line with the cases filed by other environmental activists (cf. MEHTA 2009a, 2009b), the petitions in the Debris Case and Drains Case were not filed against state-led, large-scale development projects, but rather these cases were filed as *requests for action* by different government authorities. This is an important distinction as the latter cases are dealt with differently by the courts which are able to facilitate action by state agencies rather than preventing 'development'. For example, in the Drains Case and Debris Case the state agencies were unable to join forces against the development of a certain project, but rather the state agencies mutually accused each other of being responsible. Therefore these cases developed completely differently to the case against the Commonwealth Games Village; the state agencies were rather fighting against each other in the court than jointly defending a project. While the courts played an important role in the remaking of Delhi's riverscapes and the NGT has recently facilitated action to clean the river and develop a holistic concept for its banks, their role remains ambivalent especially with regard to the poor.

The orders by the courts strongly influence the remaking of Delhi's riverscapes and further highlight the unpredictability of the judicial system. The development of a PIL is depending on a series of influencing factors including among others the personality of the involved judges, results by expert committees, public discourse and media coverage, emerging alliances, and elections or changes in the governments (cf. VÉRON 2006). Furthermore, the judicial activism on the Yamuna, including the recent cases on the protection of the floodplain and the earlier river cases focusing on pollution and flow, indicate that court-induced policies and measures do not ensure long-term action and implementation. In contrast, the enforcement of judgements and directions given by the courts is often not more effective than the implementation of already existing laws (see for similar findings DEMBOWSKI 2001: 60). Overall, the insights from this study suggest that urban environmental governance in India, as argued by SIVARAMAKRISHNAN (2015: 123), does not lack existing environmental legislations to govern the urban environment. Rather it requires a recrafting of the city-nature relationships (TIWARI et al. 2015: 175) and a redefining of the tasks and responsibility of the state agencies.

In the case of the Yamuna many ad hoc plans and measures were proposed by the court under often too short timeframes and without taking into account the complexity of the ecological system of the river. Adding to these insights, the discussed purification logics deployed by the court highlight that the courts have the powers to demarcate boundaries not only between legal and illegal, planned development and encroachment, but also between nature and society, the river and the city.

The importance of the judiciary in environmental causes is, however, an indicator for opportunities for environmental activists to participate in urban environmental governance. Other authors have even suggested that environmental activism through the courts might hinder long-term engagements between civil society organization and the state agencies in the sense of integrated urban environmental governance (cf. UPADHYAY 2001, VÉRON 2006).

Although critically engaging with middle-class environmental activism in general, this study has highlighted the important role of environmental NGOs in remaking urban environmental policies. While environmental NGOs and individuals have long campaigned for the cleaning of the river's water, only the recent development of urban mega-projects in the river zone has resulted in a wake-up call for environmentalists emphasizing on the ecological integrity of the river and its floodplain and the multiple relationships between the river and the city. The protection of the floodplain from further 'planned encroachment' became part of the already existing multiple discourses around the river. Ecological reasons ultimately prevented the 'great scheme' of channelization for the river Yamuna in the Delhi stretch. While today's local environmental NGOs did not exist in those days, Delhi-based environmental think tanks such as the Centre for Science and Environment and INTACH have been involved in the protection of the river for almost two decades. In more recent times, the constant opposition to the reclamation of land has incrementally redirected the discourse on Delhi's riverscapes towards the conservation of the floodplain.

The local NGO Yamuna Jiye Abhiyaan but also other environmental groups have been formed to lobby for the river in the course of the mega-projects along the river. By using the powers of the RTI Act to access knowledge and spatial knowledge tools like Google Earth they have been able to fulfil a "watchdog role" (MAWDSLEY 2004: 89). Overall, this rather small network of activists has constantly lobbied for the river and built up a knowledge base on the river as well as its governance structures. In contrast to the green/clean 'world-class' city discourse and the anti-poor character of bourgeois environmentalism, the work of local environmental NGOs has shown a different kind of urban environmentalism – clearly distinguishable from the institutionalized bourgeois environmentalism patronized and incorporated by the state agencies (cf. FOLLMANN 2014). The environmental NGOs working on the river have tried to collaborate with the farmers protesting against the development projects and their eviction. Yet, since all environmental NGOs are middle-class dominated, they struggled to overcome the dichotomy between middle-class conservational activism and the interests of the farmers.

Overall, the insights from this study suggest that environmental NGOs – both local action groups and larger environmental think tanks – have developed into important actors in collecting, disseminating, (re)interpreting and (re)producing environmental knowledges of Delhi's riverscapes. They have consequently emerged as key players in the social construction of environmental knowledge, the shaping of public opinion and strongly influence environmental discourses by challenging government approaches towards the river and the river-city-relationship. Nevertheless, environmentalists are challenged to 'walk a fine line' balancing between protesting, petitioning, and policy counselling.

18 CONCLUSION

City-river relationships are globally renowned to have a discrepancy between "imagined futures" and "unintended consequences" (LEIGH BONNELL 2010). The research presented in this book chose the Yamuna in Delhi to study these relationships. The problems surrounding the Yamuna in Delhi are not an isolated case. Rather, they represent issues reminiscent to other Indian urban rivers (cf. BEAR 2011, COHEN 2011, DESAI 2012b, NAGPAL & SINHA 2009, PESSINA 2012) as well as to urban river across the world (cf. CASTONGUAY & EVENDEN 2012b, DESFOR & KEIL 2000, 2004, GANDY 2006a, HOLIFIELD & SCHUELKE 2015, LEIGH BONNELL 2010, RADEMACHER 2007, 2008, 2011).

This study analyzed the relationships between the Yamuna and Delhi by explicitly focusing on Delhi's riverscapes to understand how and why urban environmental change occurs, which socio-ecological impacts are connected to this, and how and why urban environmental change has accelerated in recent decades.

The empirical chapters have presented a detailed analysis of the material and discursive remaking of Delhi's riverscapes showing that "different discursive regimes" (ESCOBAR 1996b: 326) have dominated the way the relationship between nature and society, the river and the city, have been articulated, framed and defined. The vision of the socio-ecological future of Delhi's riverscapes has clearly been shifting over time. While the planners' broader goal has always been to integrate the river into the urban fabric, the actual material outcomes of the urban development process have been at odds with this larger goal.

Both the reclamation of land and associated land-use changes (e.g. bulky and polluting infrastructure developments such as power plants, fly-ash ponds, transportation lines and depots) and the degradation of the river through increasing levels of pollution have resulted in a physical separation of the 'planned' city and the river. Delhi's riverscapes have been largely forgotten by the majority of the city's population (cf. JAIN 2011: 42). For decades they were a "neglected 'non-place'" (BAVISKAR 2011c: 45) and an "in-between space" (SHARAN 2015: 1) in the sprawling megacity dominated by slums. This changed as the 'wasted land' in the 'backyard' of the city, became the 'front yard' in the urban redevelopment and restructuring since the turn of the new millennium. The river'front' emerged as a major development frontier in the remaking of Delhi as a world-class city. The global flows of ideas and technologies (APPADURAI 1990, 1996) and the larger changes in the political and economic conditions in India, largely defined by neoliberal economic politics, have therefore had a tremendous impact on Delhi's riverscapes catapulting the riverfront onto the agenda of urban redevelopment.

The detailed reading of the planning history of Delhi's riverscapes has shown that the most recent large-scale developments along the river are rooted in long-desired urban imaginations to reclaim and develop the riverfront based on Western models like Paris or London. Basic schemes for the reclamation of ecologically sensitive areas, which long remained as 'left-over spaces', had been envisioned early on, but only the current processes of urban transformation fueled by the world-class city discourse and catalyzed by massive investments in the urban infrastructure as well as public-private real estate developments has resulted in the realization of the projects.

A multiplicity of actors embedded in complex governance structures have been involved in this remaking of Delhi's riverscapes. While a growing urban environmentalism has aimed to rejuvenate the degraded riverscapes by pressing for pollution abatement and environmental restoration, the river has remained foul smelling and black in color without any natural flow in the non-monsoonal months. The massive investments for river rejuvenation have so far completely failed. The Yamuna remains a critically threatened riverine ecosystem.

Against this background, the most recent discursive reframing that Delhi's riverscapes are the place to bring back nature into the city appears as paradox. Any such efforts will fail if the sewage of the megacity is not stopped from flowing into the river untreated and the dumping of solid waste and debris materials is allowed to continue. While different ideas for the protection of the River Zone have been discussed in the past, the river remains apparently 'unprotected' from further ecological destruction.

To date, the river is biologically dead in Delhi and the downstream areas up to the confluence with the Chambal River. Despite this, life has managed to survive on its floodplain and if the river is allowed to flow the river can be rejuvenated and reconnected to the city without channelization and concretizing its bed. No change will occur without political will. Political will can be 'created' by public awareness and pressure as shown by the actions of environmental activists lobbying for the river in Delhi. However, the rejuvenation of Delhi's riverscapes will take time – as river restoration projects generally do. The courts have pressed politicians, scientists, planners and all other involved actors into much too strict time constraints to come up with ideas, schemes and plans for Delhi's riverscapes. As a response, politicians and planners – the latter put under pressure by the former – initiated short term actions. Yet, they have largely failed to develop a long-term vision for Delhi's lifeline and action plans with feasible timeframes.

Overall, it has been shown that the existing Zonal Development Plan, which suggests at least a certain degree of protection has not provided legal protection for the floodplain. As an urban land-use plan it was actually never designed for this purpose, but rather the urban planners of the DDA instrumentalized their dominant powers over planning and land to purify the fluid and hybrid riverscapes into the fixed territories of land and legal categories for its use. In contrast to the urban planners' approach, urban environmentalists have aimed for a declaration of the socio-natural hybrid as the purified realm of the river. In doing so, both conceptualizations drew on the modern dichotomy of nature and culture, the river and the city. Until today, the fluid space, the land temporally soaked in water, has not been reconnected with the city. It has either been conceptualized as the frontline of urban expansion by the planners or as an environmentally sensitive refuge in need of protection from further urbanization and degradation. The *tale of two cities, but only one river* still remains to be rewritten as Delhi's riverscapes still represent the boundary between the materially and discursively old, traditional city versus the new 'world-class' urban India.

18.1 Riverscapes: (mega-)urban challenges and the question of governance

The riverscapes approach has stressed the historical spatial changes and related these to current socio-environmental challenges and governance processes. Current dynamic processes of urban environmental change and the reclamation of ecologically sensitive spaces like urban rivers, streams, water bodies, or wetlands are deeply connected to changing discursive framings of the role and function of these specific socio-ecological hybrids in the remaking of cities. Changing political, economic, and also socio-cultural conditions all materially and discursively affect this reinterpretation of urban nature. Through outlining the long-lasting channelization discourse for Delhi riverscapes, it has been shown that dominant discourses and their associated story-lines remain persistent over long periods of time and under completely reworked political and economic condition. This is especially the case when powerful state actors keep these discourses alive by constantly reframing the rhetoric logic as well as the proposed measures.

These insights translate to other geographical contexts as well. Therefore, while environmentally sensitive spaces in the megacities of the Global South are facing high development pressures and environmental degradation today, their remaking might have been foreseeable two or more decades ago. Taking into account the history of urban imaginaries, a discursive approach as used in this study, therefore helps to deconstruct and better understand the seemingly paradox logic of today's transformations of urban nature. A deeper understanding of the interaction of the multiplicity of actors and the global-local relations in these transformations is of great importance with regard to the manifold questions of urban sustainability as high development pressures and massive environmental degradation pose tremendous future challenges for the megacities of the Global South.

"[T]he priorities in planning and governance are currently undergoing substantial changes" (KRAAS & MERTINS 2014: 4) and the conceptualization of the urban ecology of the cities in the Global South as "inherently problematic" (ZIMMER 2015b) and urban environmental governance in these cities as completely "failed" (LAWHON et al. 2014: 9) is more and more challenged. Nevertheless, the restoration of ecologically sensitive urban spaces is far from emerging as a top priority in the remaking of the Indian cities. Often the inherent potential of urban rivers and other water bodies for a unique and environmentally-sound urban transformation remain unexplored. In contrast, a certain 'one-size-fits-all' approach is taken by many Indian cities – covering of drains and the concretization of river banks emerge as widely applied strategies to purify these hybrid spaces for urban needs that range from beautified, recreational parks to car-parking spaces.

While acknowledging the complexity of the mega-urban challenges, the development of projects such as the Golden Jubilee Park, illustrate planning incompetence and governance failures, whose deeper reasons remain opaque if discursive logic and underlying governance structure are not combined in an in-depth analysis. The analysis of the governance set-up for Delhi's riverscapes pinpoints the multiple challenges within the state system.

Due to the multiplicity of agencies the responsibilities, tasks and activities are often blurred and need to be restructured and redefined in order to develop effective ways of inter-sectoral communication, coordination and collaboration. It is evident that functional and spatial interdependencies are often not adequately reflected in still overwhelmingly sectoral divisions of governance set-ups with lacking inter-level and intra-level coordination of actions. River and water governance emerge in this context to be extraordinarily complex. Overlapping scales and responsibilities along the whole hierarchical set-up from the municipal to the federal level are involved and it is a challenging task to implement new decision-making structures and assure public participation due to the "very nature" of these governance set-ups which are "difficult to transform because the implementation of transformative decisions often implies large-scale infrastructural works" (VAN DEN BRANDELER et al. 2014: 15).

The diversity, dynamics and complexity of urban environmental change in megacities – including both land-use changes as well as environmental degradation of water and land – constitute challenges which require a massive shift in the way sectoral approaches currently deal with 'single' facets of the overall problem. Typically these are largely separated, uncoordinated and widely uncontrolled and isolated from political and societal oversight. Therefore, the governance set-ups, as evidenced for Delhi's riverscapes, are required to be reformed in such a way that clearer mandates are assigned, inter-sectoral cooperation and collaboration is strengthened and fully institutionalized, and the involved agencies (and individuals) can be held accountable.

However, as highlighted by KOOIMAN (2003: 7), interaction in governance requires an "openness to difference" in opinion of the societal problem, "a willingness to communicate" and to share the information about the societal problem, and maybe most important "a willingness to learn" by acknowledging different forms of knowledge and adapt these insights into policy-making. This willingness of interaction is not institutionalized in the Indian government environment and often remains limited to "a sensitive officer who listens" (YFW-2013-03-08) and the individual contacts within and beyond the government machinery. More often files are diverted or information is compiled in reports – often through copy and paste methods. While these reports and information are sent around, they do not appear to be read often. Even when they are read, the information often reads vague and ambiguous, and communicative action among the different state agencies is not taken to overcome these problems.

Furthermore, as these different agencies and individuals draw on multiple social constructions of the river and the city – often poles apart – coordinating structures are required to govern the complex human-nature relationships in megacities.

Moreover, due to functional and spatial interdependencies as well as scientific complexity, uncertainty and ambiguity, pressing environmental concerns in megacities of the Global South (like river pollution or the encroachment on floodplains) emerge as "persistent problems" (HOGL et al. 2012: 2). While mutual agreement is often reached for actions addressing these problems, decades go by without any measurable improvement or concrete measures of action. As a result, besides new forms of political participation and an increasing number of NGOs working on issues of the urban environment, the government retains a "substantial authority for environmental regulation" (BRIDGE & PERREAULT 2009: 481) underscoring that there is *no governance without government*.

18.2 Evaluation of the conceptual approach: do riverscapes make sense?

This study has been based on the objective of overcoming the city-centric perspective on urban rivers, focusing on the riverfront and more holistically conceptualize the multiple city-river relationships. It has been argued throughout the chapters that an analysis focusing on riverfront development and its aesthetic values for recreation or real estate development without taking into account the multiple connections between cities and rivers fails to generate a deeper understanding of urban environmental change. The limitations of the 'riverfront approach' in trying to demarcate, and spatially purify, the spheres of the city and the river are rooted in the modern ontological separation of society and nature, humans and non-human (LATOUR 2004). The combination of governance research with theoretical insights from UPE within the concept of riverscapes sets out a holistic framework to address the relationship between cities and their rivers, and contributes to a reconceptualization of the ecology of the city and the human-nature interaction in general. In doing so, it offers a more holistic approach to study urban environmental change and governance.

The conceptual framework of the hybrid riverscapes has been developed and applied in this study in order to overcome the focus on the liminal space at the water's edge, and implicitly fixed spatial boundaries between the river and the city. Taking into account both material spaces (geographic territories) and environmental narratives (discursive representations) the riverscapes approach acknowledges the assemblage of multiple non-human and human processes transforming urban rivers over space and time. The concept further rejects any pre-given 'natural' boundaries separating the river (nature) from the city (culture), but rather highlighted the discursive co-production of these categories as 'negotiated facts' between science and politics (FORSYTH 2001, HAJER 1995). The nature of the river and its hypothetical boundary with the city, the riverfront, emerge as social construction. The empirical findings illustrated that land-use planning through the exercise of master planning, and land-use zoning in particular, institutionalizes the purification of socio-natural hybrids. The riverscapes approach is therefore of great benefit for widening the analytical perspective in urban environmental governance research.

This study argues that the theoretical conceptualizations from UPE and the conceptual inspirations from science studies, offer the opportunity to challenge conventional understandings of urban environmental change in the mega-urban context of the Global South. As UPE research has deconstructed "the dynamics of hegemonic projects for socioenvironmental transformation often highlighting deployments of electoral, legislative, deliberative, and administrative processes within institutions of government and the state" (HOLIFIELD & SCHUELKE 2015: 295), the existing links between UPE and governance research are unmistakable. Particularly, the post-structuralist perspective combined with insights from governance research helps to overcome the common a-priori positionality of actors typical for earlier structural political analysis in political ecology (FORSYTH 2008: 758).

As this study has shown, the multiple processes of negotiation and bargaining between state actors and environmental NGOs are marked by fluid and varying relationships. A UPE-governance approach, methodologically rested upon the classical interpretative/hermeneutic paradigm focusing on actors and extended by a discourse analytical methodology, enables geographical research to ascertain the multiple interrelations between material changes (e.g. land-use change, environmental degradation) and discursive representations. HAJER's mid-range concepts, story-lines and discourse-coalitions (HAJER 1995, 2006), emerge here as methodological and conceptual categories allowing research in urban environmental policy-making to conceptualize the contested and changing nature of multi-actor governance.

The limitations of this approach are certainly defined by the availability of existing secondary data (e.g. in form of minutes or court documents), as well as the willingness of the involved actors to be interviewed and to explain the governance and policy-making processes. This research would not have been possible without the knowledge accumulated and documents collected by environmental activists and environmental lawyers as the retrieval of files from Indian state agencies has been a researcher's nightmare.

A further advancement of the riverscapes concept is proposed with regard to the role of the river as a non-human actor as outlined in recent work in UPE (GABRIEL 2014, GROVE 2009, HOLIFIELD & SCHUELKE 2015). In this context, the case of the Whanganui River in New Zealand, which has been declared a "legal being" as one of the first rivers in the world (cf. SALMOND 2014), is an inspiring example for both researchers and river activists to conceptualize the rights of rivers. Such a focus on rivers could 'give rivers a voice' and research could explore to what extent (urban) rivers all over the world could be declared as legal beings and specific 'rights' could be given to the rivers for the sake of their protection.

REFERENCES

Acharya, A., Lavalle, A. G., Houtzager, P. P., 2004. Civil Society Representation in the Participatory Budget and Deliberative Councils of São Paulo, Brazil. IDS Bulletin 35 (2), 40–48.
Agarwal, A., Narain, S., 1991. Third Citizens' Report. Floods, Flood Plains and Environmental Myths. Centre for Science and Environment (CSE), New Delhi.
Aggarwal, S., Butsch, C., 2011. Environmental and Ecological Threats in Indian Megacities. In: Richter, M., Weiland, U. (Eds.), Applied Urban Ecology: A Global Framework. Wiley-Blackwell, Chichester, 66–81.
Agrawal, A., Sivaramakrishnan, K., 2001. Social Nature: Resources, Representations and Rule in India. Oxford University Press, Delhi.
Agrawal, S. K., 2010. Design review in a post-colonial city: The Delhi Urban Art Commission. Cities 27 (5), 397–404.
Ahmed, W., 2011. Neoliberal Utopia and Urban Realities in Delhi. ACME: An International E-journal of Critical Geographies 10 (2), 163–188.
Allan, J. D., 2004. Landscapes and Riverscapes: The Influence of Land Use on Stream Ecosystems. Annual Review of Ecology, Evolution, and Systematics 35, 257–284.
Alley, K. D., 2002. On the banks of the Ganga. When Wastewater Meets a Sacred River. The University of Michigan Press, Ann Arbor.
Alley, K. D., 2012. The Paradigm Shift in India's River Policies: From Sacred to Transferable Waters. In: Johnston, B. R., Hiwasaki, L., Klaver, I. J., Ramos Castillo, A., Strang, V. (Eds.), Water, Cultural Diversity, and Global Environmental Change. Emerging Trends, Future Sustainability. Springer, Heidelberg, 31–48.
Alonso, A., Costa, V., 2004. The Dynamics of Public Hearings for Environmental Licensing: The case of the São Paulo Ring Road. IDS Bulletin 35 (2), 49–57.
AlSayyad, N., Roy, A. (Eds.), 2004. Urban informality: transnational perspectives from the Middle East, Latin America, and South Asia. Lexington Books, Oxford.
Altshuler, A., Luberoff, D., 2003. Mega-projects: The changing politics of urban public investment. The Brookings Institution and Lincoln Institute of Land Policy, Washington DC.
Angelo, H., Wachsmuth, D., 2014. Urbanizing Urban Political Ecology: A Critique of Methodological Cityism. International Journal of Urban and Regional Research (online first), 1–14.
Anjaria, J. S., 2009. Guardians of the Bourgeois City: Citizenship, Public Space, and Middle-Class Activism in Mumbai. City & Community 8 (4), 391–406.
Appadurai, A., 1990. Disjuncture and Difference in the Global Cultural Economy. Theory, Culture & Society 7 (2), 295–310.
Appadurai, A., 1996. Modernity at Large: Cultural Dimensions of Globalization. University of Minnesota Press, Minneapolis, London.
Arabindoo, P., 2012. Bajji on the Beach: Middle-Class Food Practices in Chennai's New Beach. In: McFarlane, C., Waibel, M. (Eds.), Urban Informalities. Ashgate, Farnham, 67–88.
Ashraf, J., 2004. Studies in historical ecology of India. Sunrise Publications, New Delhi.
Aßheuer, T., 2014. Klimawandel und Resilienz in Bangladesch. Die Bewältigung von Überschwemmungen in den Slums von Dhaka. Steiner Verlag, Stuttgart.
Aßheuer, T., Braun, B., 2011. Adaptionsfähigkeit lokaler Ökonomien an den Klimawandel. Eine institutionelle Analyse der Ziegelproduktion in Dhaka, Bangladesh. Berichte zur deutschen Landeskunde 85 (3), 293–307.
Babu, C. R., Kumar, P., Prasad, L., Agrawal, R., 2003. Valuation of Ecological Functions and Benefits: A Case Study of Wetland Ecosystems along the Yamuna River Corridors of Delhi

Region. Mumbai. http://coe.mse.ac.in/eercrep/fullrep/wetbio/WB_FR_LallanPrasad.pdf (2014-11-15).

Baccini, P., 1997. A city's metabolism: Towards the sustainable development of urban systems. Journal of Urban Technology 4 (2), 27–39.

Baghel, R., 2014. River Control in India: Spatial, Governmental and Subjective Dimensions. Springer Internalional Publishing, Cham.

Bakker, K., 2003a. Archipelagos and networks. Urbanization and water privatization in the South. Geographical Journal 169 (4), 328–341.

Bakker, K., 2003b. A political ecology of water privatization. Studies in Political Economy 70 (Spring 2003), 35–58.

Banerjee-Guha, S., 2009. Neoliberalising the 'Urban': New Geographies of Power and Injustice in Indian Cities. Economic and Political Weekly 44 (22), 95–107.

Banerjee-Guha, S., 2011. New Urbanism, Neoliberalism, and Urban Restructuring in Mumbai. In: Ahmed, W., Kundu, A., Peet, R. (Eds.), India's New Economic Policy. A Critical Analysis. Routledge, New York, 76–96.

Banerji, R., Martin, M., 1997. Yamuna: The river of death. In: Agrawal, A. (Ed.), Homicide by pesticides: what pollution does to our bodies. Centre for Science and Environment (CSE), New Delhi, 47–98.

BAPS 2005. Making of Akshardham. http://www.akshardham.com/makingofakdm/timeline.htm (2014-11-15).

Bartone, C., Bernstein, J., Leitman, J., Eigen, J., 1994. Towards Environmental Strategies for Cities. Policy Considerations for Urban Environment Management in Developing Countries. UMP Paper No. 18 Bank. World Bank, Washington DC.

Barve, S., Sen, S., 2011. Riverside restoration – city planners viewpoint: case of Mutha riverfront, Pune, India. In: Brebbia, C. A., Popov, V. (Eds.), Water Resources Management VI. WIT Press, Southampton, 165–176.

Batra, L., 2010. Out of Sight, Out of Mind: Slum Dwellers in 'World-Class' Delhi. In: Chaturvedi, B. (Ed.), Finding Delhi – Loss and Renewal in the Megacity. Penguin Books India, New Delhi, 16–36.

Batra, L., Mehra, D., 2008. Slum Demolition and Production of Neoliberal Space – Delhi. In: Mahadevia, D. (Ed.), Inside the transforming urban Asia: processes, policies and public actions. Concept Publishing Company, New Delhi, 391–414.

Baud, I., Nainan, N., 2008. "Negotiated spaces" for representation in Mumbai: ward committees, advanced locality management and the politics of middle-class activism. Environment and Urbanization 20 (2), 483–499.

Baud, I. S. A., deWit, J. (Eds.), 2008a. New Forms of Urban Governance in India. Shifts, Models, Networks and Contestations. Sage, New Delhi.

Baud, I. S. A., deWit, J., 2008b. Shifts in Urban Governance: Raising the Questions. In: Baud, I. S. A., deWit, J. (Eds.), New Forms of Urban Governance in India. Shifts, Models, Networks and Contestations. Sage, New Delhi, 1–36.

Baud, I. S. A., Dhanalakshmi, R., 2007. Governance in urban environmental management: Comparing accountability and performance in multi-stakeholder arrangements in South India. Cities 24 (2), 133–147.

Bauriedl, S., 2007. Räume lesen lernen: Methoden zur Raumanalyse in der Diskursforschung. Forum Qualitative Sozialforschung / Forum: Qualitative Social Research [Online Journal] 8 (2). http://www.qualitative-research.net/index.php/fqs/article/view/236/523.

Baviskar, A., 1995. In the Belly of the River: Tribal Conflicts over Development in the Narmada Valley. Oxford University Press, New Delhi.

Baviskar, A., 1997. Tribal Politics and Discourses of Environmentalism. Contributions to Indian Sociology 31 (2), 195–223.

Baviskar, A., 2002. The politics of the city. New Delhi. http://www.india-seminar.com /2002/ 516/516%20amita%20baviskar.htm (2014-12-01).

Baviskar, A., 2003. Between Violence and Desire: Space, Power and Identity in the Making of Metropolitan Delhi. International Social Science Journal 175, 89–98.

Baviskar, A., 2007a. Demolishing Delhi: World-Class City in the Making. In: Batra, L. (Ed.), The Urban Poor in Globalising India: Dispossession and Marginalisation. South Asian Dialogues on Ecological Democracy and Vasudhaiva Kutumbakam Publications, New Delhi, 39–44.

Baviskar, A. (Ed.), 2007b. Waterscapes. The Cultural Politics of a Natural Resource. Permanent Black, Ranikhet.

Baviskar, A., 2010. Indian Environmental Politics: An Interview. Transforming Cultures eJournal 5 (1), 169–174.

Baviskar, A., 2011a. Cows, Cars and Cycle-rickshaws: Bourgeois Environmentalism and the Battle for Delhi's Streets. In: Baviskar, A., Ray, R. (Eds.), Elite and Everyman: The Cultural Politics of the Indian Middle Classes. Routledge, New Delhi, 391–418.

Baviskar, A., 2011b. Spectacular Events, City Spaces and Citizenship: The Commonwealth Games in Delhi. In: Anjaria, J. S., McFarlane, C. (Eds.), Urban Navigations – Politics, Space and the City in South Asia. Routledge, New Delhi, 138–164.

Baviskar, A., 2011c. What the Eye Does Not See: The Yamuna in the Imagination of Delhi. Economic and Political Weekly 46 (50), 45–53.

Baviskar, A., Sinha, S., Philip, K., 2006. Rethinking Indian Environmentalism: Industrial Pollution in Delhi and Fisheries in Kerala. In: Bauer, J. (Ed.), Forging Environmentalism. Justice, Livelihood and Contested Environments. ME Sharpe, New York, 189–256.

Bear, L., 2011. Making a river of gold: Speculative state planning, informality, and neoliberal governance on the Hooghly. Focaal 2011 (61), 46–60.

Beng Huat, C., 2011. Singapore as Model: Planning Innovations, Knowledge Experts. In: Roy, A., Ong, A. (Eds.), Worlding Cities: Asian Experiments and the Art of Being Global. Blackwell, Oxford, 29–54.

Benton-Short, L., Short, J. R., 2013. Cities and Nature. Routledge, London, New York.

Bhan, G., 2009. "This is no longer the city I once knew". Evictions, the urban poor and the right to the city in millennial Delhi. Environment and Urbanization 21 (1), 127–142.

Bhan, G., 2013. Planned Illegalities. Economic and Political Weekly XLVIII (24), 58–70.

Bhan, G., Shivanand, S., 2013. (Un)Settling the City. Economic and Political Weekly XLVIII (13), 54–61.

Bhargava, D. S., 1985. Water quality variations and control technology of Yamuna River. Environmental Pollution Series A, Ecological and Biological 37 (4), 355–376.

Bhargava, D. S., 2006. Revival of Mathura's ailing Yamuna river. The Environmentalist 26 (2), 111–122.

Bharucha, R., 2006. Yamuna Gently Weeps: A Journey into the Yamuna Pushta Slum Demolitions. Sainathann Communication, New Delhi.

Bhushan, P., 2004. Supreme Court and PIL: Changing Perspectives under Liberalisation. Economic and Political Weekly 39 (18), 1770–1774.

Bhushan, P., 2009. Misplaced Priorities and Class Bias of the Judiciary. Economic and Political Weekly 44 (14), 32–37.

Bird, E. A. R., 1987. The Social Construction of Nature: Theoretical Approaches to the History of Environmental Problems. Environmental Review 11 (4), 255–264.

Birkmann, J., Garschagen, M., Kraas, F., Quang, N., 2010. Adaptive urban governance: new challenges for the second generation of urban adaptation strategies to climate change. Sustainability Science 5 (2), 185–206.

Blaikie, P., 1995. Changing Environments or Changing Views? A political Ecology for Developing Countries. Geography 80, 203–214.

Blaikie, P., 1999. A review of political ecology. Zeitschrift für Wirtschaftsgeographie 43 (3/4), 131–147.

Blaikie, P., Brookfield, P., 1987. Land Degradation and Society. Methuen, London.

Blaikie, P., Muldavin, J. S. S., 2004. Upstream, Downstream, China, India: The Politics of Environment in the Himalayan Region. Annals of the Association of American Geographers 94 (3), 520–548.

Blok, A., Jensen, T. E., 2011. Bruno Latour: Hybrid Thoughts in a Hybrid World. Routledge, New York.

Blomley, N., 2008. Simplification is complicated: property, nature, and the rivers of law. Environment and Planning A 40 (8), 1825–1842.

Bogner, A., Littig, B., Menz, W. (Eds.), 2009a. Interviewing Experts. Palgrave Macmillan, Basingstoke.

Bogner, A., Littig, B., Menz, W., 2009b. Introduction: Expert Interviews – An Introduction to a New Methodological Debate. In: Bogner, A., Littig, B., Menz, W. (Eds.), Interviewing Experts. Palgrave Macmillan, Basingstoke, 1–16.

Bon, B., 2015. A new megaproject model and a new funding model. Travelling concepts and local adaptations around the Delhi metro. Habitat International 45, Part 3, 223–230.

Bose, P. S., 2013. Bourgeois environmentalism, leftist development and neoliberal urbanism in the City of Joy. In: Roshan Samara, T., He, S., Chen, G. (Eds.), Locating Right to the City in the Global South. Routledge, London, New York, 127–151.

Brand, P., Thomas, M., 2005. Urban Environmentalism: Global Change and the Mediation of Local Conflict. Routledge, London.

Braun, B., 2009. Nature. In: Castree, N., Demeritt, D., Liverman, D., Rhoads, B. (Eds.), A Companion to Environmental Geography. Wiley-Blackwell, Chichester, 19–36.

Braun, B., Schulz, C., Soyez, D., 2003. Konzepte und Leitthemen einer ökologischen Modernisierung der Wirtschaftsgeographie. Zeitschrift für Wirtschaftsgeographie 47 (3–4), 231–248.

Braun, B., Wainwright, J., 2001. Nature, Poststructuralism, and Politics. In: Castree, N., Braun, B. (Eds.), Social Nature – Theory, Practice, and Politics. Blackwell, Oxford, 41–63.

Brenner, N., Theodore, N., 2002. Cities and the Geographies of "Actually Existing Neoliberalism". Antipode 34 (3), 349–379.

Bridge, G., Perreault, T., 2009. Environmental Governance. In: Castree, N., Demeritt, D., Liverman, D., Rhoads, B. (Eds.), A Companion to Environmental Geography. Wiley-Blackwell, Chichester, 475–497.

Brosius, C., 2010. India's Middle Class: New Forms of Urban Leisure, Consumption and Prosperity. Routledge, New Delhi.

Bryant, R. L., 1992. Political ecology: An emerging research agenda in Third-World studies. Political Geography 11 (1), 12–36.

Bryant, R. L., 1998. Power, knowledge and political ecology in the third world: a review. Progress in Physical Geography 22 (1), 79–94.

Bryant, R. L., Bailey, S., 1997. Third World Political Ecology. Routledge, London.

Budds, J., 2008. Whose scarcity? The hydrosocial cycle and the changing waterscape of La Ligua river basin, Chile. In: Goodman, M., Boykoff, M., Evered, K. (Eds.), Contentious Geographies: Environmental Knowledge, Meaning, Scale. Ashgate, Aldershot, 59–78.

Budds, J., 2009. Contested H2O: Science, policy and politics in water resources management in Chile. Geoforum 40 (3), 418–430.

Bulkeley, H., 2005. Reconfiguring environmental governance: Towards a politics of scales and networks. Political Geography 24 (8), 875–902.

Bulkeley, H., Betsill, M., 2003. Cities and climate change: urban sustainability and global environmental governance. Routledge, London.

Bunce, S., Desfor, G., 2007. Introduction to "Political ecologies of urban waterfront transformations". Cities 24 (4), 251–258.

Castonguay, S., Evenden, M., 2012a. Introduction. In: Castonguay, S., Evenden, M. (Eds.), Urban Rivers: Remaking Rivers, Cities, and Space in Europe and North America. University of Pittsburgh Press, Pittsburgh, 1–13.

Castonguay, S., Evenden, M. (Eds.), 2012b. Urban Rivers: Remaking Rivers, Cities, and Space in Europe and North America. University of Pittsburgh Press, Pittsburgh.
Castree, N., 1995. The nature of produced nature: Materiality and knowledge construction in Marxism. Antipode 27 (1), 12–48.
Castree, N., 2002. False Antitheses? Marxism, Nature and Actor-Networks. Antipode 34 (1), 111–146.
Castree, N., 2005. Nature. Routledge, New York.
Castree, N., 2008. The Production of Nature. In: Sheppard, E., Barnes, T. J. (Eds.), A Companion to Economic Geography. Blackwell Publishing Ltd, Oxford, 275–289.
Castree, N., Braun, B. (Eds.), 2001. Social Nature – Theory, Practice, and Politics. Blackwell, Oxford.
Castree, N., Macmillan, T., 2001. Dissolving Dualism: Actor-networks and the Reimagination of Nature. In: Castree, N., Braun, B. (Eds.), Social Nature - Theory, Practice, and Politics. Blackwell, Oxford, 208–244.
Castro, J. E., 2007. Water governance in the twentieth-first century. Ambiente & sociedade 10, 97–118.
Census of India (Government of India - Ministry of Home Affairs), 2011a. Census of India 2011 - Provisional Population Totals Paper 1 of 2011: NCT of Delhi. http://www.censusindia gov.in/2011-prov-results/prov_data_products_delhi.html (2013-03-03).
Census of India (Government of India - Ministry of Home Affairs), 2011b. Census of India 2011 - Provisional Population Totals. Urban Agglomerations and Cities. http://censusindia.gov.in/2011-prov-results/paper2/data_files/India2/1.%20Data%20Highlight.pdf (2013-03-03).
Census of India (Government of India - Ministry of Home Affairs), 2011c. Census of India 2011- Rural Urban Distribution of Population. http://censusindia.gov.in/2011-prov-results/paper2/data_files/india/Rural_Urban_2011.pdf (2012-03-12).
CGWB (Central Groundwater Board, Ministry of Water Resources, Government of India), 2012. Ground Water Year Book 2011–12 National Capital Territory of Delhi. http://www.down toearth.org.in/dte/userfiles/images/report_20130422.pdf (2014-12-01).
Chakrabarty, B., Pandey, K. R., 2008. Indian Government and Politics. Sage, New Delhi.
Chang, T. C., Huang, S., 2011. Reclaiming the City: Waterfront Development in Singapore. Urban Studies 48 (10), 2085–2100.
Chaplin, S. E., 1999. Cities, sewers and poverty: India's politics of sanitation. Environment and Urbanization 11 (1), 145–158.
Chatterjee, I., 2009. Social conflict and the neoliberal city: a case of Hindu–Muslim violence in India. Transactions of the Institute of British Geographers 34 (2), 143–160.
Chatterjee, I., 2011a. From Red Tape to Red Carpet? Violent Narratives of Neoliberalizing Ahmedabad. In: Ahmed, W., Kundu, A., Peet, R. (Eds.), India's New Economic Policy. A Critical Analysis. Routledge, New York, 154–178.
Chatterjee, I., 2011b. Governance as 'Performed', Governance as 'Inscribed': New Urban Politics in Ahmedabad. Urban Studies 48 (12), 2571–2590.
Chatterjee, P., 2004. The Politics of the Governed. Columbia University Press, New York.
Chatterjee, R., Gupta, B. K., Mohiddin, S. K., Singh, P. N., Shekhar, S., Purohit, R., 2009. Dynamic groundwater resources of National Capital Territory, Delhi: assessment, development and management options. Environmental Earth Sciences 59 (3), 669–686.
Chattopadhyay, P., 2005. River Interlinking. Economic and Political Weekly 40 (11), 998.
Chilla, T., 2005a. 'Stadt-Naturen' in der Diskursanalyse. Konzeptionelle Hintergründe und empirische Möglichkeiten. Geographische Zeitschrift 93 (3), 183–196.
Chilla, T., 2005b. Stadt und Natur – Dichotomie, Kontinuum, soziale Konstruktion? Raumforschung und Raumordnung 63 (3), 179–188.
Coelho, K., Raman, N., 2010. Salvaging and Scapegoating: Slum Evictions on Chennai's Waterways. Economic and Political Weekly 45 (21), 19–23.

Coelho, K., Raman, N., 2013. From the Frying Pan to the Floodplain: Negotiating Land, Water, and Fire in Chennai's Development. In: Rademacher, A. M., Sivaramakrishnan, K. (Eds.), Ecologies of Urbanism in India: Metropolitan Civility and Sustainability. Hong Kong University Press, Hong Kong, 145–168.

Cohen, B., 2011. Modernising the Urban Environment: The Musi River Flood of 1908 in Hyderabad, India. Environment and History 17 (3), 409–432.

Colopy, C., 2012. Dirty, Sacred Rivers. Confronting South Asia's Water Crisis. Oxford University Press, Oxford.

Colten, C., 2012. Fluid Geographies: Urbanizing River Basins. In: Castonguay, S., Evenden, M. (Eds.), Urban Rivers: Remaking Rivers, Cities, and Space in Europe and North America. University of Pittsburgh Press, Pittsburgh, 201–218.

Cook, J., Oviatt, K., Main, D., Kaur, H., Brett, J., 2014. Re-conceptualizing urban agriculture: an exploration of farming along the banks of the Yamuna River in Delhi, India. Agriculture and Human Values 32 (2), 265–279.

Cornwall, A., 2002. Locating Citizen Participation. IDS Bulletin 33 (2), i–x.

Cornwall, A., 2004. Introduction: New Democratic Spaces? The Politics and Dynamics of Institutionalised Participation. IDS Bulletin 35 (2), 1–10.

Cosgrove, D., 1984. Social Formation and Symbolic Landscape. Croom Helm, London.

Cosgrove, D., 1990. An elemental division: water control and engineered landscape. In: Cosgrove, D., Petts, G. (Eds.), Water, engineering and landscape: water control and landscape transformation in the modern period. Belhaven Press, London, 1–11.

Cosgrove, D., Daniels, S., 1988. The Iconography of Landscape: Essays on the Symbolic Representation, Design and Use of Past Environments. Cambridge University Press, Cambridge.

Cosgrove, D., Petts, G. (Eds.), 1990. Water, engineering and landscape: water control and landscape transformation in the modern period. Belhaven Press, London.

CPCB (Central Polution Control Board), 2004. Status of Sewerage and Sewage Treatment Plants in Delhi. New Delhi, http://www.cpcb.nic.in/newitems/13.pdf (2012-10-06).

CPCB (Central Pollution Control Board), 2006. Water Quality Status of Yamuna River (1999–2005). http://www.cpcb.nic.in/newitems/11.pdf (2012-06-06).

CPCB (Central Pollution Control Board), 2008. CPCB functions. http://www.cpcb.nic.in/Functions.php (2014-11-04).

CPCB (Central Pollution Control Board), 2010. Status of Water Quality in India - 2010. http://cpcb.nic.in/WQSTATUS_REPORT2010.pdf (2013-06-06).

Cronon, W., 1992. Nature's Metropolis. A. A. Norton, New York.

Crutzen, P. J., 2002. Geology of mankind. Nature 415, 23.

Crutzen, P. J., Stoermer, E. F., 2000. The "Anthropocene". IGBP Newsletter 41, 17–18.

CSE (Centre for Science and Environment), 1999. Fifth Citizens' Report. State of India's Environment. New Delhi.

CSE (Centre for Science and Environment), 2004. Master violators. http://www.downtoearth.org.in/node/11770 (29 January 2015).

CSE (Centre for Science and Environment), 2007. Sewage Canal: How to Clean the Yamuna. New Delhi.

CSE (Centre for Science and Environment, River Pollution Unit), 2009. Review of the interceptor plan for the Yamuna. http://www.cseindia.org/userfiles/CSE_interceptor_analysis.pdf (2015-03-01).

CSE (Centre for Science and Environment), 2012. Excreta Matters (Vol.2). New Delhi.

Cullet, P., Madhav, R., 2009. Water Law Reforms in India: Trends and Prospects. In: Iyer, R. R. (Ed.), Water and the laws in India. Sage, New Delhi, 511–534.

Cullet, P., Paranjape, S., Thakkar, H., Vani, M. S., Joy, K. J., Ramesh, M. K., 2012. Water Conflict in India. Towards a New Legal and Institutional Framework. Forum for Policy Dialogue on

Water Conflicts in India. Pune. http://soppecom.org/pdf/3Water%20conflicts%20in%20 India.pdf (2013-05-05).

CWC (Central Water Commission, Ministry of Water Resources, GoI), 2014. Water Resources Information System of India: Eastern Yamuna Canal Project JI01889. http://india-wris.nrsc.gov.in/wrpinfo/index.php?title=Eastern_Yamuna_Canal_Project_JI01889 (2014-11-15).

CWPRS (Central Water and Power Research Station), 1993. Hydraulics Model Studies for Channelisation of River Yamuna at Delhi. Pune.

CWPRS (Central Water and Power Research Station, Ministry of Water Resources, GoI), 2007. Hydraulic model studies for assessing the effect of Akshardham Bund on the flow conditions in the river Yamuna at Delhi. Technical Report No: 4428. Pune.

D'Souza, R., 2003. Damming the Mahanadi river: The emergence of multi-purpose river valley development in India (1943–46). Indian Economic & Social History Review 40 (1), 81–105.

D'Souza, R., 2006a. Drowned and Dammed. Colonial Capitalism and Flood Control in Eastern India. Oxford, New Delhi.

D'Souza, R., 2006b. Water in British India: The Making of a 'Colonial Hydrology'. History Compass 4 (4), 621–628.

D'Souza, R., Nagendra, H., 2011. Changes in public commons as a consequence of urbanization: the Agara lake in Bangalore, India. Environ Manage 47 (5), 840–50.

Dalrymple, W., 2009. The Last Mughal. The Fall of Delhi 1857. Bloomsbury Publishing, London.

Daniell, K. A., Barreteau, O., 2014. Water governance across competing scales: Coupling land and water management. Journal of Hydrology 519, Part C (0), 2367–2380.

Das, D., 2015. Hyderabad: Visioning, restructuring and making of a high-tech city. Cities 43 (0), 48–58.

Datta, A., 2012. The Illegal City: Space, law and gender in a Delhi Squatter Settlement. Ashgate, Farnham.

Davidson, D. J., Frickel, S., 2004. Understanding Environmental Governance: A Critical Review. Organization & Environment 17 (4), 471–492.

Davis, M., 2006. Planet of Slums. Verso, London, New York.

DDA - Special Project Cell (Delhi Development Authority, Special Project Cell Yamuna Riverfront), 1998. Development of Yamuna River. New Delhi.

DDA (Delhi Development Authority), 1962. Delhi Master Plan. New Delhi.

DDA (Delhi Development Authority - City Planning Development Wing), 1986. Development of Urban Villages in Delhi (1986–2001). New Delhi, republished by R.G. Gupta, City/Policy Planner, UPS Campus, Block-A, Preet Vihar, Delhi-92, http://www.rgplan.org/delhi/DEVELOPMENT%20OF%20URBAN%20VILLAGES%20IN%20DELHI%20-2013.docx (2014-12-13).

DDA (Delhi Development Authority), 1990. Master Plan for Delhi. Published in Gazette of India, Part II of Section 3, Sub-section (ii), Extraordinary, dated 01.08.1990.

DDA (Delhi Development Authority), 1998. Draft Zonal Development Plan for River Yamuna Area (Zone O and part of Zone P), March 1998. New Delhi.

DDA (Delhi Development Authority), 2006a. Annual Administration Report 2005–2006, Delhi Development Authority, New Delhi.

DDA (Delhi Development Authority), 2006b. Construction of Common Wealth Games Village. Form-1 and Form 1A. Report submitted to Ministry of Environment and Forests for Environmental Clearance. September 2006. Materials accessed by YJA by RTI. New Delhi.

DDA (Delhi Development Authority), 2006c. Draft Master Plan for Delhi - 2021. New Delhi, http://dda.org.in/planning/draft_master_plans.htm (2014-11-15).

DDA (Delhi Development Authority), 2006d. Draft Zonal Development Plan for River Yamuna Area (Zone O and part Zone 'P') - May 2006. New Delhi, http://dda.org.in/planning/docs/MODIFIED%20zonal%20development%20plan%20report.doc (2014-12-01).

DDA (Delhi Development Authority), 2007. Master Plan for Delhi 2021. New Delhi.

DDA (Delhi Development Authority), [2007] 2010. Master Plan 2021 (Reprint dated May 2010 of the version from 7th February 2007). http://dda.org.in/tendernotices_docs/jan12/reprint%20 mpd2021.pdf (2014-11-15).

DDA (Delhi Development Authority), 2008. Draft Zonal Development Plan for River Yamuna/River Front, Zone O. July 2008. New Delhi, http://dda.org.in/planning/docs/O-Zone-Rep.pdf (2014-10-08).

DDA (Delhi Development Authority), 2010a. Proposal for Yamuna River Front development Zone O. DDA Public Notice dated 26th April 2010. New Delhi, http://www.dda.org.in /tendernotices_docs/may10/River%20Front%20Development.pdf (2014-11-15).

DDA (Delhi Development Authority), 2010b. Zonal Development Plan for River Yamuna / River Front, Zone O. New Delhi, http://dda.org.in/tendernotices_docs/august10/Report%20of% 20ZDP%20of%20Zone%20O%20Notified%20on%2010Aug10.pdf (2014-10-08).

DDA (Delhi Development Authority), 2011. Annual Administration Report 2009-2010. New Delhi, http://www.dda.org.in/ddanew/pdf/rti/april11/DDA%20Annual%20Report%202009-10%20 Eng.pdf (2014-10-08).

DDA (Delhi Development Authority), 2013a. Annual Report 2011-12. New Delhi, http://dda.org.in/ ddanew/pdf/Annual_Reports/Final%20Annual%20Report%20-%20English%20part%20-1.pdf (2014-10-08).

DDA (Delhi Development Authority), 2013b. Public Notice, 28th September 2013: Public Consultations regarding Re-delineation and Rezoning of the Zone O. File No.: F.20(12)/2013-MP. New Delhi, http://dda.org.in/tendernotices_docs/sep13/public_noticeZone%20O%20 Redelineation.pdf and http://dda.org.in/tendernotices_docs/sep13/draft%20public%20notice %20for%20zone%20O.doc (2014-12-01).

DDA (Delhi Development Authority), 2015. Master Plan for Delhi 2021 (Incorporating modifications up to January 2015). New Delhi, http://dda.org.in/ddanew/planning.aspx (2015-03-15).

de Mello-Théry, N., Bruno, L., Dupont, V., Zérah, M.-H., Correia, B. O., Saglio-Yatzimirsky, M.-C., Ribeiro, W., 2013. Public Policies, Environment and Social Exclusion. In: Saglio-Yatzimirsky, M.-C., Landy, F. (Eds.), Megacity Slums, 213–256.

Dean, M., 2010. Governmentality. Power and Rule in Modern Society. Sage Publications, London.

Deeke, A., 1995. Experteninterviews - ein methodologisches und forschungspraktisches Problem. Einleitende Bemerkungen und Fragen zum Workshop. In: Brinkmann, C., Deeke, A., Völkel, B. (Eds.), Experteninterviews in der Arbeitsmarktforschung. Diskussionsbeiträge zu methodischen Fragen und praktischen Erfahrungen. Bundesanstalt für Arbeit, Nürnberg, 7–22.

del Cerro Santamaría, G. (Ed.), 2013. Urban Megaprojects: A Worldwide View. Research in Urban Sociology, 13. Emerald, Bingley.

Delhi Town Planning Committee, 1913. Final Report on the Town Planning of the New Imperial Capital. Delhi, Superintendent of Government Printing.

Dembowski, H., 1999. Courts, Civil Society and Public Sphere: Environmental Litigation in Calcutta. Economic and Political Weekly 34 (1/2), 49–56.

Dembowski, H., 2001. Taking the State to Court. Public Interest Litigation and the Public Sphere in Metropolitan India. Asia House, Cologne.

Demeritt, D., 1998. Science, social constructivism and nature. In: Braun, B., Castree, N. (Eds.), Remaking Reality: Nature at the Millennium. Routledge, New York, 172–192.

Demeritt, D., 2002. What is the 'social construction of nature'? A typology and sympathetic critique. Progress in Human Geography 26 (6), 767–790.

Desai, R., 2012a. Entrepreneurial Urbanism in the Time of Hindutva: City Imagineering, Place Marketing, and Citizenship in Ahmedabad. In: Desai, R., Sanyal, R. (Eds.), Urbanizing Citizenship. Contested Spaces in Indian Cities. Sage, New Delhi, 31–57.

Desai, R., 2012b. Governing the urban poor: Riverfront development, slum resettlement and the politics of inclusion in Ahmedabad. Economic and Political Weekly 47 (2), 49–56.

Desfor, G., Keil, R., 1999. Contested and polluted terrain. Local Environment 4 (3), 331–352.

Desfor, G., Keil, R., 2000. Every river tells a story: the Don River (Toronto) and the Los Angeles River (Los Angeles) as articulating landscapes. Journal of Environmental Policy and Planning 2 (1), 5–23.

Desfor, G., Keil, R., 2004. Nature and the City. Making Environmental Policy in Toronto and Los Angeles. The University of Arizona Press, Tuscon.

deWit, J., Nainan, N., Palnitkar, S., 2008. Urban Decentralisation in Indian Cities: Assessing the Performance of Neighbourhood Level Wards Committees. In: Baud, I. S. A., deWit, J. (Eds.), New Forms of Urban Governance in India. Shifts, Models, Networks and Contestations. Sage, New Delhi, 65–83.

Dhawale, M., 2010. Narratives of the Environment of Delhi. A Ramble in the Archives and Reminiscences. Indian National Trust for Art and Cultural Heritage (INTACH), New Delhi.

Dias, A., 1994. Judicial activism in the development and enforcement of environmental law: some comparative insights from the Indian experience. Journal of Environmental Law 6 (2), 243–262.

Dittrich, C., 2003. Bangalore. Polarisierung und Fragmentierung in Indiens Hightech-Kapitale. Geographische Rundschau 55 (10), 40–45.

Dittrich, C., 2007. Die südindische Hightech-Metropole Bangalore im Zeichen wirtschaftlicher Globalisierung. Handbuch des Geographieunterrichts 8,2, 87–94.

Divan, S., Rosencranz, A., 2001. Environmental law and policy in India. Cases Materials and Statutes. Oxford University Press, New Delhi.

DJB (Delhi Jal Board GNCTD), 2014a. Delhi Jal Board - About Us. http://www.delhi.gov.in/wps/wcm/connect/doit_djb/DJB/Home/About+Us (2015-01-15).

DJB (Delhi Jal Board, GNCTD), 2014b. Sewerage Master Plan for for Delhi - 2031. Final Report. Prepared by DJB in collaboration with AECOM and WAPCOS. New Delhi.

Doshi, S., 2013. The Politics of the Evicted: Redevelopment, Subjectivity, and Difference in Mumbai's Slum Frontier. Antipode 45 (4), 844–865.

DoUD (Department of Urban Development, Government of NCT of Delhi), 2006. City Development Plan Delhi. New Delhi, prepared by IL&FS Ecosmart Limited, New Delhi, http://jnnurm.nic.in/wp-content/uploads/2010/12/CDP_Delhi.pdf (2014-11-15).

DoUD (Department of Urban Development; Government of NCT of Delhi), 2010. Report of the High Powered Committee on Yamuna River Development. New Delhi.

Dovey, K., 2005. Fluid City: Transforming Melbourne's Urban Waterfront University of New South Wales Press, Sydney.

Dowling, R., 2010. Power, subjectivity, and Ethics in Qualitative Research. In: Hay, I. (Ed.), Qualitative Research Methods in Human Geography. Oxford University Press, Oxford, 26-39.

Dryzek, J. S., 2005. The politics of the earth: environmental discourses, 2nd edition. Oxford University Press, Oxford.

Dubey, C., Mishra, B., Shukla, D., Singh, R., Tajbakhsh, M., Sakhare, P., 2012. Anthropogenic arsenic menace in Delhi Yamuna Flood Plains. Environmental Earth Sciences 65 (1), 131–139.

Duncan, J., 1993. Landscapes of the self/landscapes of the other(s): cultural geography 1991-92. Progress in Human Geography 17 (3), 367–377.

Dunn, K., 2010. Interviewing. In: Hay, I. (Ed.), Qualitative Research Methods in Human Geography. Oxford University Press, Oxford, 101–138.

Dupont, V., 2000. Spatial and Demographic Growth of Delhi since 1947 and Main Migration Flows. In: Dupont, V., Tarlo, E., Vidal, D. (Eds.), Delhi. Urban Space and Human Destinies. Manohar-CSH, New Delhi, 229–239.

Dupont, V., 2004. Socio-spatial differentiation and residential segregation in Delhi: a question of scale? Geoforum 35 (2), 157–175.

Dupont, V. (Ed.), 2005a. Peri-Urban Dynamics: Population, Habitat and Environment on the Peripheries of Large Indian Metropolises. A review of concepts and general issues. CSH Occasional Paper N° 14 / 2005. Centre de Sciences Humaines, New Delhi.

Dupont, V., 2005b. Residential Prectices, Creation and Use of Space: Unauthorized Colonies in Delhi. In: Hust, E., Mann, M. (Eds.), Urbanization and governance in India. Manohar, New Delhi, 311–342.

Dupont, V., 2007. Conflicting stakes and governance in the peripheries of large Indian metropolises – An introduction. Cities 24 (2), 89–94.

Dupont, V., 2008. Slum Demolitions in Delhi since the 1990s: An Appraisal. Economic and Political Weekly 43 (28), 89–94.

Dupont, V., 2011. The Dream of Delhi as a Global City. International Journal of Urban and Regional Research 35 (3), 533–554.

Dupont, V., Ramanathan, U., 2008. The Courts and the Squatter Settlements in Delhi - Or the Intervention of the Judiciary in Urban 'Governance'. In: Baud, I. S. A., deWit, J. (Eds.), New Forms of Urban Governance in India. Shifts, Models, Networks and Contestations. Sage, New Delhi, 312–343.

Dupont, V., Tarlo, E., Vidal, D. (Eds.), 2000. Delhi. Urban Space and Human Destinies. Manohar-CSH, New Delhi.

DUSIB (Delhi Urban Shelter Improvement Board), 2014. JJ Clusters Details. http://delhishelterboard.in/main/?page_id=3644 (2014-10-22).

Dutta, R., (PEACE Institute Charitable Trust), 2009. The Unquiet River. An Overview of Select decisions of the Courts on the River Yamuna. New Delhi, http://www.peaceinst.org/publication/book-let/The%20Unquiet%20River%20An%20Overview%20of%20Select%20Decisions%20of%20the%20Courts%20on%20the%20River%20Yamuna.pdf (2014-12-12).

Dwivedi, R., 2001. Environmental Movements in the Global South: Issues of Livelihood and Beyond. International Sociology 16 (1), 11–31.

Dwyer, R., 2004. The Swaminarayan Movement. In: Jacobsen, K., Pratap Kumar, P. (Eds.), South Asians in the Diaspora: Histories and Traditions Brill, Leiden, 180–199.

Economy, E. C., 2004. The River Runs Black: The Environmental Challenge to China's Future. Cornell University Press, Ithaca, NY.

Eden, S., Tunstall, S., 2006. Ecological versus social restoration? How urban river restoration challenges but also fails to challenge the science? policy nexus in the United Kingdom. Environment and Planning C: Government and Policy 24 (5), 661–680.

Eden, S., Tunstall, S., Tapsell, S., 2000. Translating nature: river restoration as nature-culture. Environment and Planning D 18, 257–273.

Edensor, T., Jayne, M. (Eds.), 2012. Urban Theory beyond the West. Routledge, London, New York.

Ehlers, E., 2008. Das Anthropozän: Die Erde im Zeitalter des Menschen. Wissenschaftliche Buchgesellschaft, Darmstadt.

Ellis, R., 2011. The Politics of the Middle: Re-centering Class in the Postcolonial. ACME: An International E-journal of Critical Geographies 10 (1), 69–81.

Ellis, R., 2012. "A World Class City of Your Own!": Civic Governmentality in Chennai, India. Antipode 44 (4), 1143–1160.

Engineers of India, CH2M Hill (India) PVT LTD, 2008. Laying of Interceptor Sewer along Najafgarh, Supplementary and Shahdara Drain for Abatement of Pollution in River Yamuna. Detailed Project Report. December 2008. Prepared for DJB, GNCTD. New Delhi, http://www.indiaenvironmentportal.org.in/files/Revised%20DFR.pdf (2015-03-02).

ENVIS Centre on NGO and Parliament (Ministry of Environment and Forests, operated by WWF India), 2014. NGOs and the Environment. http://www.wwfenvis.nic.in/ViewMajorActivity.aspx?Id=3566&Year=2014 (2015-03-04).

EQMS (EQMS India Pvt. Ltd.), 2006. Environmental Impact Assessment Study of Commonwealth Games Village. Submitted to Delhi Development Authority. New Delhi.

Erlewein, A., 2013. Disappearing rivers — The limits of environmental assessment for hydropower in India. Environmental Impact Assessment Review 43, 135–143.

Escobar, A., 1996a. Constructing Nature. Elements for a poststructural political ecology. In: Peet, R., Watts, M. (Eds.), Liberation ecologies. Environment, development, social movements. Routledge, London, 46–68.

Escobar, A., 1996b. Construction nature. Futures 28 (4), 325–343.

Escobar, A., 1998. Whose knowledge, Whose nature? Biodiversity, Conservation, and the Political Ecology of Social Movements. Journal of Political Ecology 5, 53–82.

Escobar, A., 2010. Postconstructivist political ecologies. In: Redclift, M., Woodgate, G. (Eds.), The international handbook of environmental sociology. Edward Elgar, Cheltenham, 91–105.

Escudero, C. N., 2001. Urban environmental governance. Comparing air quality managemnet in London and Mexico city. Ashgate, Aldershot.

Etzold, B., 2013. The Politics of Street Food. Contested Governance and Vulnerabilities in Dhaka. Franz Steiner Verlag, Stuttgart.

Evans, J. P., 2007. Wildlife Corridors: An Urban Political Ecology. Local Environment 12 (2), 129–152.

Expert Committee MoEF, 2014. Restoration and Conservation of River Yamuna (Final Report) submitted to the National Green Tribunal with reference to Main Application no. 06 of 2012, Expert Committee, Ministry of Environment and Forests comprising of Prof. C. R. Babu (Delhi University, Delhi), Prof. A. K. Gosain (IIT-Delhi), Prof. Brij Gopal (Jaipur), New Delhi.

Fainstein, S. S., 2008. Mega-projects in New York, London and Amsterdam. International Journal of Urban and Regional Research 32 (4), 768–785.

Farago, M. E., Mehra, A., Banerjee, D. K., 1989. A preliminary investigation of pollution in the River Yamuna, Delhi, India: metal concentrations in river bank soils and plants. Environmental Geochemistry and Health 11 (3–4), 149–156.

Faria, S. C. d., Bessa, L. F. M., Tonet, H. C., 2009. A theoretical approach to urban environmental governance in times of change. Management of Environmental Quality: An International Journal 20 (6), 638–648.

Fernandes, L., 2006. India's New Middle Class: Democratic Politics State Power and the Restructuring of Urban Space in India. University of Minnesota Press, Minneapolis.

Fischer, F., 2000. Citizens, Experts, and the Environment: The Politics of Local Knowledge. Duke University Press, Durham.

Fischer, F., 2003. Reframing Public Policy: Discursive Politics and Deliberative Practices. Oxford University Press, Oxford.

Fischer, F., Forester, J. (Eds.), 1993. The Argumentative Turn in Policy Analysis and Planning. Duke University Press, Durham.

Fitzsimmons, M., 1989. The Matter of Nature. Antipode 21 (2), 106–120.

Flick, U., 2004. Triangulation in Qualitative Research. In: Flick, U., von Kardorff, E., Steinke, I. (Eds.), A Companion to Qualitative Research. Sage, London, 178–183.

Flick, U., 2014a. An Introduction to Qualitative Research. Sage, Los Angeles.

Flick, U., 2014b. Mapping the Field. In: Flick, U. (Ed.), The SAGE Handbook of Qualitative Data Analysis. Sage, Los Angeles, 3–18.

Flybbjerg, B., 1998. Rationality and Power: Democracy in Practice. University of Chicago Press, Chicago.

Flyvbjerg, B., Bruzelius, N., Rothengatter, W., 2003. Megaprojects and risk: An anatomy of ambition. Cambridge University Press, New York.

Follmann, A., 2014. Delhi's changing riverfront: bourgeois environmentalism and the reclamation of Yamuna's floodplain for a world-class city in the making. In: Joshi, H., Viguier, A. (Eds.), Ville et Fleuve en Asie du Sud: regards croisés. Inalco, Paris, 153–176.

Follmann, A., 2015. Urban mega-projects for a 'world-class' riverfront – The interplay of informality, flexibility and exceptionality along the Yamuna in Delhi, India. Habitat International 45, Part 3, 213–222.

Follmann, A., Trumpp, T., 2013. Armutsbekämpfung oder Bekämpfung der Armen. Weltstadtvisionen und Slum-Räumungen in Delhi. Geographische Rundschau 65 (10), 4–11.

Forsyth, T., 2001. Environmental Social Movements in Thailand: How Important is Class? Asian Journal of Social Sciences 29 (1), 35–51.
Forsyth, T., 2003. Critical political ecology. The politics of environmental science. Routledge, London.
Forsyth, T., 2008. Political ecology and the epistemology of social justice. Geoforum 39 (2), 756–764.
Foucault, M., 1984. Truth and Power. In: Rabinow, P. (Ed.), Foucault Reader. Pantheon Books, New York, 109–133.
Frey, K., 2008. Development, good governance, and local democracy. Brazilian Political Science Review. http://socialsciences.scielo.org/scielo.php?script=sci_arttext&pid=S1981-38212008000100007&lng=en&nrm=iso (2015-01-16).
Friedmann, J., 1986. The World City Hypothesis. Development and Change 17 (1), 69–83.
Gabriel, N., 2014. Urban Political Ecology: Environmental Imaginary, Governance, and the Non-Human. Geography Compass 8 (1), 38–48.
Gadgil, M., Guha, R., 1994. Ecological Conflicts and Environmental Movement in India. Development and Change 25, 101–136.
Gandy, M., 2002. Concrete and Clay: Reworking Nature in New York City. MIT Press, Cambridge.
Gandy, M., 2004. Rethinking urban metabolism: water, space and the modern city. City 8 (3), 363–379.
Gandy, M., 2006a. Riparian Anomie: Reflections on the Los Angeles River. Landscape Research 31 (2), 135–145.
Gandy, M., 2006b. Urban nature and the ecological imaginary. In: Heynen, N., Kaïka, M., Swyngedouw, E. (Eds.), In the Nature of Cities. Urban political ecology and the politics of urban metabolism. Routledge, New York, 63–74.
Gandy, M., 2008. Landscapes of disaster: water, modernity, and urban fragmentation in Mumbai. Environment and Planning A 40 (1), 108–130.
Gandy, M., 2013. Marginalia: Aesthetics, Ecology, and Urban Wastelands. Annals of the Association of American Geographers 103 (6), 1301–1316.
Gandy, M., 2014. The Fabric of Space: Water, Modernity, and the Urban Imagination. MIT Press, Cambridge.
Gangai, I. P. D., Ramachandran, S., 2010. The role of spatial planning in coastal management—A case study of Tuticorin coast (India). Land Use Policy 27 (2), 518–534.
Garschagen, M., 2014. Risky change? Vulnerability and adaptation between climate change and transformation dynamics in Can Tho City, Vietnam. Steiner Verlag, Stuttgart.
Gautam, S. K., Sharma, D., Tripathi, J. K., Ahirwar, S., Singh, S., 2013. A study of the effectiveness of sewage treatment plants in Delhi region. Applied Water Science 3 (1), 57–65.
Gebhardt, H., Mattisek, A., Reuber, P., 2007. Neue Kulturgeographie? Perspektiven, Potentiale und Probleme. Geographische Rundschau 59 (7/8), 12–20.
Gellert, P. K., Lynch, B. D., 2003. Mega-projects as displacements. International Social Science Journal 55 (1), 15–25.
Geological Survey of India, no date. Geology and mineral resources of the Northern Region States. http://36igc.org/files/GSI_Northern_Region_States.pdf (2014-10-05).
Ghertner, D. A., 2008. Analysis of New Legal Discourse behind Delhi's Slum Demolitions. Economic and Political Weekly 43 (20), 57–66.
Ghertner, D. A., 2010. Calculating without numbers: aesthetic governmentality in Delhi's slums. Economy and Society 39 (2), 185–217.
Ghertner, D. A., 2011a. Gentrifying the State, Gentrifying Participation: Elite Governance Programs in Delhi. International Journal of Urban and Regional Research 35 (3), 504–532.
Ghertner, D. A., 2011b. Green evictions: environmental discourse of a "slum-free" Delhi. In: Peet, R., Robbins, P., Watts, M. (Eds.), Global Political Ecology. Routledge, London, 145–165.

Ghertner, D. A., 2011c. The Nuisance of Slums: Environmental Law and the Production of Slum Illegality in India. In: Anjaria, J. S., McFarlane, C. (Eds.), Urban Navigations – Politics, Space and the City in South Asia. Routledge, New Delhi, 23–49.

Ghertner, D. A., 2011d. Nuisance Talk and the Propriety of Property: Middle Class Discourses of a Slum-Free Delhi. Antipode 44 (4), 1161–1187.

Ghertner, D. A., 2011e. Rule by Aesthetics: World-Class City Making in Delhi. In: Roy, A., Ong, A. (Eds.), Worlding Cities: Asian Experiments and the Art of Being Global. Blackwell, Oxford, 279–306.

Ghertner, D. A., 2013. Nuisance Talk: Middle-Class Discourses of a Slum-Free Delhi. In: Rademacher, A. M., Sivaramakrishnan, K. (Eds.), Ecologies of Urbanism in India: Metropolitan Civility and Sustainability. Hong Kong University Press, Hong Kong, 249–276.

Ghosh, A., Kennedy, L., Ruet, J., Tawa Lama-Rewal, S., Zerah, M.-H., 2009. Comparing urban governance in Delhi, Hyderabad, Kolkata, and Mumbai. In: Ruet, J., Tawa Lama-Rewal, S. (Eds.), Governing India's Metropolises. Routledge, New Delhi, 24–54.

Ghosh, N., 2007. Air pollution in India: The post-liberalisation era. In: Shaw, A. (Ed.), Indian Cities In Transition. Orient Longman, Chennai, 125–159.

Gibbs, G., 2007. Analyzing Qualitative Data. Sage, London.

Giddens, A., 1984. The Constitution of Society. University of California Press, Berkeley.

Gidwani, V., 2013. Value Struggles: Waste Work and Urban Ecology in Delhi. In: Rademacher, A. M., Sivaramakrishnan, K. (Eds.), Ecologies of Urbanism in India: Metropolitan Civility and Sustainability. Hong Kong University Press, Hong Kong, 169–200.

Gill, G. N., 2012. Human Rights and the Environment in India: Access through Public Interest Litigation. Environmental Law Review 14 (3), 200–218.

Gill, G. N., 2013. Access to environmental justice in India with special reference to National Green Tribunal: a step in the right direction. OIDA International Journal of Sustainable Development 6 (4), 25–36.

Gill, G. N., 2014. Environmental protection and developmental interests: A case study of the River Yamuna and the Commonwealth Games, Delhi, 2010. International Journal of Law in the Built Environment 6 (1/2), 69–90.

Ginther, K., Denters, E., De Waart, P. J. I. M. (Eds.), 1995. Sustainable Development and Good Governance. Nijhoff Publishers, Dordrecht.

Glasze, G., Mattissek, A., 2009. Diskursforschung in der Humangeographie: Konzeptionelle Grundlagen und empirische Operationalisierungen. In: Glasze, G., Mattissek, A. (Eds.), Handbuch Diskurs und Raum. Theorien und Methoden für die Humangeographie sowie die sozial- und kulturwissenschaftliche Raumforschung. transcript, Bielefeld, 11–59.

GNCTD (Government of NCT of Delhi, coordinated by Irrigation and Flood Control Department), 2014. Flood Control Order. http://delhi.gov.in/wps/wcm/connect/71171b804778e229 bccaff92d4a674f7/Flood+Control+Order-2011.pdf?MOD=AJPERES (2014-11-14).

Goldar, B., Banerjee, N., 2004. Impact of informal regulation of pollution on water quality in rivers in India. Journal of Environmental Management 73 (2), 117–130.

Goldman, M., 2011a. Speculating on the Next World City. In: Roy, A., Ong, A. (Eds.), Worlding Cities: Asian Experiments and the Art of Being Global. Blackwell, Oxford, 229–258.

Goldman, M., 2011b. Speculative Urbanism and the Making of the Next World City. International Journal of Urban and Regional Research 35 (3), 555–581.

Gonsalves, C., 2010. Has the judiciary abandoned the environment? In: Benjamin, V. M. (Ed.), Has the judiciary abandoned the environment? Human Rights Law Network (HRLN), New Delhi, 3–38.

Gopal, B., Chauhan, M., 2007. River Yamuna from Source to Delhi: Human Impacts and Approaches to Conservation. In: Martin, P., Gopal, B., Southey, C. (Eds.), Restoring River Yamuna: Concepts, Strategies and Socio-Economic Considerations. National Institute of Ecology, New Delhi, 45–69.

Gopal, B., Goel, U., Chauhan, M., Bansal, R., Chandra Kuman, S., 2002. Regulation of Human Activities Along Rivers and Lakes. A Background Document for the Proposed Notification on River Regulation Zone. Prepared for the National River Conservation Directorate, Ministry of Environment and Forests, Government of India. National Institute of Ecology, New Delhi, December 2002. http://www.downtoearth.org.in/dte/userfiles/images/RRZ-Background-Report-Dec2002.pdf (2014-03-22).

Govt. of U.P. (National Capital Region Planning Cell, Housing and Urban Planning Department, Government of Uttar Pradesh), 2013. U. P. Sub Regional Plan (Approved in 33rd Meeting of NCR Planning Board held on 01.07.2013 under the provisions of National Capital Region Planning Board Act, 1985). Ghaziabad, http://ncrup.up.nic.in/SubRegionalPlan.aspx (2014-11-11).

Grove, K., 2009. Rethinking the nature of urban environmental politics: Security, subjectivity, and the non-human. Geoforum 40 (2), 207–216.

Guha, R., 1989. The Unquiet Woods. Ecological Change and Peasant Resistance in the Hilamalya. Oxford University Press, New Delhi.

Guha, R., Martinez-Alier, J., 1997. Varieties of Environmentalism: Essays North and South. Earthscan, London.

Gupta, N., 1981. Delhi Between Two Empires, 1803–1931: Society, Government, and Urban Growth. Oxford University Press, Delhi.

Gupta, N., 2012. Role of NGOs in Environmental Protection: A Case Study of Ludhiana City in Punjab. JOAAG 7 (2), 9–18.

Gupta, R. G., 1995. Shelter for Poor in the Fourth World (Vol. 2). Shipra Publications, New Delhi.

Gururani, S., 2013. Flexible Planning: The Making of India's 'Millenium City', Gurgaon. In: Rademacher, A. M., Sivaramakrishnan, K. (Eds.), Ecologies of Urbanism in India: Metropolitan Civility and Sustainability. Hong Kong University Press, Hong Kong, 119–144.

Haberman, D. L., 2006. River of Love in an Age of Pollution. The Yamuna River of Northern India. University of California Press, Berkeley.

Haberman, D. L., 2011. How Long is the Life of Yamuna-ji? Yale - TERI Workshop on the Yamuna River, New Delhi, 3–5 January 2011, http://fore.research.yale.edu/yamuna-river-conference/ (2014-06-12).

Hackenbroch, K., 2013. The Spatiality of Livelihoods – Negotiations of Access to Public Space in Dhaka, Bangladesh. Franz Steiner Verlag, Stuttgart.

Hagerman, C., 2007. Shaping neighborhoods and nature: Urban political ecologies of urban waterfront transformations in Portland, Oregon. Cities 24 (4), 285–297.

Haidvogl, G., 2012. The Channelization of the Danube and Urban Spatial Development in Vienna in the Nineteenth and Early Twentieth Centuries. In: Castonguay, S., Evenden, M. (Eds.), Urban Rivers: Remaking Rivers, Cities, and Space in Europe and North America. University of Pittsburgh Press, Pittsburgh, 113–129.

Hajer, M., 1993. Discourse Coalitions and the Institutionalisation of Practice: The Case of Acid Rain in Great Britain. In: Fischer, F., Forester, J. (Eds.), The Argumentative Turn in Policy Analysis and Planning. Duke University Press, Durham, 43–76.

Hajer, M., 1995. The Politics of Environmental Discourse. Ecological Modernization and the Policy Process. Clarendon Press, Oxford.

Hajer, M., 2002. Discourse Analysis and the Study of Policy Making. European Political Science 2 (1), 61–65.

Hajer, M., 2006. Doing discourse analysis: coalitions, practices, meaning. In: van den Brink, M., Metze, T. (Eds.), Words matter in policy and planning. Netherlands Graduate School of Urban and Regional Research, Utrecht, 65–76.

Hajer, M., 2008. Diskursanalyse in der Praxis: Koalitionen, Praktiken und Bedeutung. In: Janning, F., Toens, K. (Eds.), Die Zukunft der Policy-Forschung. VS Verlag für Sozialwissenschaften, Wiesbaden, 211–222.

Hajer, M., Versteeg, W., 2005. A Decade of Discourse Analysis of Environmental Politics: Achievements, Challenges, Perspectives. Journal of Environmental Policy & Planning 7 (3), 175–184.
Hansjürgens, B., Heinrichs, D., 2014. Megacities and Climate Change: Early Adapters, Mainstream Adapters, and Capacities. In: Kraas, F., Aggarwal, S., Coy, M., Mertins, G. (Eds.), Megacities: Our Global Urban Future. Springer, Dordrecht, 9–24.
Haraway, D., 1991. Simians, Cyborgs, and Women. The Reinvention of Nature. Routledge, New York.
Hardoy, J. E., Satterthwaite, D., 1991. Environmental problems of third world cities: A global issue ignored? Public Administration and Development 11 (4), 341–361.
Harkness, T., Sinha, A., 2004. Taj Heritage Corridor: Intersections between History and Culture on the Yamuna Riverfront. Places 6 (2), 62–69.
Harris, L. M., 2006. Irrigation, gender, and social geographies of the changing waterscapes of southeastern Anatolia. Environment and Planning D: Society and Space 24 (2), 187–213.
Harris, N., 2003. Globalisation and the Management of Indian Cities. Economic and Political Weekly 38 (25), 2535–2543.
Harris, N., 2007. Globalisation and the management of Indian cities. In: Shaw, A. (Ed.), Indian Cities In Transition. Orient Longman, Chennai, 1–28.
Hartmann, T., 2013. Land policy for German Rivers: making Space for the Rivers. In: Warner, J. F., van Buuren, A., Edelenbos, J. (Eds.), Making Space for the River. Governance experiences with multifunctional river flood management in the US and Europe. IWA, London, 121–134.
Harvey, D., 1989. From manageralism to entrepreneuralism: The Transformation of Urban Governance in Late Capitalism. Geografiska Annaler, Series B, Human Geography 71 (1), 3–17.
Harvey, D., 1993. The Nature of Environment: Dialectics of Social and Environmental Change. Socialist Register 29, 1–51.
Harvey, D., 1996a. Cities or urbanization? City 1 (1–2), 38–61.
Harvey, D., 1996b. Justice, Nature and the Geography of Difference. Blackwell, Cambridge, MA.
Hashmi, S., 2010. Death of a River. Yamuna.Elbe Seminar: Contemporary flows, fluid times, Max Mueller Bhawan, New Delhi, also published under http://kafila.org/2011/02/04/death-of-a-river-yamuna-jamuna-delhi/ (2014-11-15).
Hashmi, S., 2013. Dillis without the Yamuna. Down to Earth, Aug 15–31, Centre for Science and Environment (CSE), New Delhi. http://www.downtoearth.org.in/content/rivers-close-and-personal (2014-11-15).
Haslam, S. M., 2008. The Riverscape and the River. Cambridge University Press, Cambridge.
Hazards Centre, 2004a. Pollution, Pushta and Prejudice, Report No. 29, Hazards Centre, New Delhi.
Hazards Centre, 2004b. A Report on Pollution of the Yamuna at the Pushta in Delhi. Report No. 27, Hazards Centre, New Delhi.
Healey, P., 2013. Circuits of Knowledge and Techniques: The Transnational Flow of Planning Ideas and Practices. International Journal of Urban and Regional Research 37 (5), 1510–1526.
Heikkila, E. J., 2011. Environmentalism with Chinese Characteristics? Urban River Revitalization in Foshan. Planning Theory & Practice 12 (1), 33–55.
Heinelt, H., 2002. Achieving Sustainable and Innovative Policies through Participatory Governance. In: Heinelt, H., Getimis, P., Kafkalas, G., Smith, R., Swyngedouw, E. (Eds.), Participatory Governance in Multi-Level Context. Springer, Wiesbaden, 17–32.
Heinrichs, D., Krellenberg, K., Hansjürgens, B., 2012. Introduction: Megacities in Latin America as Risk Habitat. In: Heinrichs, D., Krellenberg, K., Hansjürgens, B., Martínez, F. (Eds.), Risk Habitat Megacity. Springer, Berlin, 3–17.
Hewitt de Alcántara, C., 1998. Uses and abuses of the concept of governance. International Social Science Journal 50 (155), 105–113.
Heynen, N., 2006. Green urban political ecologies: toward a better understanding of inner-city environmental change. Environment and Planning A 38 (3), 499–516.

Heynen, N., 2014. Urban political ecology I: The urban century. Progress in Human Geography 38 (4), 598–604.

Heynen, N., Kaïka, M., Swyngedouw, E. (Eds.), 2006a. In the Nature of Cities. Urban political ecology and the politics of urban metabolism. Routledge, New York.

Heynen, N., Kaïka, M., Swyngedouw, E., 2006b. Urban Political Ecology. Politicizing the production of urban natures. In: Heynen, N., Kaïka, M., Swyngedouw, E. (Eds.), In the Nature of Cities. Urban political ecology and the politics of urban metabolism. Routledge, New York, 1–20.

Hill, C., 2008. South Asia. An Environmental History. ABC-CLIO, Santa Barbara.

Hinchliffe, S., 2008. Reconstituting nature conservation: Towards a careful political ecology. Geoforum 39 (1), 88–97.

Hirsch, P., 2006. Water Governance Reform and Catchment Management in the Mekong Region. The Journal of Environment & Development 15 (2), 184–201.

HLC (High Level Committee for the Commonwealth Games, 2010), 2011. Commonwealth Games Village - Second Report of HCL. New Delhi, http://www.archive.india.gov.in/high_level/reports.htm (2014-05-05).

Hochstetler, K., 2002. After the Boomerang: Environmental Movements and Politics in the La Plata River Basin. Global Environmental Politics 2 (4), 35–57.

Hodson, M., Marvin, S., 2009. 'Urban Ecological Security': A New Urban Paradigm? International Journal of Urban and Regional Research 33 (1), 193–215.

Hogl, K., Kvarda, E., Nordbeck, R., Pregernig, M., 2012. Legitimacy and effictiveness of environmental governance – concepts and perspectives. In: Hogl, K., Kvarda, E., Nordbeck, R., Pregernig, M. (Eds.), Environmental Governance: The Challenge of Legitimacy and Effectiveness. Edward Elgar, Chelterham, 1–28.

Hohn, U., Neuer, B., 2006. New urban governance: Institutional change and consequences for urban development. European Planning Studies 14 (3), 291–298.

Holifield, R., Schuelke, N., 2015. The Place and Time of the Political in Urban Political Ecology: Contested Imaginations of a River's Future. Annals of the Association of American Geographers 105 (2), 294–303.

Homm, S., 2014. Global Players – Local Struggles. Spatial Dynamics of Industrialisation and Social Change in Peri-urban Chennai, India. Franz Steiner, Stuttgart.

Homm, S., Bohle, H., 2012. "India's Shenzhen" - a miracle? Critical reflections on New Economic Geography, with empirical evidence from peri-urban Chennai. Erdkunde 66 (4), 281–294.

Hopf, C., 2004. Qualitative Interviews: An Overview. In: Flick, U., von Kardorff, E., Steinke, I. (Eds.), A companion to qualitative research. Sage, London, 209–213.

Hordijk, M., Baud, I., 2011. Inclusive Adaptation: Linking Participatory Learning and Knowledge Management to Urban Resilience. Local Sustainability 1, 111–121.

Hosagrahar, J., 2005. Indigenous Modernities: Negotiating Architecture and Urbanism. Routledge, London.

Hosagrahar, J., 2011. Landscapes of Water in Delhi: Negotiating Global Norms and Local Cultures. In: Sorensen, A., Okata, J. (Eds.), Megacities: Urban Form, Governance and Sustainability. Springer, Tokyo, 111–132.

Hust, E., 2005. Introduction: Problems of Urbanization and Urban Governance in India. In: Hust, E., Mann, M. (Eds.), Urbanization and governance in India. Manohar, New Delhi, 1–26.

Hust, E., Mann, M. (Eds.), 2005. Urbanization and governance in India. Manohar, New Delhi.

IDS (Knowledge, Technology and Society Team, Institute of Development Studies), 2006. Understanding policy processes. A review of IDS research on the environment. Brighton, http://r4d.dfid.gov.uk/pdf/ThematicSummaries/Understanding_Policy_Processes.pdf (2014-04-01).

IFCD (Department of Irrigation and Flood Control, Govt. NCT of Delhi), 2010. Flood Problem due to River Yamuna. http://delhi.gov.in/wps/wcm/connect/DoIT_Irrigation/irrigation+and+flood+control/extra2/flood+problem+due+to+river+yamuna (2014-10-01).

References

IPCC (Intergovernmental Panel on Climate Change), 2014. Climate Change 2014: Impacts, Adaptation, and Vulnerability. Part B: Regional Aspects. Contribution of Working Group II to the Fifth Assessment Report of the Intergovernmental Panel on Climate Change [Barros, V.R., C.B. Field, D.J. Dokken, M.D. Mastrandrea, K.J. Mach, T.E. Bilir, M. Chatterjee, K.L. Ebi, Y.O. Estrada, R.C. Genova, B. Girma, E.S. Kissel, A.N. Levy, S. MacCracken, P.R. Mastrandrea, and L.L. White (eds.)]. Cambridge University Press, Cambridge, http://www.ipcc.ch/pdf/assessment-report/ar5/wg2/WGIIAR5-PartB_FINAL.pdf (2015-03-03).

Israel, M., Hay, I., 2006. Research Ethics for Social Scientists. Sage, London.

Iyer, R. R., 2001a. Water: Charting a Course for the Future: I. Economic and Political Weekly 36 (13), 1115–1122.

Iyer, R. R., 2001b. Water: Charting a Course for the Future: II. Economic and Political Weekly 36 (14/15), 1235–1245.

Iyer, R. R., 2002. Linking of Rivers: Judicial Activism or Error? Economic and Political Weekly 37 (46), 4595–4596.

Iyer, R. R., 2003a. Interlinking of Rivers. Economic and Political Weekly 38 (28), 2910–2991.

Iyer, R. R., 2003b. Water - perspectives, issues, concerns. Sage, New Delhi.

Iyer, R. R., 2008. Floods, Himalayan Rivers, Nepal: Some Heresies. Economic and Political Weekly 43 (46), 37–40.

Iyer, R. R. (Ed.), 2009a. Water and the laws in India. Sage, New Delhi.

Iyer, R. R., 2009b. The Yamuna. Need for Inclusive Debate. eSocialScience, Navi Mumbai. www.esocialsciences.org/Download/repecDownload.aspx?fname=Document11462009580.7573358.doc&fcategory=Articles&AId=2056&fref=repec (2015-01-01).

Jacobs, K., 2006. Discourse Analysis and its Utility for Urban Policy Research. Urban Policy and Research 26 (1), 39–52.

Jagmohan, M., 2005. Soul and Structure of Governance in India. Allied Publishers, New Delhi.

Jagmohan, M., 2009. River Pollution. A.P.H. Publishing Corporation, New Delhi.

Jain, A. G., Rosencranz, A., 2014. The Indian Supreme Court Promotes Interlinking of India's Rivers: Judicial Overreach? Environmental Law Reporter 44 (5). http://ssrn.com/abstract=2431762 (2014-03-06).

Jain, A. K., 1989. Delhi—planning and growth. International Journal of Environmental Studies 34 (1–2), 65–77.

Jain, A. K., 1990. The Making of a Metropolis: Planning and Growth of Delhi. National Book Organization, New Delhi.

Jain, A. K., 2009a. River Pollution. APH Publishing Corporation, New Delhi.

Jain, C. K., 2004. Metal fractionation study on bed sediments of River Yamuna, India. Water Research 38 (3), 569–578.

Jain, M., Siedentop, S., 2014. Is spatial decentralization in National Capital Region Delhi, India effective? An intervention-based evaluation. Habitat International 42 (0), 30–38.

Jain, M., Siedentop, S., Taubenböck, H., Namperumal, S., 2013. From Suburbanization to Counterurbanization? Investigating Urban Dynamics in the National Capital Region Delhi, India. Environment and Urbanization Asia 4 (2), 247–266.

Jain, P., 2009b. Sick Yamuna, Sick Delhi. Searching a Correlation. PEACE Institute Charitable Trust, New Delhi, http://www.peaceinst.org/publication/book-let/Sick%20Yamuna%20Sick%20 Delhi.pdf (2015-02-01).

Jain, S., 2011. In search of Yamuna. Reflections on a River Lost. Vitasta, New Delhi.

Jasanoff, S., 1997. NGOs and the Environment: From Knowledge to Action. Third World Quarterly 18 (3), 579–594.

Jenkins, R., 1999. Democratic Politics and Economic Reforms in India. Cambridge, Cambridge University Press.

Jessop, B., 1997. Capitalism and Its Future: Remarks on Regulation, Government and Governance. Review of International Political Economy 4 (3), 561–581.

Jessop, B., 1998. The rise of governance and the risks of failure: the case of economic development. International Social Science Journal 50 (155), 29–45.

Jessop, B., 2002. Liberalism, Neoliberalism, and Urban Governance: A State–Theoretical Perspective. Antipode 34 (3), 452–472.

Jodhka, S. S., 2002. Nation and Village: Images of Rural India in Gandhi, Nehru and Ambedkar. Economic and Political Weekly 37 (32), 3343–3353.

Johnson, D. A., 2010. Land Acquisition, Landlessness, and the Building of New Delhi. Radical History Review (108), 91–116.

Jolly, U. S., 2010. Challenges for a Mega City: Delhi, a Planned City with Unplanned Growth. Concept Publishing Company, New Delhi.

Jungk, R., Müllert, N., 1987. Future workshops: How to Create Desirable Futures. Institute for Social Inventions, London.

Kaïka, M., 2005. City of Flows. Modernity, Nature and the City. Routledge, London.

Kaïka, M., 2006a. Dams as Symbols of Modernization: The Urbanization of Nature Between Geographical Imagination and Materiality. Annals of the Association of American Geographers 96 (2), 276–301.

Kaïka, M., 2006b. The political ecology of water scarcity: the 1989–1991 Athenian drought. In: Heynen, N., Kaïka, M., Swyngedouw, E. (Eds.), In the Nature of Cities. Urban political ecology and the politics of urban metabolism. Routledge, New York, 157–172.

Karpouzoglou, T., 2011. 'Our power rests with numbers'. Re-visiting the divide between 'expert' and 'lay' knowledge in contemporary wastewater governance: New Delhi, India University of Sussex (PhD). Sussex.

Karpouzoglou, T., Zimmer, A., 2012. Closing the Gap between 'Expert' and 'Lay' Knowledge in the Governance of Wastewater: Lessons and Reflections from New Delhi. IDS Bulletin 43 (2), 59–68.

Kaushik, C. P., Sharma, H. R., Jain, S., Dawra, J., Kaushik, A., 2008. Pesticide residues in river Yamuna and its canals in Haryana and Delhi, India. Environmental Monitoring and Assessment 144 (1–3), 329–40.

Keeley, J., Scoones, I., 1999. Understanding Policy Processes. A Review. IDS Working Paper 89, Brighton. http://www.ids.ac.uk/files/Policy_Processes06.pdf (2014-03-20).

Keeley, J., Scoones, I., 2003. Understanding Environmental Policy Processes: Cases from Africa. Earthscan, London.

Keil, R., 2003. Urban political ecology. Urban Geography 24 (8), 723–738.

Keil, R., 2005. Progress report - Urban political ecology. Urban Geography 26 (7), 640–651.

Keil, R., Desfor, G., 2003. Ecological Modernisation in Los Angeles and Toronto. Local Environment 8 (1), 27–44.

Keil, R., Graham, J., 1998. Reasserting Nature. Constructing urban environments after Fordism. In: Castree, N., Braun, B. (Eds.), Remaking Reality. Nature at the Millenium. Routledge, New York, 100–125.

Keller, R., 2013. Doing Discourse Research. Sage, Los Angeles.

Kelman, A., 2003. A River and Its City: The Nature of Landscape in New Orleans. University of Carlifornia Press, Berkeley.

Kennedy, C., Pincetl, S., Bunje, P., 2011. The study of urban metabolism and its applications to urban planning and design. Environmental Pollution 159 (8–9), 1965–1973.

Kennedy, L., 2007. Regional industrial policies driving peri-urban dynamics in Hyderabad, India. Cities 24 (2), 95–109.

Kennedy, L., 2009. Large Scale Economic and Infrastructure Projects in India's Metropolitan Cities. New Policies and Practices among Competing Subnational States. Rijswijk. http://newurbanquestion.ifou.org/proceedings/7%20The%20New%20Metropolitan%20Region/E017_Kenn_Reviewed_corrLK_.pdf (2014-12-03).

Kennedy, L., 2014. The Politics of Economic Restructuring in India. Economic Governance and State Spatial Rescaling. Routledge, London, New York.

Kennedy, L., 2015. The politics and changing paradigm of megaproject development in metropolitan cities. Habitat International 45, Part 3, 163–168.
Kennedy, L., Zérah, M.-H., 2008. The Shift to City-Centric Growth Strategies: Perspectives from Hyderabad and Mumbai. Economic and Political Weekly 43 (39), 110–117.
Kersting, N., Caulfield, J., Nickson, R. A., Olowu, D., Wollmann, H., 2009. Local Governance Reform in Global Perspective. VS Verlag für Sozialwissenschaften, Wiesbaden.
Khagram, S., Nicholas, K., Bever, D., Warren, J., Richards, E., Oleson, K., Kitzes, J., Katz, R., Hwang, R., Goldman, R., Funk, J., Brauman, K., 2010. Thinking about knowing: conceptual foundations for interdisciplinary environmental research. Environmental Conservation 37 (4), 388–397.
Kohli, K., Manju, M., 2005. Eleven Years of the Environment Impact Assessment Notification, 1994: How Effective Has It Been? Kalpavriksh Environmental Action Group and the Just Environment Trust and Environmental Justice Initiative, New Delhi.
Kooiman, J., 2003. Governing as Governance. Sage, London.
Kraas, F., 2003. Megacities as Global Risk Areas. Petermanns Geographische Mitteilungen 147 (4), 6–15.
Kraas, F., 2007. Megacities and global change: key priorities. Geographical Journal 173 (1), 79–82.
Kraas, F., Aggarwal, S., Coy, M., Mertins, G. (Eds.), 2014. Megacities: Our Global Urban Future. Springer, Dordrecht.
Kraas, F., Mertins, G., 2014. Megacities and global change. In: Kraas, F., Aggarwal, S., Coy, M., Mertins, G. (Eds.), Megacities: Our Global Urban Future. Springer, Dordrecht, 1–5.
Krings, T., 2008. Politische Ökologie. Grundlagen und Arbeitsfelder eines geographischen Ansatzes der Mensch-Umwelt-Forschung. Geographische Rundschau 60 (12), 4–9.
Krueger, R., Gibbs, D. (Eds.), 2007. The Sustainable Development Paradox Urban Political Economy in the United States and Europe. Guilford Press, New York.
Kudva, N., 2009. The everyday and the episodic: the spatial and political impacts of urban informality. Environment and Planning A 41 (7), 1628–1641.
Kulshrestha, S. K., 2012. Urban and Regional Planning in India: A Handbook for Professional Practice. Sage, New Delhi.
Kumar, M., 2014. Erstwhile villages in urban India. Development in Practice 25 (1), 124–132.
Kumar, S., 2012. Clean air, Dirty Logic? Environmentral Activis, Citizenship, and the Public Sphere in Delhi. In: Desai, R., Sanyal, R. (Eds.), Urbanizing Citizenship. Contested Spaces in Indian Cities. Sage, New Delhi, 135–160.
Kumar, S., Jha, P., Baier, K., Jha, R., Azzam, R., 2012. Pollution of Ganga River Due to Urbanization of Varanasi: Adverse Conditions Faced by the Slum Population. Environment and Urbanization Asia 3 (2), 343–352.
Kundu, A., 2003. Urbanisation and Urban Governance: Search for a Perspective beyond Neo-Liberalism. Economic and Political Weekly 38 (29), 3079–3087.
Kvale, S., 2007. Doing Interviews. Sage, London.
Lahiri-Dutt, K., 2000. Imagining rivers. Economic and Political Weekly 32 (27), 2395–2400.
Lahiri-Dutt, K., 2014. Beyond the water-land binary in geography: Water/lands of Bengal re-visioning hybridity. ACME 13 (3), 505–529.
Lascoumes, P., Le Gales, P., 2007. Introduction: Understanding Public Policy through Its Instruments—From the Nature of Instruments to the Sociology of Public Policy Instrumentation. Governance 20 (1), 1–21.
Latour, B., 1993. We Have Never Been Modern. Harvester Wheatsheaf, London.
Latour, B., 1996. On actor-network theory: a few clarifications. Soziale Welt 47 (4), 369–381.
Latour, B., 2004. Politics of Nature: How to Bring Science into Democracy. Harvard University Press, Cambridge, MA.
Lavau, S., 2011. Curious Indeed, or Curious in Deed? Some peculiarities of post-settlement relations with an antipodean river. Australian Geographer 42 (3), 241–256.

Lawhon, M., Ernstson, H., Silver, J., 2014. Provincializing Urban Political Ecology: Towards a Situated UPE Through African Urbanism. Antipode 46 (2), 497–516.

Lawhon, M., Murphy, J. T., 2012. Socio-technical regimes and sustainability transitions: Insights from political ecology. Progress in Human Geography 36 (3), 354–378.

Le Galès, P., 1998. Regulations and Governance in European Cities. International Journal of Urban and Regional Research 22 (3), 482–506.

Leach, M., Bloom, G., Ely, A., Nightingale, P., Scoones, I., Shah, E., Smith, A., 2007. Understanding governance: Pathways to sustainability. Steps Centre, Brighton. http://steps-centre.org/wp-content/uploads/final_steps_governance.pdf (2014-03-13).

Leach, M., Mearns, R., 1996. Environmental Change and Policy In: Leach, M., Mearns, R. (Eds.), The Lie of the Land: Challenging Received Wisdom on the African Environment. James Currey, Oxford, 1–33.

Lees, L., 2004. Urban geography: discourse analysis and urban research. Progress in Human Geography 28 (1), 101–107.

Lefebvre, H., 1991 [1974]. The Production of Space. Blackwell, Oxford.

Lefebvre, H., 1996. Writings on Cities. Blackwell, Oxford.

Leftwich, A., 1993. Governance, democracy and development in the Third World. Third World Quarterly 14 (3), 605–624.

Legg, S., 2006a. Governmentality, congestion and calculation in colonial Delhi. Social & Cultural Geography 7 (5), 709–729.

Legg, S., 2006b. Post-Colonial Developmentalities: From the Delhi Improvement Trust to the Delhi Development Authority. In: Raju, S., Kumar, M. S., Corbridge, S. (Eds.), Colonial and Postcolonial Geographies of India. Sage, New Delhi, 182–204.

Legg, S., 2007. Spaces of Colonialism. Delhi's Urban Governmentalities. Blackwell, Malden.

Leigh Bonnell, J., 2010. Imagined Futures and Unintended Consequences: An Environmental History of Toronto's Don River Valley. Department of Theory and Policy Studies in Education, Ontario Institute for Studies in Education, University of Toronto, Toronto. https://tspace.library.utoronto.ca/bitstream/1807/24690/1/Bonnell_Jennifer_201006_PhD_thesis.pdf (2014-05-05).

Lele, S., Dubash, N. K., Dixit, S., 2010. A Structure for Environment Governance: A Perspective. Economic and Political Weekly 45 (6), 13–16.

Lemos, M. C., Agrawal, A., 2006. Environmental Governance. Annual Review of Environment and Resources 31 (1), 297–325.

Littig, B., 2009. Interviewing the Elite - Interviewing Experts: Is there a Difference? In: Bogner, A., Littig, B., Menz, W. (Eds.), Interviewing Experts. Palgrave Macmillan, Basingstoke, 98–116.

Loftus, A., 2006. The metabolic processes of capital accumulation in Durban's waterscape. In: Heynen, N., Kaïka, M., Swyngedouw, E. (Eds.), In the Nature of Cities. Urban political ecology and the politics of urban metabolism. Routledge, New York, 173–190.

Loftus, A., 2007. Working the Socio-Natural Relations of the Urban Waterscape in South Africa. International Journal of Urban and Regional Research 31 (1), 41–59.

Loftus, A., 2009. Rethinking Political Ecologies of Water. Third World Quarterly 30 (5), 953–968.

Loftus, A., 2012. Everyday Environmentalism. Creating an Urban Political Ecology. University of Minnesota Press, Minneapolis.

Loftus, A., Lumdsen, F., 2008. Reworking hegemony in the urban waterscape. Transactions of the Institute of British Geographers 33, 109–126.

Lord, A., 2014. Making a 'Hydropower Nation': Subjectivity, Mobility, and Work in the Nepalese Hydroscape. Himalaya, the Journal of the Association for Nepal and Himalayan Studies 34 (2), Article 13.

Lorenzen, G., Sprenger, C., Taute, T., Pekdeger, A., Mittal, A., Massmann, G., 2010. Assessment of the potential for bank filtration in a water-stressed megacity (Delhi, India). Environmental Earth Sciences 61 (7), 1419–1434.

Lübken, U., 2012. Rivers and Risk in the City: The Urban Floodplain as a Contested Space. In: Castonguay, S., Evenden, M. (Eds.), Urban Rivers: Remaking Rivers, Cities, and Space in Europe and North America. University of Pittsburgh Press, Pittsburgh, 130–144.

Mahadevia, D., 2003. Globalisation, urban reforms und metropolitan response: India. Manak, New Delhi.

Mahadevia, D., Narayanan, H., 2008a. Shanghaing Mumbai: Politics of Evictions and Resistance in Slum Settlements. In: Mahadevia, D. (Ed.), Inside the transforming urban Asia: processes, policies and public action. Concept Publishing Company, New Delhi, 549–589.

Mahadevia, D., Narayanan, H., 2008b. Slumbay to Shanghai: Envisioning Renewal or Take Over. In: Mahadevia, D. (Ed.), Inside the transforming urban Asia: processes, policies and public action. Concept Publishing Company, New Delhi, 94–131.

Maitra, S., 1991. Housing in Delhi: DDA's Controversial Role. Economic and Political Weekly 26 (7), 344–346.

Malard, F., Trockner, K., Ward, J., 2000. Physicochemical heterogeneity in a glacial riverscape. Landscape Ecology 15, 679–695.

Mandal, P., Upadhyay, R., Hasan, A., 2010. Seasonal and spatial variation of Yamuna River water quality in Delhi, India. Environ Monit Assess 170 (1–4), 661–70.

Mann, M., 2005. Geschichte Indiens. Vom 18. bis zum 20. Jahrhundert. Schöningh UTB, Paderborn.

Mann, M., 2006. Metamorphosen einer Metropole: Delhi 1911–1977. In: Ahuja, R., Brosius, C. (Eds.), Mumbai Delhi Kolkata - Annäherungen an die Megastädte Indiens. Draupadi, Heidelberg, 127–142.

Mann, M., 2007. Delhi's Belly: On the Management of Water, Sewage and Excreta in a Changing Urban Environment during the Nineteenth Century. Studies in History 23 (1), 1–31.

Mann, M., Sehrawat, S., 2009. A City With a View: The Afforestation of the Delhi Ridge, 1883–1913. Modern Asian Studies 43 (2), 543–570.

Maria, A., 2006. The role of groundwater in Delhi's water supply. Interaction Between Formal and Informal Development of the Water System, and Possible Scenarios of Evolution. In: Tellam, J. H., Rivett, M. O., Israfilov, R. G., Herringshaw, L. G. (Eds.). Springer Netherlands, 459–470.

Maria, A., 2008. Urban water crisis in Delhi. Stakeholder responses and potential scenarios of evolution, Paris.

Marston, S. A., 2000. The social construction of scale. Progress in Human Geography 24 (2), 219–242.

Martin, A., McGuire, S., Sullivan, S., 2013. Global environmental justice and biodiversity conservation. Geographical Journal 179 (2), 122–131.

Mathur, N., 2012. On the Sabarmati Riverfront. Urban Planning as Totalitarian Governance in Ahmedabad. Economic and Political Weekly 47 (47/48), 64–75.

Mattissek, A., 2007. Diskursanalyse in der Humangeographie – „State of the Art". Geographische Zeitschrift 95 (1/2), 37–55.

Mattissek, A., Reuber, P., 2004. Die Diskursanalyse als Methode in der Geographie – Ansätze und Potentiale. Geographische Zeitschrift 92 (4), 227–242.

Mawdsley, E., 2004. India's Middle Classes and the Environment. Development and Change 35 (1), 79–103.

Mawdsley, E., 2009. 'Environmentality' in the neoliberal city: attitudes, governance and social justice. In: Lange, H., Meier, L. (Eds.), The New Middle Classes. Globalizing Lifestyles, Consumerism and Environmental Concern. Springer, Heidelberg, 253–268.

Mawdsley, E., Mehra, D., Beazley, K., 2009. Nature Lovers, Picnickers and Bourgeois Environmentalism. Economic and Political Weekly 44 (11), 49–59.

Mayntz, R., 2003. New challenges to governance theory. In: Bang, H. P. (Ed.), Governance as social and political communication. Manchester University Press, Manchester, 27–40.

Mayntz, R., 2004. Governance im modernen Staat. In: Benz, A. (Ed.), Governance - Regieren in komplexen Regelsystemen. VS Verlag, Wiesbaden.

Mayntz, R., 2005. Governance Theory als fortentwickelte Steuerungstheorie? In: Schuppert, G. F. (Ed.), Governance-Forschung. Vergewisserung über Stand und Entwicklungslinien. Nomos, Baden-Baden, 11–20.

Mayntz, R., 2009a. Governancetheorie: Erkenntnisinteresse und offene Fragen. In: Grande, E., May, S. (Eds.), Perspektiven der Governance-Forschung. Nomos, Baden-Baden, 9–19.

Mayntz, R., 2009b. Über Governance. Institutionen und Prozesse politischer Regelung. Campus, Frankfurt a. M.

Mayntz, R., Scharpf, F. (Eds.), 1995. Gesellschaftliche Selbstregulierung und politische Steuerung. Schriften des Max-Planck-Instituts für Gesellschaftsförderung Band 23. Campus, Frankfurt, New York.

McCann, E. J., Ward, K., 2011. Introduction. Urban Assemblages: Territories, Relations, Practices, and Power. In: McCann, E. J., Ward, K. (Eds.), Mobile Urbanism. University of Minnesota Press, Minneapolis, xiii–xxxv.

McCully, P., 1996. Silent Rivers: The Ecology and Politics of Large Dams. Zed Books, London.

McFarlane, C., 2008. Governing the Contaminated City: Infrastructure and Sanitation in Colonial and Post-Colonial Bombay. International Journal of Urban and Regional Research 32 (2), 415–435.

McFarlane, C., 2012. Rethinking Informality: Politics, Crisis, and the City. Planning Theory & Practice 13 (1), 89–108.

McFarlane, C., Waibel, M. (Eds.), 2012. Urban Informalities. Ashgate, Farnham.

McGranahan, G., Satterthwaite, D., 2000. Environmental health or ecological sustainability? Reconciling the Brown and Green agendas in urban development. In: Pugh, C. (Ed.), Sustainable Cities in Developing Countries. Earthscan, London, 73–90.

McKinsey, 2003. Vision Mumbai. Transforming Mumbai into a world-class city. A Bombay First – McKinsey Report. Mumbai.

McKinsey Global Institute, 2010. India's Urban Awakening: building inclusive cities, sustaining economic growth. San Francisco.

Mehra, A., Farago, M. E., Banerjee, D. K., 1998. Impact of Fly Ash from Coal-Fired Power Stations in Delhi, with Particular Reference to Metal Contamination. Environ Monit Assess 50 (1), 15–35.

Mehra, D., 2013. Planning Delhi ca. 1936–1959. South Asia: Journal of South Asian Studies 36 (3), 354–374.

Mehta, L., Karpouzoglou, T., 2015. Limits of policy and planning in peri-urban waterscapes: The case of Ghaziabad, Delhi, India. Habitat International 48 (0), 159–168.

Mehta, M. C., 2009a. In the Public Interest. Landmark Judgements and Orders of the Supreme Court of India on Environment & Human Rights - Volume I. Prakriti Publications, New Delhi.

Mehta, M. C., 2009b. In the Public Interest. Landmark Judgements and Orders of the Supreme Court of India on Environment & Human Rights - Volume II. Prakriti Publications, New Delhi.

Mehta, P. B., 2006. India's Judiciary: The Promise of Uncertainty. In: Chopra, P. (Ed.), The Supreme Court Versus the Constitution: A Challenge to Federalism. Sage, New Delhi, 155–177.

Mels, T., 2009. Analysing Environmental Discourses and Representations. In: Castree, N., Demeritt, D., Liverman, D., Rhoads, B. (Eds.), A Companion to Environmental Geography. Willey-Blackwell, Chichester, 385–399.

Menon-Sen, K., Bhan, G., 2008. Swept off the Map. Surviving Eviction and Resettlement in Delhi. Yoda Press, New Delhi.

Menon, M., Kohli, K., 2009. From Impact Assessment to Clearance Manufacture. Economic and Political Weekly 44 (23), 20–23.

Menon, M., Rodriguez, S., Sridhar, A., 2007. Coastal Zone Management: Better or Bitter Fare? Economic and Political Weekly 42 (38), 3838–3840.

Meuser, M., Nagel, U., 2009. The Expert Interview and Changes in Knowledge Production. In: Bogner, A., Littig, B., Menz, W. (Eds.), Interviewing Experts. Palgrave Macmillan, Basingstoke, 17–42.

Miraftab, F., 2004. Invited and Invented Spaces of Participation: Neoliberal Citizenship and Feminists' Expanded Notion of Politics. Wagadu 1 (Spring 2004), 1–7.

Miraftab, F., 2009. Insurgent Planning: Situating Radical Planning in the Global South. Planning Theory 8 (1), 32–50.

Mirza, M. M. Q., Ahmed, A. U., Ahmad, Q. K. (Eds.), 2008. Interlinking of Rivers in India: Issues and Concerns. Taylor & Francis, Boca Raton.

Mishra, D. K., 2009. Floods and Some Legal Concerns. In: Iyer, R. R. (Ed.), Water and the laws in India. Sage, New Delhi, 309–328.

Misra, A. K., 2010a. A River about to Die: Yamuna. Journal of Water Resource and Protection 2 (5), 489–500.

Misra, M., 2010b. Dreaming of a Blue Yamuna. In: Chaturvedi, B. (Ed.), Finding Delhi - Loss and Renewal in the Megacity. Penguin Books India, New Delhi, 71–86.

Misra, M., 2015. NGT Orders Maily se Nirmal Yamuna - will this lead to a rejuvenated Yamuna? SANDRP: Dams, Rivers & People 13 (1–2). https://sandrp.files.wordpress.com/2015/03/drp-feb-march-2015.pdf (2015-03-30).

Mistelbacher, J., 2005. Urbanisierungsdynamik in Indien: Das Beispiel Delhi. Geographische Rundschau 57 (10), 20–29.

Mittal, N., Petrarulo, L., Perera, N., 2015. Urban Environmental Governance: Bangladesh as a Case in Point. In: Leal Filho, W. (Ed.), Climate Change in the Asia-Pacific Region. Springer International Publishing, 119–142.

MoEF (Ministry of Environment and Forests, Government of India), 2006. EIA Regulations, Notification S.O.1533, 14 September 2006. http://www.moef.nic.in/legis/eia/eia-2006.htm (2014-05-05).

MoEF (Ministry of Environment and Forests, GoI), 2014a. About the Ministry. http://envfor.nic.in/about-ministry/about-ministry (2014-11-05).

MoEF (Ministry of Environment and Forests, GoI), 2014b. Environmental Impact Assessment. http://envfor.nic.in/major-initiatives/environmental-clearances (2014-05-05).

MoEF (Ministry of Environment and Forests, GoI), 2014c. National Green Tribunal (NGT). http://www.moef.nic.in/rules-regulations/national-green-tribunal-ngt (2014-11-05).

MoEF (Ministry of Environment and Forests, GoI), 2014d. National River Conservation Directorate (NRCD). http://envfor.nic.in/division/national-river-conservation-directorate-nrcd (2014-05-05).

Mohan, D., 2008. Mythologies, Metro Rail Systems and Future Urban Transport. Economic and Political Weekly 43 (03), 41–53.

Mohapatra, P. K., Singh, R. D., 2003. Flood Management in India. In: Mirza, M. M., Dixit, A., Nishat, A. (Eds.), Flood Problem and Management in South Asia. Springer Netherlands, 131–143.

Mol, A. P. J., 2002. Ecological Modernization and the Global Economy. Global Environmental Politics 2 (2), 92–115.

Molle, F., 2007. Scales and power in river basin management: the Chao Phraya River in Thailand. Geographical Journal 173 (4), 358–373.

Molle, F., 2009. River-basin planning and management: The social life of a concept. Geoforum 40 (3), 484–494.

Molle, F., Mollinga, P., Wester, P., 2009. Hydraulic bureaucracies and the hydraulic mission: Flows of water, flows of power. Water Alternatives 2 (3), 328–349.

Monstadt, J., 2009. Conceptualizing the political ecology of urban infrastructures: insights from technology and urban studies. Environment and Planning A 41 (8), 1924–1942.

Mookherjee, D., Geyer, H. S., Hoerauf, E., 2014. Dynamics of an Evolving City-Region in the Developing World: The National Capital Region of Delhi Revisited. International Planning Studies, 1–15.

Moseley, W., Perramond, E., Hapke, H., Laris, P., 2014. An Introduction to Human-Environmental Geography. Local Dynamics and Global Processes. Wiley-Blackwell, Chichester.

MoUAE (Government of India, Ministry of Urban Affairs and Employment, Department or Urban Development (Delhi Division)), 1999. Notification No. K-20013/21/93-DDIB. Published in the Gazette of India Part III Section 3, sub-Section (ii), 21st September 1999. New Delhi.

MoUD (Government of India, Ministry of Urban Development (Delhi Division)), 2006. Notification S.O. 1321 (E.) Published in the Gazette of India: Extraordinary, Part II, Section 3 (ii), 18th August 2006. New Delhi.

MoUEPA, MoUD (Ministry of Urban Employment and Poverty Alleviation, GoI and Ministry of Urban Development, GoI), 2005. Jawaharlal Nehru National Urban Renewal Mission - Overview. http://jnnurm.nic.in/wp-content/uploads/2011/01/UIGOverview.pdf (2014-11-05).

MoWR (Ministry of Water Resources, GoI), 1987. National Water Policy. http://cgwb.gov.in/documents/nwp_1987.pdf (2014-11-05).

MoWR (Ministry of Water Resources, GoI), 2002. National Water Policy. New Delhi, April 2002, http://cgwb.gov.in/documents/nwp_2002.pdf (2014-11-05).

MoWR (Ministry of Water Resources, GoI), 2012. National Water Policy. http://wrmin.nic.in/writereaddata/NationalWaterPolicy/NWP2012Eng6495132651.pdf (2014-11-05).

Murdoch, J., 1997. Inhuman/nonhuman/human: actor-network theory and the prospects for a nondualistic and symmetrical perspective on nature and society. Environment and Planning D: Society and Space 15 (6), 731–756.

Murdoch, J., 1998. The spaces of actor-network theory. Geoforum 29 (4), 357–374.

Murdoch, J., 2006. Post-structuralist geography: a guide to relational space. Sage, London.

Myers, G. A., 2008. Sustainable Development and Environmental Justice in African Cities. Geography Compass 2 (3), 695–708.

Nagpal, S., Sinha, A., 2009. The Gomti Riverfront in Lucknow, India: Revitalization of a Cultural Heritage Landscape. Journal of Urban Design 14 (4), 489–506.

Naib, S., 2013. The right to information in India. Oxford University Press, New Delhi.

Nanda, V. P., 2006. The "Good Governance" Concept Revisited. The ANNALS of the American Academy of Political and Social Science 603 (1), 269–283.

Nandi, S., Gamkhar, S., 2013. Urban challenges in India: A review of recent policy measures. Habitat International 39 (0), 55–61.

Nanson, G. C., Croke, J. C., 1992. A genetic classification of floodplains Geomorphology 4 (6), 459–486.

Narain, V., Nischal, S., 2007. The peri-urban interface in Shahpur Khurd and Karnera, India. Environment and Urbanization 19 (1), 261–273.

Narula, K. K., Wendland, F., Rao, D. D. B., Bansal, N. K., 2001. Water resources development in the Yamuna river basin in India. Journal of Environmental Studies and Policy 4 (1), 21–33.

Natter, W., Zierhofer, W., 2002. Political ecology, territoriality and scale. GeoJournal 58 (4), 225–231.

NCRPB (National Capital Region Planning Board), 1988. Regional Plan 2001 - National Capital Region (approved by the Board on November 3, 1988). New Delhi.

NCRPB (National Capital Region Planning Board), 2005. Regional Plan-2021. New Delhi, http://ncrpb.nic.in/regionalplan2021.php (2014-11-05).

NCRPB (National Capital Region Planning Board), 2013a. Draft Revised Regional Plan 2021 (Approved in 33rd Meeting of the NCR Planning Board held on 1st July, 2013). New Delhi, http://ncrpb.nic.in/pdf_files/Draft%20Revised%20Regional%20Plan%202021/4-in-1.pdf (2014-12-05).

NCRPB 2013b. Regional Plan 2021, National Capital Region Planning Board, Ministry of Urban Development, Government of India, New Delhi.

NEERI (National Environmental Engineering Research Institute), 1999. Initial Environmental Examination of Development Plan in Yamuna River Stretch between New Railway Bridge and Proposed ILFS Bridge - Phase I Report. January 1999. Nagpur.

NEERI (National Environmental Engineering Research Institute), 2000. Environmental Management Plan for Rejuvenation of River Yamuna in NCT. April 2000. Nagpur.

NEERI (National Environmental Engineering Research Institute), 2005. Environmental Management Plan for Rejuvenation of River Yamuna in NCT. October 2005. Nagpur.

NEERI (National Environmental Engineering Research Institute), 2008. Commonwealth Games Village Complex by Delhi Development Authority in Pocket III of Sub-Zone 6 of Yamuna River. Report prepared on request of DDA. January 2008. Nagpur.

Negi, R., 2011. Neoliberalism, Environmentalism, and Urban Politics in Delhi. In: Ahmed, W., Kundu, A., Peet, R. (Eds.), India's New Economic Policy. A Critical Analysis. Routledge, New York, 179–198.

Neumann, R. P., 2008. Political ecology: theorizing scale. Progress in Human Geography 33 (3), 398–406.

Neumann, R. P., 2009a. Political Ecology. In: Kitchin, R., Thrift, N. (Eds.), International Encyclopedia of Human Geography. Elsevier, Oxford, 228–233.

Neumann, R. P., 2009b. Political ecology II: theorizing region. Progress in Human Geography 34 (3), 368–374.

Neumann, R. P., 2011. Political ecology III: Theorizing landscape. Progress in Human Geography 35 (6), 843–850.

Nijman, J., 2007. Mumbai since liberalisation: The space-economy of India's gateway city. In: Shaw, A. (Ed.), Indian Cities In Transition. Orient Longman, Chennai, 238–259.

Nissel, H., 2001. Auswirkungen von Globalisierung und New Economic Policy im urbanen System Indiens. Mitteilungen der Österreichischen Geographischen Gesellschaft 143, 63–90.

Nissel, H., 2009. Contesting Urban Space. Megacities and Globalization in India. Geographische Rundschau / International edition 5 (1), 40–46.

Njeru, J., 2006. The urban political ecology of plastic bag waste problem in Nairobi, Kenya. Geoforum 37 (6), 1046–1058.

Nüsser, M. (Ed.), 2014. Large Dams in Asia. Contested Environments between Technological Hydroscapes and Social Resistance. Springer, Dordrecht.

Nüsser, M., Baghel, R., 2010. Discussing Large Dams in Asia after the World Commission on Dams: Is a Political Ecology Approach the Way Forward? Water Alternatives 3 (2), 231–248.

Olwig, K. R., 1996. Recovering the Substantive Nature of Landscape. Annals of the Association of American Geographers 86 (4), 630–653.

Ong, A., 2011. Introduction. Worlding Cities, or the Art of Being Global. In: Roy, A., Ong, A. (Eds.), Worlding Cities: Asian Experiments and the Art of Being Global. Blackwell, Oxford, 1–26.

Otto-Zimmermann, K. (Ed.), 2011. Resilient Cities 1. Cities and Adaptation to Climate Change - Proceedings of the Global Forum 2010. Springer, Dordrecht.

Otto-Zimmermann, K. (Ed.), 2012. Resilient Cities 2. Cities and Adaptation to Climate Change - Proceedings of the Global Forum 2011. Springer, Dordrecht.

Pandey, G., 2001. Remembering Partition. Violence, Nationalism and History in India. Cambridge University Press, Cambridge.

Panigrahi, J. K., Amirapu, S., 2012. An assessment of EIA system in India. Environmental Impact Assessment Review 35 (0), 23–36.

Panwar, H. S., 2009. Reviving River Yamuna. An Actionable Blue Print for a BLUE RIVER. Peace Institute Charitable Trust, New Delhi, http://www.peaceinst.org/publication/booklet/Reviving%20River%20Yamuna%20An%20Actionable%20Blue%20Print%20for%20a%20Blue%20River.pdf (2014-10-09).

Parker, D. J., 1995. Floodplain development policy in England and Wales. Applied Geography 15 (4), 341–363.

Parmar, D. L., Keshari, A. K., 2012. Sensitivity analysis of water quality for Delhi stretch of the River Yamuna, India. Environ Monit Assess 184 (3), 1487–508.

Parnell, S., Oldfield, S. (Eds.), 2014. The Routledge Handbook on Cities of the Global South. Routledge, New York.

Parnell, S., Robinson, J., 2012. (Re)theorizing Cities from the Global South: Looking Beyond Neoliberalism. Urban Geography 33 (4), 593–617.

Parnell, S., Simon, D., Vogel, C., 2007. Global environmental change: conceptualising the growing challenge for cities in poor countries. Area 39 (3), 357–369.

Peace Institute Charitable Trust, 2010. On the brink ... Water Governance in the Yamuna River Basin in Haryana. New Delhi, http://www.yamunajiyeabhiyaan.com/Admin/downloadsfile/dfiles/5_Impact%20Assessment%20of%20Bridges%20and%20Barrages%20on%20River%20Yamuna.pdf (2014-11-12).

PEACE Institute Charitable Trust & Environics Trust, 2009. Impact Assessment of Bridges and Barrages on River Yamuna (Wazirabad - Okhla Section). New Delhi, http://www.yamunajiyeabhiyaan.com/Admin/downloadsfile/dfiles/5_Impact%20Assessment%20of%20Bridges%20and%20Barrages%20on%20River%20Yamuna.pdf (2014-11-12).

Peck, J., Tickell, A., 2002. Neoliberalizing Space. Antipode 34 (3), 380–404.

Peck, L., 2005. Delhi. A Thousand Years of Building. Roli Books, New Delhi.

Peet, R., Watts, M. (Eds.), 1996. Liberation ecologies. Environment, development, social movements. Routledge, London.

Pessina, G., 2012. Trying to be "sustainable" and "global": Ahmedabad and the Sabarmati River Front Development Project. AESOP 26th Annual Congress, Ankara, 11–15 July 2012, http://www.arber.com.tr/aesop2012.org/arkakapi/cache/absfilAbstractSubmissionFullContent 1074.pdf (2014-09-13).

Pierre, J. E., 1998. Partnerships in Urban Governance: European and American Experiences. MacMillan, London.

Pierre, J. E., Peters, G., 2000. Governance, Politics and the State. St. Martin's Press, New York.

Pincetl, S., 2012. Nature, urban development and sustainability – What new elements are needed for a more comprehensive understanding? Cities 29, Supplement 2 (0), S32–S37.

Planning Commission, (GoI), 2009. Delhi Development Report. New Delhi, http://planningcommission.nic.in/plans/stateplan/sdr/sdr_delhi1909.pdf (2014-11-12).

Planning Commission, (GoI), 2011. Report of working group on flood management and region specific issues for XII plan. New Delhi, http://planningcommission.gov.in/aboutus/committee/wrkgrp12/wr/wg_flood.pdf (2014-11-12).

Prashad, V., 2001. The technology of sanitation in colonial Delhi. Modern Asian Studies 35, 113–155.

Prime Minister's Office (Government of India), 2007. Subject: Development of river Yamuna. http://ud.delhigovt.nic.in/PMO-1.pdf (2014-12-01).

Punjab Government, 1912. A Gazetteer of Delhi (1912). Compiled and published under the authority of the Punjab Government, http://revenueharyana.gov.in/html/gazeteers/delhi%20gazeetter%201912.pdf (2014-10-29).

Raco, M., 2009. Governance, Urban. In: Kitchin, R., Thrift, N. (Eds.), International Encyclopedia of Human Geography. Elsevier, Oxford, 622–627.

Rademacher, A. M., 2005. Culturing urban ecology: development, statemaking, and river restoration in Kathmandu. Yale University (PhD thesis). Yale.

Rademacher, A. M., 2007. Farewell to the Bagmati Civilisation: Losing Riverscape and Nation in Kathmandu. National Identities 9 (2), 127–142.

Rademacher, A. M., 2008. Fluid City, Solid State: Urban Environmental Territory in a State of Emergency, Kathmandu. City & Society 20 (1), 105–129.

Rademacher, A. M., 2009. When Is Housing an Environmental Problem? Reforming Informality in Kathmandu. Current Anthropology 50 (4), 513–533.

References

Rademacher, A. M., 2011. Reigning the River: Urban Ecologies and Political Transformation in Kathmandu. Duke University Press, Durham.

Rademacher, A. M., Sivaramakrishnan, K. (Eds.), 2013a. Ecologies of Urbanism in India: Metropolitan Civility and Sustainability. Hong Kong University Press, Hong Kong.

Rademacher, A. M., Sivaramakrishnan, K., 2013b. Introduction: Ecologies of Urbanism in India. In: Rademacher, A. M., Sivaramakrishnan, K. (Eds.), Ecologies of Urbanism in India: Metropolitan Civility and Sustainability. Hong Kong University Press, Hong Kong, 1–41.

Rai, R. K., Upadhyay, A., Ojha, C. S. P., Singh, V. P., 2012. The Yamuna River Basin. Water Resources and Environment. Springer Netherlands.

Rajamani, L., 2007. Public Interest Environmental Litigation in India: Exploring Issues of Access, Participation, Equity, Effectiveness and Sustainability. Journal of Environmental Law 19 (3), 293–321.

Ramachandran, R., 1989. Urbanization and Urban Systems in India. Oxford University Press, New Delhi.

Ramanathan, U., 2006. Illegality and the Urban Poor. Economic and Political Weekly 41 (29), 3193–3197.

Ramsar Convention Secretariat 2015. About the Ramsar Convention.

Ranganathan, M., 2015. Storm Drains as Assemblages: The Political Ecology of Flood Risk in Post-Colonial Bangalore. Antipode 47 (5), 1300-1320.

Razzaque, J., 2004. Public Interest Environmental Litigation in India, Pakistan and Bangladesh. Kluwer Law International, The Hague.

Razzaque, J., 2013. Environmenal Governance in Europe and Asia. Routledge, New York.

Reed, M. G., Bruyneel, S., 2010. Rescaling environmental governance, rethinking the state: A three-dimensional review. Progress in Human Geography 34 (5), 646–653.

Ren, X., Weinstein, L., 2013. Urban governance, mega-projects and scalar transformations in China and India. In: Roshan Samara, T., He, S., Chen, G. (Eds.), Locating Right to the City in the Global South. Routledge, London, New York, 107–126.

Renaud, F. G., Kuenzer, C. (Eds.), 2012. The Mekong Delta System. Interdisciplinary Analyses of a River Delta. Springer, Dordrecht.

Rhodes, R. A. W., 1996. The New Governance: Governing without Government. Political Studies 44 (4), 652–667.

Rhodes, R. A. W., 1997. Understanding Governance. Open University Press, Buckingham.

Richardson, T., Jensen, O. B., 2003. Linking Discourse and Space. Towards a Cultural Sociology of Space in Analysing Spatial Policy Discourses. Urban Studies 40 (1), 7–22.

Robbins, P., 2012. Political Ecology: A Critical Introduction. Wiley-Blackwell, Chichester.

Robinson, J., 2002. Global and World Cities: A View off the Map. International Journal of Urban and Regional Research 26 (3), 531–554.

Robinson, J., 2006. Ordinary cities: between modernity and development. Routledge, London.

Robinson, J., 2011. The Spaces of Circulating Knowledge: City Strategies and Global Urban Governmentality. In: McCann, E. J., Ward, K. (Eds.), Mobile Urbanism. Cities and Policymaking in the Global Age. University of Minnesota Press, Minneapolis, 15–40.

Rocheleau, D. E., 2008. Political ecology in the key of policy: From chains of explanation to webs of relation. Geoforum 39 (2), 716–727.

Rocheleau, D. E., Thomas-Slayter, B., Wangari, E., 1996. Feminist Political Ecology: Global issues and local experiences. Routledge, London.

Rosenau, J., Czempiel, D., 1992. Governance without Government. Order and Change in Worlds Politics. Cambridge University Press, Cambridge.

Roy, A., 2003. Requiem, Calcutta: Gender and the Politics of Poverty. University of Minnesota Press, Minneapolis.

Roy, A., 2004. The Gentlemen's City: Urban Informality in Calcutta of New Communism. In: Roy, A., AlSayyad, N. (Eds.), Urban informality: transnational perspectives from the Middle East, Latin America, and South Asia. Lexington Books, Oxford, 147–170.

Roy, A., 2005. Urban Informality. Towards an Epistemology of Planning. Journal of the American Planning Association 71 (2), 147–158.

Roy, A., 2009a. The 21st-Century Metropolis: New Geographies of Theory. Regional Studies 43 (6), 819–830.

Roy, A., 2009b. Why India Cannot Plan Its Cities: Informality, Insurgency and the Idiom of Urbanization. Planning Theory 8 (1), 76–87.

Roy, A., 2012. Urban Informality: The Production of Space and Practice of Planning. In: Weber, R., Crane, R. (Eds.), The Oxford Handbook of Urban Planning. Oxford University Press, Oxford, 691–705.

Rubin, H., Rubin, I., 2012. Qualitative interviewing: the art of hearing data. Sage, Thousand Oaks.

Rubin, M., 2013. Courting change: the role of apex courts and court cases in urban governance: a Delhi-Johannesburg comparison. Ph.D. Thesis. University of the Witwatersrand Johannesburg.

Ruet, J., Tawa Lama-Rewal, S. (Eds.), 2009. Governing India's Metropolises. Routledge, New Delhi.

Rydin, Y., 1998. Land Use Planning and Environmental Capacity: Reassessing the Use of Regulatory Policy Tools to Achieve Sustainable Development. Journal of Environmental Planning and Management 41 (6), 749–765.

Salmond, A., 2014. Tears of Rangi. Water, power, and people in New Zealand. Hau: Journal of Ethnographic Theory 4 (3), 285–309.

Sami, N., 2013. From Farming to Development: Urban Coalitions in Pune, India. International Journal of Urban and Regional Research 37 (1), 151–164.

Sandercock, L., Dovey, K., 2002. Pleasure, Politics and the "Public Interest" - Melbourne's Riverscape Revitalization. Journal of the American Planning Association 68 (2), 151–164.

Sandhu, C., Grischek, T., Kumar, P., Ray, C., 2010. Potential for Riverbank filtration in India. Clean Technologies and Environmental Policy 13 (2), 295–316.

Sassen, S., 1991. The global city. Princeton University Press, New York.

Savage, V. R., Huang, S., Chang, T. C., 2004. The Singapore River thematic zone: sustainable tourism in an urban context. Geographical Journal 170 (3), 212–225.

Scott, J. C., 1998. Seeing Like a State: How Certain Schemes to Improve the Human Condition have Failed. Yale University Press, New Haven.

Sehgal, M., Garg, A., Suresh, R., Dagar, P., 2012. Heavy metal contamination in the Delhi segment of Yamuna basin. Environ Monit Assess 184 (2), 1181–96.

Sehgal, M., Suresh, R., Sharma, V. P., Kumar Gautam, S., 2014. Study on heavy metal contamination of agricultural soils and vegetables grown in the Delhi segment of Yamuna river basin. Global Health Perspectives 2014 (2).

Selbach, V., 2009. Wasserversorgung und Verwundbarkeit in der Megastadt Delhi/Indien. University of Cologne Cologne. http://kups.ub.uni-koeln.de/2860/ (2014-11-01).

Sharan, A., 2002. Claims on cleanliness. Environment and justice in contemporary Delhi. In: Vasudevan, R., Sundaram, R., Bagchi, J., Narula, M., Lovink, G., Sengupta, S. (Eds.), Sarai Reader 02 - The Cities of Everyday Life. Sarai, CSDS, New Delhi, 31–37.

Sharan, A., 2013. One Air, Two Interventions: Delhi in the Age of Environment. In: Rademacher, A. M., Sivaramakrishnan, K. (Eds.), Ecologies of Urbanism in India: Metropolitan Civility and Sustainability. Hong Kong University Press, Hong Kong, 71–92.

Sharan, A., 2014. In the City, out of Place. Nuisance, Pollution, and Dwelling in Delhi, c. 1850–2000. Oxford University Press, New Delhi.

Sharan, A., 2015. A river and the riverfront: Delhi's Yamuna as an in-between space. City, Culture and Society (in press). doi:10.1016/j.ccs.2014.12.001.

Sharma, D., Kansal, A., 2011a. Current condition of the Yamuna River - an overview of flow, pollution load and human use. Yale - TERI Workshop on the Yamuna River, New Delhi, January 3–5, 2011, http://fore.research.yale.edu/yamuna-river-conference/ (2014-06-06).

Sharma, D., Kansal, A., 2011b. The status and effects of the Yamuna Action Plan (YAP). Yale - TERI Workshop on the Yamuna River, New Delhi, January 3–5, 2011, http://fore.research.yale.edu/yamuna-river-conference/ (2014-06-12).

Sharma, D., Kansal, A., 2011c. Water quality analysis of River Yamuna using water quality index in the national capital territory, India (2000–2009). Applied Water Science 1 (3–4), 147–157.

Sharma, D., Singh, R. K., 2009. DO-BOD modeling of River Yamuna for national capital territory, India using STREAM II, a 2D water quality model. Environ Monit Assess 159 (1–4), 231–40.

Shatkin, G., 2007. Global cities of the South: Emerging perspectives on growth and inequality. Cities 24 (1), 1–15.

Shatkin, G., 2011. Planning Privatopolis: Representation and Contestation in the Development of Urban Integrated Mega-Projects. In: Roy, A., Ong, A. (Eds.), Worlding Cities: Asian Experiments and the Art of Being Global. Blackwell, Oxford, 77–97.

Shatkin, G., 2014. Contesting the Indian City: Global Visions and the Politics of the Local. International Journal of Urban and Regional Research 38 (1), 1–13.

Shatkin, G., Vidyarthi, S., 2013. Introduction. Contesting the Indian City: Global Visions and the Politics of the Local. In: Shatkin, G., Vidyarthi, S. (Eds.), Contesting the Indian City. Global Visions and the Politics of the Local. Wiley Blackwell, Chichester, 1–38.

Shaw, A. (Ed.), 2007. Indian Cities In Transition. Orient Longman, Chennai.

Shaw, A. (Ed.), 2012. Indian Cities. Oxford University Press, New Delhi.

Shaw, A., Satish, M. K., 2007. Metropolitan restructuring in post-liberalized India: Separating the global and the local. Cities 24 (2), 148–163.

Shekhar, S., Prasad, R. K., 2009. The groundwater in the Yamuna flood plain of Delhi (India) and the management options. Hydrogeology Journal 17 (7), 1557–1560.

Sheppard, J. W., 2006. The Paradox of Urban Environmentalism: Problem and Possibility. Ethics, Place & Environment 9 (3), 299–315.

Sherman, T. C., 2013. From 'Grow More Food' to 'Miss a Meal': Hunger, Development and the Limits of Post-Colonial Nationalism in India, 1947–1957. South Asia: Journal of South Asian Studies 36 (4), 571–588.

Shutkin, W., 2001. The land that could be: environmentalism and democracy in the twenty first century. MIT Press, Cambridge.

Siddiqui, K. (Ed.), 2004. Megacity governance in South Asia. A comparative study. The University Press Limited, Dhaka.

Siddiqui, K., Ranjan, N., Kapuria, S., 2004. Delhi. In: Siddiqui, K. (Ed.), Megacity governance in South Asia. A comparative study. The University Press Limited, Dhaka.

Siemiatycki, M., 2006. Message in a Metro: Building Urban Rail Infrastructure and Image in Delhi, India. International Journal of Urban and Regional Research 30 (2), 277–292.

Simone, A., 2010. City Life from Jakarta to Dakar: Movements at the Crossroads. Routledge, New York.

Singh, A. P., Ghosh, S. K., Sharma, P., 2006. Water quality management of a stretch of river Yamuna: An interactive fuzzy multi-objective approach. Water Resources Management 21 (2), 515–532.

Singh, P., 1989. The Nine Cities of Delhi. In: Singh, P., Dhamija, R. (Eds.), Delhi - the deepening urban crisis. Sterling Publishers, New Delhi, 13–18.

Singh, R., 2003. Interlinking of Rivers. Economic and Political Weekly 38 (19), 1885–1886.

Singh, R., 2007. Urban Planning of the Heritage City of Varanasi (India) and its role in Regional Development. In: Schrenk, M., Popovich, V. V., Benedikt, J. (Eds.), REAL CORP 007: To Plan Is Not Enough. Strategies, Concepts, Plans, Projects and their Successful Implementation in Urban, Regional and Real Estate Development. Proceedings of 12th International Conference on Urban Planning and Spatial Development in the Information Society, Wien/Schwechat, 259–268.

Singh, R., 2009. Wastewater problems and social vulnerability in Megacity Delhi / India. Institute of Geography, University of Cologne (Dissertation). Cologne.

Singh, R. B. (Ed.), 2015. Urban Development Challenges, Risks and Resilience in Asian Mega Cities. Springer, Tokyo.

Singh, S., Kumar, M., 2006. Heavy metal load of soil, water and vegetables in peri-urban Delhi. Environ Monit Assess 120 (1–3), 79–91.

Singh, U., 2006. Delhi Acient History. Social Science Press, New Delhi.

Sinha, A., Harkness, T., 2009. Views of the Taj - Figure in the Landscape. Landscape Journal 28 (2), 198–217.

Sinha, A., Ruggles, D. F., 2004. The Yamuna Riverfront, India: A Comparative Study of Islamic And Hindu Traditions In Cultural Landscapes. Landscape Journal 23 (2), 141–152.

Sivaramakrishnan, K. C., 2011. Re-visioning Indian Cities. The Urban Renewal Mission. Sage, New Delhi.

Sivaramakrishnan, K. C., 2015. Governance of Megacities: Fractured Thinking, Fragmented Setup. Oxford University Press, Oxford.

Sivaramakrishnan, K. C., Kundu, A., Singh, B. N., 2007. Handbook of Urbanization in India. An Analysis of Trends and Processes. Oxford University Press, New Delhi.

Smith, N., 1984. Uneven Development. Nature, Capital and the Production of Space. Blackwell, Oxford.

Smith, N., 2006. Foreword. In: Heynen, N., Kaïka, M., Swyngedouw, E. (Eds.), In the Nature of Cities. Urban political ecology and the politics of urban metabolism. Routledge, New York, xi–xv.

Soni, C., 2006. Killing Delhi's Lifeline. Tehelka, puplication date: 2006-08-19, http://archive.tehelka.com/story_main18.asp?filename=Cr081906Killing_Delhi.asp, (2015-01-10).

Soni, V., 2007. Three waters - An evaluation of urban groundwater resource in Delhi. Current Science 93 (6), 760–761.

Soni, V., Gosain, A. K., Datta, P. S., Singh, D., 2009. A new scheme for large-scale natural water storage in the floodplains: the Delhi Yamuna floodplains as a case study. Current Science 96 (10), 1338–1342.

Soni, V., Shekhar, S., Singh, D., 2014. Environmental flow for the Yamuna river in Delhi as an example of monsoon rivers in India. Current Science 106 (4), 558–564.

Soni, V., Singh, D., 2013. Floodplains: self-recharging and self-sustaining aquifers for city water. Current Science 104 (4), 420–422.

Sonnenfeld, D. A., Mol, A. P. J., 2006. Environmental Reform in Asia: Comparisons, Challenges, Next Steps. The Journal of Environment & Development 15 (2), 112–137.

Sorensen, A., 2011. Megacity Sustainability: Urban Form, Development, and Governance. In: Sorensen, A., Okata, J. (Eds.), Megacities: Urban Form, Governance and Sustainability. Springer, Tokyo, 397–418.

Sorensen, A., Okata, J. (Eds.), 2011. Megacities: Urban Form, Governance and Sustainability. Springer, Tokyo.

Sreedharan, E., 2009. Restrict Yamuna with walls and develop low-lying areas. The Times of India, puplication date: 2009-05-20.

Sridharan, E., 2011. The Growth and Sectoral Composition of India's Middle Classes: Their Impact on the Politics of Economic Liberalization. In: Baviskar, A., Ray, R. (Eds.), Elite and Everyman: The Cultural Politics of the Indian Middle Classes. Routledge, New Delhi, 27–57.

Sridharan, N., 2008. New Forms of Contestation and Cooperation in Indian Urban Governance In: Baud, I. S. A., deWit, J. (Eds.), New Forms of Urban Governance in India. Shifts, Models, Networks and Contestations. Sage, New Delhi, 291–311.

Srirangan, K., 1997. Land policies in Delhi: Their contribution to unauthorised land development. University of London, Development Planning Unit London.

Srivastava, S., 2009. Urban Spaces, Disney-Divinity and Moral Middle Classes in Delhi. Economic and Political Weekly 44 (26–27), 338–345.

Stang, F., 2002. Indien. Wissenschaftliche Buchgesellschaft Darmstadt, Darmstadt.

References

Stoker, G., 1998. Governance as theory: five propositions. International Social Science Journal 50 (155), 17–28.

Stott, P., Sullivan, S., 2000. Political Ecology: Science, Myth and Power. Arnold, London.

Sullivan, S. M., Watzin, M., Keeton, W., 2007. A riverscape perspective on habitat associations among riverine bird assemblages in the Lake Champlain Basin, USA. Landscape Ecology 22 (8), 1169–1186.

Survey of India, 1913a. Delhi & Vicinity, Coloured to Show Land Acquisition Proposals, Nos 53 H/2 & H/6. Survey of India prepared by Delhi Town Planning (Swinton) Committee 1912–13: Final report on the town planning of the new imperial capital. Simla, Delhi.

Survey of India, 1913b. Map for the Delhi Town Planning Committee. Final report on the town planning of the new imperial capital (1912–13). Shimla, Delhi.

Survey of India, 1922. Delhi and vicinity. Scale 1:15,840. 1 map on 4 sheets: col.; 134 x 156 cm, sheets 76 x 101 cm. Calcutta.

Survey of India, 1933. Delhi guide map, 6th Edition (revised). Dehra Dun.

Survey of India, 1945. Delhi guide map [surveyed 1939–1942]. Dehra Dun.

Suryanarayanan, N., 1997. National water policy in India. In: Biswas, A. K., Toledo, H. G., Velasco, H. G., Quiroz, C. T. (Eds.), National Water Master Plans for Developing Countries. Oxford University Press, Calcutta, 142–181.

Swyngedouw, E., 1996. The city as a hybrid: On nature, society and cyborg urbanization. Capitalism Nature Socialism 7 (2), 65–80.

Swyngedouw, E., 1997. Power, nature, and the city. The conquest of water and the political ecology of urbanization in Guayaquil, Ecuador: 1880–1990. Environment and Planning A 29 (2), 311–332.

Swyngedouw, E., 1999. Modernity and Hybridity: Nature, Regeneracionismo, and the Production of the Spanish Waterscaspe, 1890–1930. Annals of the Association of American Geographers 89 (3), 443–465.

Swyngedouw, E., 2004. Social Power and the Urbanization of Water. Oxford University Press, Oxford.

Swyngedouw, E., 2005a. Dispossessing H 2 O: the contested terrain of water privatization. Capitalism Nature Socialism 16 (1), 81–98.

Swyngedouw, E., 2005b. Governance innovation and the citizen: The Janus face of governance-beyond-the-state. Urban Studies 42 (11), 1991–2006.

Swyngedouw, E., 2006a. Circulations and metabolisms. (Hybrid) Natures and (Cyborg) cities. Science as Culture 15 (2), 105–121.

Swyngedouw, E., 2006b. Metabolic urbanization: the making of cyborg cities. In: Heynen, N., Kaïka, M., Swyngedouw, E. (Eds.), In the Nature of Cities. Urban political ecology and the politics of urban metabolism. Routledge, New York, 21–40.

Swyngedouw, E., 2009. The Political Economy and Political Ecology of the Hydro-Social Cycle. Journal of Contemporary Water Research & Education (142), 56–60.

Swyngedouw, E., 2014. 'Not A Drop of Water...': State, Modernity and the Production of Nature in Spain, 1898–2010. Environment and History 20 (1), 67–92.

Swyngedouw, E., Heynen, N. C., 2003. Urban political ecology, justice and the politics of scale. Antipode 35 (5), 898–918.

Swyngedouw, E., Kaïka, M., 2000. The Environment of the City ... or the Urbanization of Nature. In: Bridge, G., Watson, S. (Eds.), A Companion to the City. Blackwell, Oxford, 567–580.

Swyngedouw, E., Kaïka, M., 2013. The Urbanization of Nature: Great Promises, Impasse, and New Beginnings. In: Bridge, G., Watson, S. (Eds.), The New Blackwell Companion to the City. Wiley / Blackwell, Oxford, 96–107.

Swyngedouw, E., Moulaert, F., Rodriguez, A., 2002. Neoliberal Urbanization in Europe: Large-Scale Urban Development Projects and the New Urban Policy. Antipode 34 (3), 542–577.

Talyan, V., Dahiya, R. P., Sreekrishnan, T. R., 2008. State of municipal solid waste management in Delhi, the capital of India. Waste Management 28 (7), 1276–1287.

Tarlo, E., 2000. Welcome to History: A Resettlement Colony in the Making. In: Dupont, V., Tarlo, E., Vidal, D. (Eds.), Delhi. Urban Space and Human Destinies. Manohar-CSH, New Delhi, 51–72.

Tarlo, E., 2003. Unsettling Memories. Narratives of the Emergency in Delhi. University of California Press, Berkeley.

Tawa Lama-Rewal, S., Ghosh, A., 2005. Democratization in Progress. Women and Local Politics in Urban India. Tulika Books, New Delhi.

Taylor, M., 2007. Community Participation in the Real World: Opportunities and Pitfalls in New Governance Spaces. Urban Studies 44 (2), 297–317.

Thakur, M. K., 2009. Global Discourses and Local Mobilisations: The Case of 'Barh Mukti Abhiyan'. Sociological Bulletin 58 (2), 212–228.

Tiwari, P., Nair, R., Ankinapalli, P., Hingorani, P., Gulati, M., 2015. India's Reluctant Urbanization. Palgrave Macmilan, New York.

Town Planning Organisation, 1956. Interim General Plan for Greater Delhi. Government of India. Delhi, http://ncrpb.nic.in/pdf_files/Interim%20General%20Plan%20for%20Greater%20Delhi.PDF (2014-11-12).

Trepl, L., 1996. City and ecology. Capitalism Nature Socialism 7 (2), 85–94.

Trivedi, R. C., 2010. Water quality of the Ganga River – An overview. Aquatic Ecosystem Health & Management 13 (4), 347–351.

Truelove, Y., Mawdsley, E., 2011. Discourses of citizenship and criminality in clean, green Delhi. In: Clark-Deces, I. (Ed.), A Companion to the Anthropology of India. Wiley-Blackwell, Chichester, 407–425.

Trumpp, T., Kraas, F., 2015. Urban Cultural Heritage in Delhi, India: An Asset for the Future or a Neglected Resource? Asien 134 (Januar 2015), 9–29.

UC (Unauthorized Colonies Cell, Department of Urban Development. Government of NCT of Delhi), 2014, New Delhi. http://www.delhi.gov.in/wps/wcm/connect/doit_udd/Urban+Development/Our+Services/Unauthorized+Colonies+Cells+%28UC%29/ (2014-12-10).

UN (United Nations), 2011. On-line Data: Urban and Rural Population. World Population Prospects: The 2010 Revision and World Urbanization Prospects: The 2011 Revision http://esa.un.org/unup/unup/index_panel1.html (2014-03-03).

UN (United Nations - Department of Economic and Social Affairs), 2014. World Urbanization Prospects. The 2014 Revision. Highlights. New York, http://esa.un.org/unpd/wup/Highlights/WUP2014-Highlights.pdf (2014-10-03).

UN Habitat (United Nations Human Settlements Programme), 2002. The Global Campaign on Urban Governance. Concept Paper, 2nd Edition, March 2002. Nairobi, http://ww2.unhabitat.org/cdrom/transparency/html/docs/y010.pdf (2014-03-03).

UN Habitat (United Nations Human Settlements Programme), 2010. State of the World's Cities Report 2010/2011. Bridging the Urban Divide. Nairobi.

UNESCAP (United Nations Economic and Social Commission for Asia and Pacific), 2005. Urban Environmental Governance. For Sustainable Development in Asia and the Pacific: A Regional Overview. Bangkok.

Upadhyay, V., 2001. Forests, People and Courts: Utilising Legal Space. Economic and Political Weekly 36 (24), 2131–2134.

Uppal, V., 2009. The Impact of Commonwealth Games 2010 on Urban Development of Delhi. Theoretical and Empirical Researches in Urban Management 1 (10), 7–29.

Uppal, V., Ghosh, D., 2006. The Impact of Commonwealth Games 2010 on Urban Development of Delhi. New Delhi. http://ideas.repec.org/p/ess/wpaper/id2456.html (2014-10-06).

UYRB (Upper Yamuna River Board), no date. Upper Yamuna River Board (UYRB) Ministry of Water Resources, Government of India. http://uyrb.gov.in/ (2014-10-06).

Vaidya, C., 2009. Urban Issues, reforms and way forward in India. Department of Economic Affairs, Ministry of Finance, Government of India, Working Paper No.4/2009-DEA - July 2009, New Delhi. http://indiagovernance.gov.in/files/urban_issues_reforms.pdf (2014-10-06).

van den Brandeler, F., Hordijk, M., von Schönfeld, K., Sydenstricker-Neto, J., 2014. Decentralization, participation and deliberation in water governance: a case study of the implications for Guarulhos, Brazil. Environment and Urbanization.

van Dijk, T., 2011. Networks of Urbanization in Two Indian Cities. Environment and Urbanization Asia 2 (2), 303–319.

Varma Pakalapati, U., 2010. Hi-tech Hyderabad and the Urban Poor: Reformed Out of the System. In: Banerjee-Guha, S. (Ed.), Accumulation by dispossession: transformative cities in the new global order. Sage, New Delhi, 125–150.

Véron, R., 2006. Remaking urban environments the political ecology of air pollution in Delhi. Environment and Planning A 38 (11), 2093–2109.

Viehoff, V., 2009. Engineering Modernity: The provision of water for Tangier 1840–1956. Department of Geography, University College London (Doctoral thesis). London.

Vijay, R., Sargoankar, A., Gupta, A., 2007. Hydrodynamic simulation of river Yamuna for riverbed assessment: a case study of Delhi region. Environ Monit Assess 130 (1–3), 381–7.

Wachsmuth, D., 2012. Three Ecologies: Urban Metabolism and the Society-Nature Opposition. The Sociological Quarterly 53 (4), 506–523.

Walia, A., Mehra, N. K., 1998a. A Seasonal Assessment of the Impact of Coal Fly Ash Disposal on the River Yamuna, Delhi, II. Biology. Water, Air, and Soil Pollution 103 (1–4), 315–339.

Walia, A., Mehra, N. K., 1998b. A Seasonal Assessment of the Impact of Coal Fly Ash Disposal on the River Yamuna, Delhi. I. Chemistry. Water, Air, and Soil Pollution 103 (1–4), 277–314.

Walker, P. A., 2005. Political ecology: where is the ecology? Progress in Human Geography 29 (1), 73–82.

Walker, P. A., 2006. Political ecology: where is the policy? Progress in Human Geography 30 (3), 382–395.

Walker, P. A., 2007. Political ecology: where is the politics? Progress in Human Geography 31 (3), 363–369.

WAPCOS (Water and Power Consultancy Services), 2003. Pilot Study on Water Use Efficiency for Western Yamuna Canal. Planning Commission, GoI. New Delhi, http://planning commission.nic.in/reports/sereport/ser/index.php?repts=stdy_wjc.htm (2014-10-09).

Watson, V., 2009. 'The planned city sweeps the poor away…': Urban planning and 21st century urbanisation. Progress in Planning 72 (3), 151–193.

Watts, M., 2008. Political Ecology. In: Sheppard, E., Barnes, T. J. (Eds.), A Companion to Economic Geography. Blackwell, Oxford, 181–197.

Weinstein, L., 2009. Redeveloping Dharavi: Toward a political economy of slums and slum redevelopment in globalizing Mumbai. University of Chicago (PhD dissertation, Doctor of Philosophy). Chicago.

Wescoat Jr, J., White, G., 2003. Water for Life. Water Management and Environmental Policy. Cambridge University Press, Cambridge.

Whatmore, S., 1999. Hybrid geographies: rethinking the human in human geography. In: Massey, D., Allen, J., Sarre, P. (Eds.), Human Geography Today. Polity Press, Cambridge, 24–39.

Whatmore, S., 2002. Hybrid geographies: natures; cultures, spaces. Sage, London.

Whatmore, S., Boucher, S., 1993. Bargaining with Nature: The Discourse and Practice of 'Environmental Planning Gain'. Transactions of the Institute of British Geographers 18 (2), 166–178.

Whatmore, S. J., 2009. Mapping knowledge controversies: science, democracy and the redistribution of expertise. Progress in Human Geography 33 (5), 587–598.

White, R., 1995. The Organic Machine: The Remaking of the Columbia River. Hill and Wang, New York.

Whitehead, M., 2013. Neoliberal Urban Environmentalism and the Adaptive City: Towards a Critical Urban Theory and Climate Change. Urban Studies 50 (7), 1348–1367.

Wiens, J., 2005. Riverine landscapes: taking landscape ecology into the water. Freshwater Biology 47, 501–515.

Williams, G., 2009. Governance, Good. In: Kitchin, R., Thrift, N. (Eds.), International Encyclopedia of Human Geography. Elsevier, Oxford, 606–614.

Williams, G., Mawdsley, E., 2006a. India's Evolving Political Ecologies. In: Raju, S., Kumar, M. S., Corbridge, S. (Eds.), Colonial and Postcolonial Geographies of India. Sage, New Delhi, 261–278.

Williams, G., Mawdsley, E., 2006b. Postcolonial environmental justice: Government and governance in India. Geoforum 37 (5), 660–670.

Williams, R., 1983. Keywords: a vocabulary of culture and society. Flamingo, London.

Winchester, H., Kong, L., Dunn, K., 2003. Landscapes - Ways of Imagining the World. Routledge, New York.

Winchester, H., Rofe, M., 2010. Qualitative Research and its Place in Human Geography. In: Hay, I. (Ed.), Qualitative Research Methods in Human Geography. Oxford University Press, Oxford, 3–25.

Wisner, B., Blaikie, P., Cannon, T., Davis, I., 2004. At Risk. Natural hazards, people's vulnerability and disasters. Routledge, London, New York.

Witzel, A., 2000. The Problem-Centered Interview. Forum Qualitative Sozialforschung / Forum: Qualitative Social Research 1 (1). http://www.qualitative-research.net/index.php/fqs/article/view/1132/2521 (2014-05-05).

Witzel, A., Reiter, H., 2012. The Problem-Centred Interview. Sage, London.

WOAT (World Organization against Torture, Habitat International Coalition-Housing and Land Rights Network), 2004. Joint Urgent Action Appeal, 'Over 300,000 people to be forcefully evicted from Yamuna Pushta in Delhi: 40,000 homes demolished so far', Case IND-FE050504 (Delhi, Geneva, Cairo, 5 May 2004). http://www.omct.org/escr/urgent-interventions/india/2004/05/d2319/ (2014-05-05).

Wolman, A., 1965. The metabolism of cities. Scientific American 213, 179–190.

World Meteorological Organization 2015. World Weather Information Service - New Delhi India (based on India Meteorological Department). http://worldweather.wmo.int/en/city.html?cityId=224 (2015-04-01).

Worster, D., [1985] 1992. Rivers of empire: water, aridity, and the growth of the American West, Oxford.

Wylie, J., 2007. Landscape. Routledge, London.

Yang, G., Calhoun, C., 2007. Media, Civil Society, and the Rise of a Green Public Sphere in China. China Information 21 (2), 211–236.

Yates, J. S., Gutberlet, J., 2011. Reclaiming and recirculating urban natures: integrated organic waste management in Diadema, Brazil. Environment and Planning A 43 (9), 2109–2124.

YRDA (Yamuna River Development Authority), 2007. First Three-Monthly Action Taken Report (August–November 2007). Unpublished material accessed via RTI filed by Yamuna Jiye Abhiyaan.

Zalasiewicz, J., Williams, M., Steffen, W., Crutzen, P., 2010. The New World of the Anthropocene. Environmental Science & Technology 44 (7), 2228–2231.

Zérah, M.-H., 2000. Water: Unreliable Supply in Delhi. Manohar Publishers, New Delhi.

Zérah, M.-H., 2009. Participatory Governance in Urban Management and the Shifting Geometry of Power in Mumbai. Development and Change 40 (5), 853–877.

Zérah, M.-H., Dupont, V., Tawa Lama-Rewal, S. (Eds.), 2011. Urban Policies and the Right to the City in India: Rights, Responsibilities and Citizenship. United Nations Educational, Scientific and Cultural Organization (UNESCO) & Centre de Sciences Humaines (CSH), New Delhi.

Zérah, M.-H., Landy, F., 2013. Nature and urban citizenship redefined: The case of the National Park in Mumbai. Geoforum 46 (0), 25–33.

Zérah, M. H., 2007. Conflict between green space preservation and housing needs: The case of the Sanjay Gandhi National Park in Mumbai. Cities 24 (2), 122–132.

Zierhofer, W., 1999. Geographie der Hybriden. Erdkunde 53 (1), 1–13.

Zimmer, A., 2010. Urban Political Ecology. Theoretical concepts, challenges, and suggested future directions. Erdkunde 64 (4), 343–354.
Zimmer, A., 2012a. Enumerating the Semi-Visible: The Politics of Regularising Delhi's Unauthorised Colonies. Economic and Political Weekly 47 (30), 89–97.
Zimmer, A., 2012b. Everyday governance of the waste waterscapes. A Foucauldian analysis of Delhi's informal settlements. Universität zu Bonn, Bonn. http://hss.ulb.uni-bonn.de/2012/2956/2956.pdf.
Zimmer, A., 2015a. Urban Political Ecology in Megacities: The Case of Delhi's Waste Water In: Singh, R. B. (Ed.), Urban Development Challenges, Risks and Resilience in Asian Mega Cities. Springer, Tokyo, 119–140.
Zimmer, A., 2015b. Urban political ecology 'beyond the West': engaging with South Asian urban studies In: Bryant, R. L. (Ed.), The International Handbook of Political Ecology. Edward Elgar, Cheltenham, UK & Northampton MA, USA, 591–603.
Zimmer, A., Sakdapolrak, P., 2012. The Social Practices of Governing: Analysing Waste Water Governance in a Delhi Slum. Environment and Urbanization Asia 3 (2), 325–341.
Zimmerer, K. S., 2009. Biodiversity. In: Castree, N., Demeritt, D., Liverman, D., Rhoads, B. (Eds.), A Companion to Environmental Geography. Wiley-Blackwell, Chichester, 50–65.
Zimmerer, K. S., Bassett, K., 2003. Approaching Political Ecology. Society, Nature, and Scale im Human-Environment Studies. In: Zimmerer, K. S., Bassett, K. (Eds.), Political Ecology: An Integrative Approach to Geography and Environment-Development Studies. Guilford Press, New York, 1–28.
Zukin, S., 1991. Landscapes of Power. University of California Press, Berkeley.

APPENDIX

Appendix I: List of interviews

No.	Code	Date	Actor	Length	Transcript*
1	DDA-01/	2009-12-02	DDA Planning Unit	-	mm
2	NGO-01/	2010-02-16	NGO	62 min	ft
3	NGO-02/	2010-02-17	NGO	45 min	ft
4	NGO-03/	2010-03-26	NGO	52 min	ft
5	NGO-04/	2010-04-01	NGO	67 min	ft
6	MCD-01/	2010-09-17	MCD Slum Department	45 min	mm
7	NGO-04/	2010-09-18	NGO	44 min	ft
8	NGO-05/	2010-09-18	NGO	65 min	ft
9	MCD-02/	2010-09-20	MCD Slum Department	-	mm
10	DDA-01/	2010-09-21	DDA Planning Unit	69 min	ft
11	DTTDC/	2010-09-21	DTTDC	32 min	mm
12	DDA-02/	2010-09-23	DDA (retired)	-	ft
13	ARCH-01/	2010-09-23	-	22 min	ft
14	NGO-06/	2010-09-24	NGO	42 min	ft
15	IFCD/	2010-09-27	IFCD	45 min	ft
16	CEMDE-01/	2011-03-01	CEMDE	40 min	ft
17	NGO-07-	2011-03-02	NGO	92 min	ft
18	DDA-03/	2011-03-03	DDA Landscape Unit	90 min	ft
19	SCIENTIST-01/	2011-03-04	IIT Delhi	34 min	ft
20	NGO-04/	2011-03-05	NGO	127 min	tt
21	NGO-05/	2011-03-05	NGO	42 min	ft
22	CEMDE-02/	2011-03-09	CEMDE	49 min	ft
23	FARM-01/	2011-03-09	Farmer East Delhi, Delhi Peasants	50 min	ft
24	NGO-03/	2011-11-04	NGO	100 min	ft
25	NGO-04/	2011-11-07	NGO	107 min	ft
26	SCIENTIST-02/	2011-11-08	former CGWB	90 min	ft
27	NGO-08/	2011-11-12	NGO	62 min	ft
28	NGO-09/	2011-11-12	NGO	62 min	ft
29	NGO-10/	2011-11-14	NGO	39 min	ft
30	CGWB/	2011-11-16	CGWB	27 min	ft
31	CPCB/	2011-11-16	CPCB	48 min	ft
32	ARCH-02/	2011-11-17	-	72 min	ft
33	NGO-11/	2011-11-18	NGO	41 min	ft
34	DDA-04/	2011-11-18	DDA Landscape Unit	82 min	ft

Appendix

No.	Code	Date	Actor	Length	Transcript*
35	NGO-12/	2011-11-20	NGO	69 min	ft
36	DJB/	2011-11-21	DJB	-	mm
37	DDA-05/	2011-11-22	DDA Planning Unit	-	mm
38	DDA-06/	2011-11-22	DDA Planning Unit	52 min	ft
39	DDA-07/	2011-11-23	DDA (retired)	85 min	ft
40	NGO-13/	2011-11-24	NGO	87 min	ft
41	SCIENTIST-03/	2011-11-25	former CPCB	84 min	ft
42	NIDRM-/	2011-11-28	NIDRM	35 min	ft
43	FARM-02/	2011-12-04	Farmer Bela Estate (Hindi)	106 min	tn
44	FARM-03/	2011-12-04	Farmer Bela Estate (Hindi)	60 min	tn
45	FARM-04/	2011-12-04	Farmer Bela Estate (Hindi)	14 min	tn
46	FARM-05/	2011-12-04	Farmer Bela Estate (Hindi)	20 min	tn
47	FARM-06/	2011-12-04	Farmer Bela Estate (Hindi)	32 min	tn
48	FARM-07/	2011-12-04	Farmer Bela Estate (Hindi)	10 min	tn
49	DDA-H/	2011-12-05	DDA Horticulture Unit, Bela Estate	15 min	mm
50	LG/	2011-12-05	Lieutenant Governor of Delhi	-	mm
51	MoUD/	2011-12-05	MoUD	43 min	mm
52	FARM-01/	2011-12-06	Farmer East Delhi, Delhi Peasants	59 min	ft
53	NGO-04/	2011-12-06	NGO	92 min	ft
54	DDA-08/	2011-12-07	DDA	-	ft
55	DDA-09/	2011-12-07	DDA Engineering Unit	40 min	mm
56	NGO-14/	2011-12-07	NGO	50 min	ft
57	NGO-15/	2011-12-08	NGO	75 min	ft
58	NGO-02/	2011-12-09	NGO	49 min	ft
59	ARCH-01/	2011-12-09	Architect	48 min	ft
60	CEMDE-03/	2011-12-10	CEMDE	-	mm
61	DDA-10/	2013-02-08	DDA (retired)	55 min	ft
62	NGO-16/	2013-02-10	NGO	60 min	ft
63	NGO-04/	2013-02-11	NGO	55 min	ft
64	NGO-17/	2013-02-11	NGO	66 min	ft
65	NGO-18/	2013-02-13	NGO	38 min	ft
66	NGO-19/	2013-02-13	NGO	-	mm
67	LAW-01/	2013-02-14	-	57 min	ft
68	IFCD/	2013-02-15	IFCD	38 min	ft
69	FARM-08/	2013-03-06	Farmer Bela Estate (Hindi)	20 min	tn
70	LAW-02/	2013-03-06	-	22 min	ft
71	FARM-02/	2013-03-06	Farmer Bela Estate (Hindi)	51 min	tn
72	NGO-20/	2013-03-06	NGO	51 min	mm

*ft = full transcript; mm = memory minutes; tn = notes, translated by field assistant

Appendix II: List of minutes of the Yamuna Standing Committee (33–83)

No.	Date	Code	No.	Date	Code
33	23.05.1975	YSC/33/1975-05-23	59	19.04.1999	YSC/59/1999-04-19
34	16.12.1976	YSC/34/1976-12-16	60	27.06.2000	YSC/60/2000-06-27
35	25.04.1978	YSC/35/1978-04-25	61	08.08.2001	YSC/61/2001-08-08
36	28.10.1978	YSC/36/1978-10-28	62	31.07.2002	YSC/62/2002-07-31
37	26.04.1979	YSC/37/1979-04-26	63	06.06.2003	YSC/63/2003-06-06
38	20.07.1979	YSC/38/1979-07-20	64	29.10.2003	YSC/64/2003-10-29
39	26.09.1979	YSC/39/1979-09-26	65	27.04.2004	YSC/65/2004-04-27
40	01.10.1980	YSC/40/1980-10-01	66	23.08.2004	YSC/66/2004-08-23
41	03.10.1981	YSC/41/1981-10-03	67	27.09.2004	YSC/67/2004-09-27
42	05.06.1982	YSC/42/1982-06-05	68	16.03.2005	YSC/68/2005-03-16
43	10.12.1982	YSC/43/1982-12-10	69	08.09.2005	YSC/69/2005-09-08
44	18.05.1984	YSC/44/1984-05-18	70	11.11.2005	YSC/70/2005-11-11
45	28.08.1985	YSC/45/1985-08-28	71	28.07.2006	YSC/71/2006-07-28
46	22.09.1988	YSC/46/1988-09-22	72	08.01.2007	YSC/72/2006-01-08
47	21.12.1988	YSC/47/1988-12-21	73	18.03.2008	YSC/73/2008-03-18
48	17.02.1989	YSC/48/1989-02-17	74	15.07.2008	YSC/74/2008-07-15
49	25.05.1989	YSC/49/1989-05-25	75	23.02.2009	YSC/75/2009-02-23
50	26.02.1990	YSC/50/1990-02-26	76	07.08.2009	YSC/76/2009-08-07
51	25.06.1991	YSC/51/1991-06-25	77	08.01.2010	YSC/77/2010-01-08
52	10.05.1993	YSC/52/1993-05-10	78	17.06.2011	YSC/78/2011-06-17
53	23.08.1994	YSC/53/1994-08-23	79	27.01.2012	YSC/79/2012-01-27
54	26.06.1995	YSC/54/1995-06-26	80	06.03.2012	YSC/80/2012-03-06
55	15.07.1996	YSC/55/1996-07-15	81	14.12.2012	YSC/81/2012-12-14
56	08.04.1997	YSC/56/1997-04-08	82	19.07.2013	YSC/82/2013-07-19
57	18.11.1997	YSC/57/1997-11-18	83	27.09.2013	YSC/83/2013-09-27
58	17.02.1998	YSC/58/1998-02-17			

Minutes accessed by Yamuna Jiye Abhiyaan

Appendix III: Minutes Meetings of TAG, YRDA

No.	Date	Code
1st	22.11.2007	TAG-YRDA/1/2007-11-22
2nd	07.03.2008	TAG-YRDA/2/2008-03-07
3rd	23.05.2008	TAG-YRDA/3/2008-05-23
4th	23.09.2008	TAG-YRDA/4/2008-09-23
5th	14.01.2009	TAG-YRDA/5/2009-01-14
6th	29.04.2009	TAG-YRDA/6/2009-04-29
7th	30.06.2009	TAG-YRDA/7/2009-06-30
8th	26.08.2009	TAG-YRDA/8/2009-08-26
9th	10.11.2009	TAG-YRDA/9/2009-11-10
10th	04.12.2009	TAG-YRDA/10/2009-12-04
11th	21.01.2010	TAG-YRDA/11/2010-01-21
12th	24.02.2010	TAG-YRDA/12/2010-02-24
13th	11.03.2010	TAG-YRDA/13/2010-03-11
14th	08.04.2010	TAG-YRDA/14/2010-04-08

Minutes accessed via RTI, Yamuna Jiye Abhiyaan

Appendix IV: Zonal Development Plan (ZDP) for River Zone (Zone O)

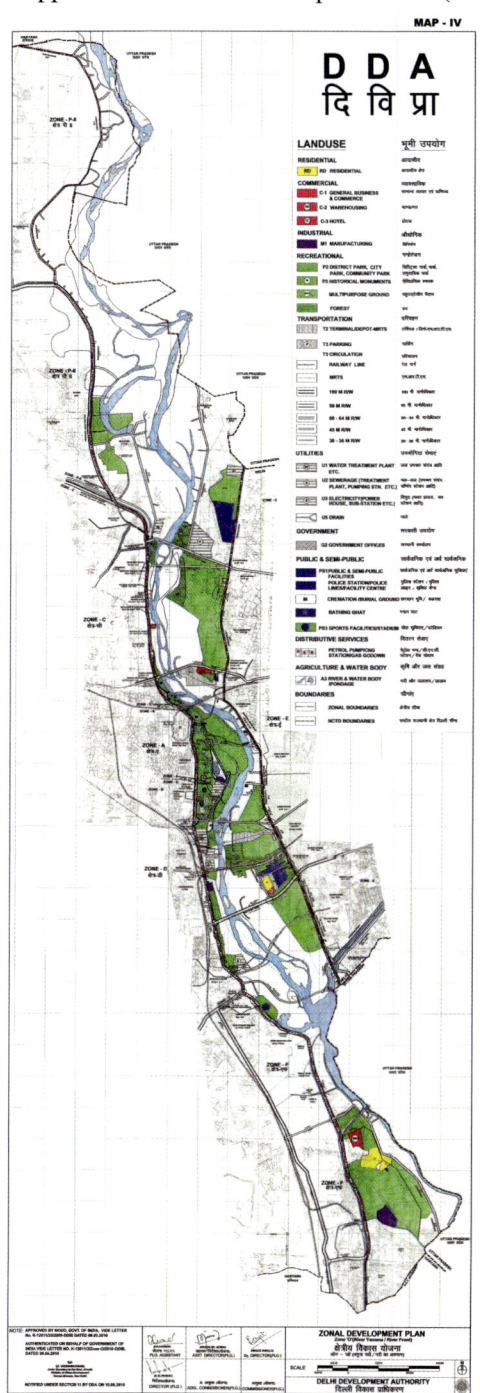

Source: DDA (Delhi Development Authority) 2010. Land Use Plan - Zonal Development Plan for River Yamuna / River Front, Zone O (Map IV). New Delhi, http://dda.org.in/tender-notices_docs/august10/Landuse%20Plan%20of%20Zone%20O%20Notified%20on%2010Aug10.jpg (28 November 2012).

382   Appendix

Appendix V: Master Plan for Delhi 1962 (MPD)

Source: DDA (1962),
Online available: http://dda.org.in/planning/map_mpd_1962.htm (last access 15th April 2015)

Appendix VI: Delhi Master Plan 2001 (MPD-2001)

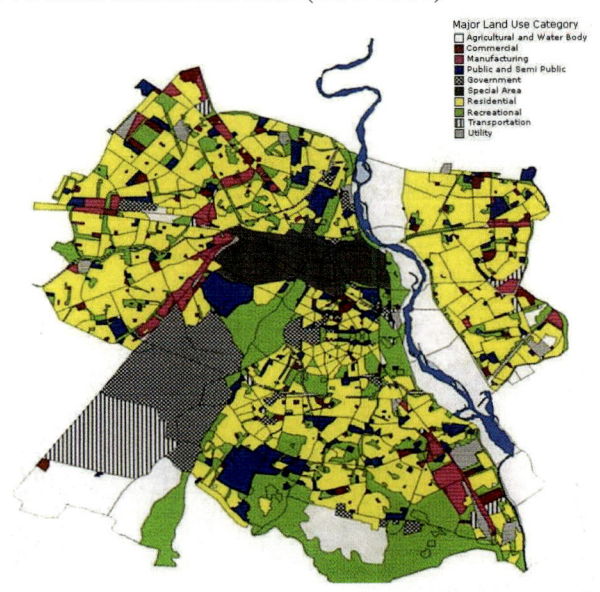

Source: DDA (1990),
Online available: https://dda.org.in/planning/map_mpd_2001.htm (last access 15th April 2015)

Appendix VII: Proposed Structure Plan for Biodiversity Zone

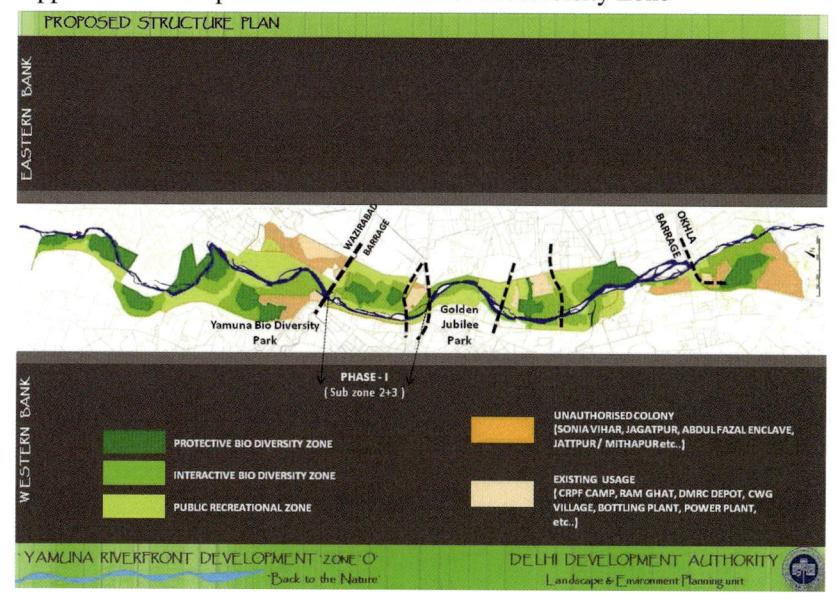

Source: Public Notice DDA for Yamuna Riverfront Development, April 2010 (DDA 2010a)

Appendix VIII: The riverfront promenade – public recreation zone

Source: DDA Landscape Unit 2011
(published also in the Times of India, 4th October 2011: 'Holiday by the Yamuna in 2015')

384   Appendix

Appendix IX: Draft ZDP for O-Zone prepared by DDA 2006

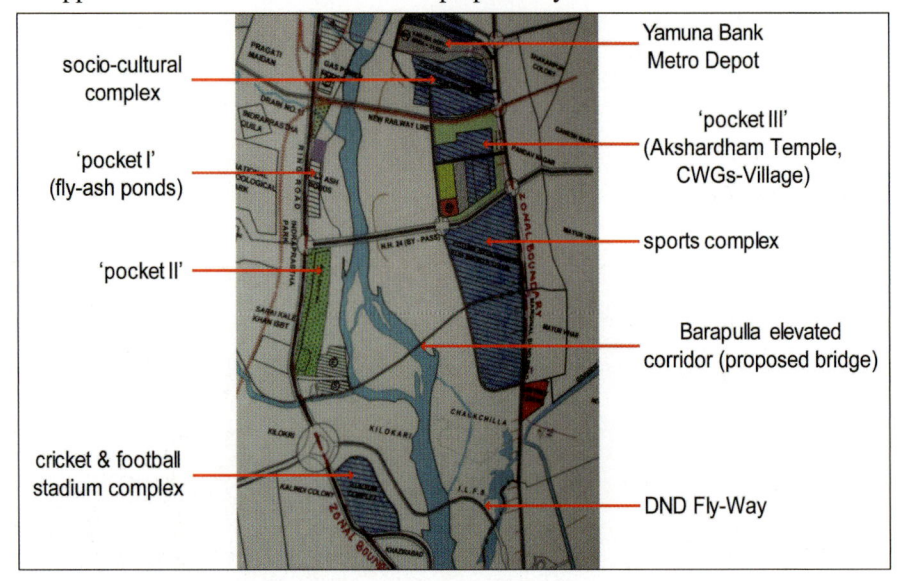

Source: adapted from photography of the draft ZDP-2006 (DDA 2006d), retrieved from presentation by Yamuna Jiye Abhiyaan dated 3rd December 2007 (online available at http://www.peaceinst.org/publication/book-let/Yamuna %20Jiye%20Abhiyaan%20-%203.12.2007.pdf, last access 7th April 2015); for legend see Appendix V: Zonal Development Plan (ZDP) for River Zone (Zone O)

Appendix X: Model for channelization of Yamuna River (ca. 1986)

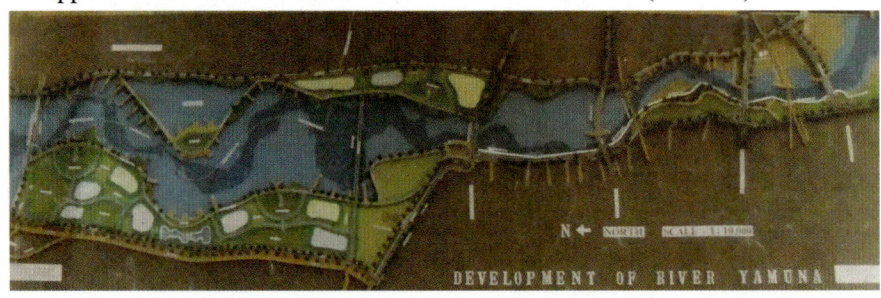

Source: Project report Yamuna Board DDA, prepared 1992–1994 (materials provided by R.G. Gupta)

MAPS

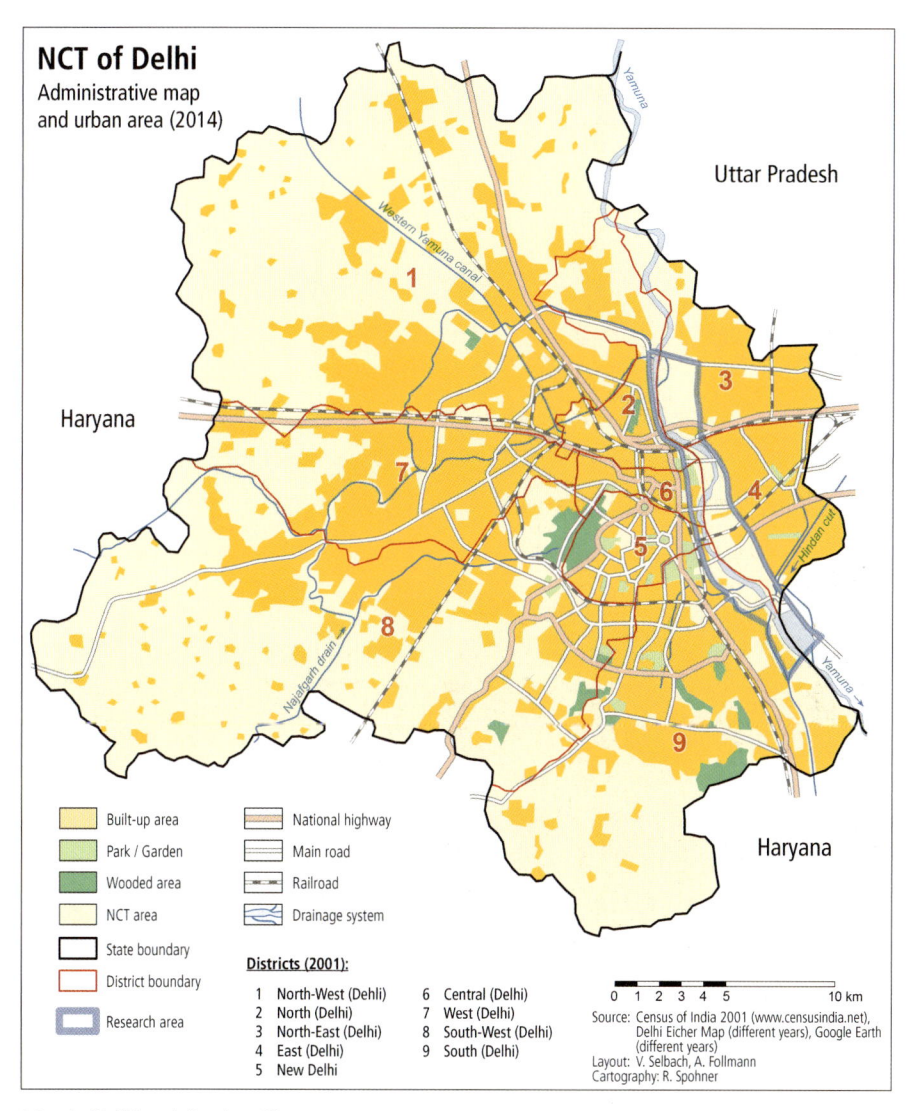

Map 1: Delhi and the river Yamuna

386 Maps

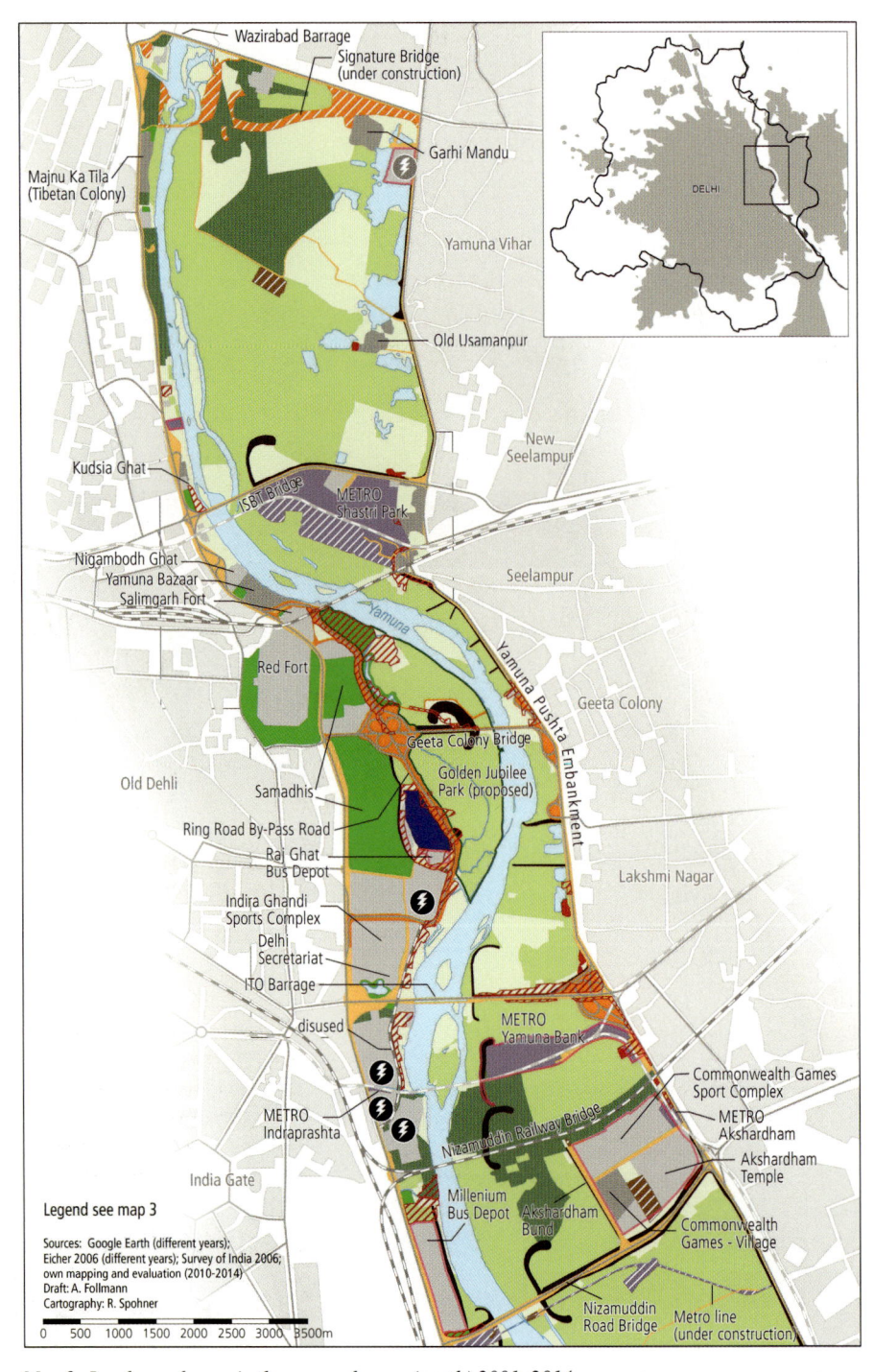

Map 2: Land-use change in the research area (north) 2001–2014

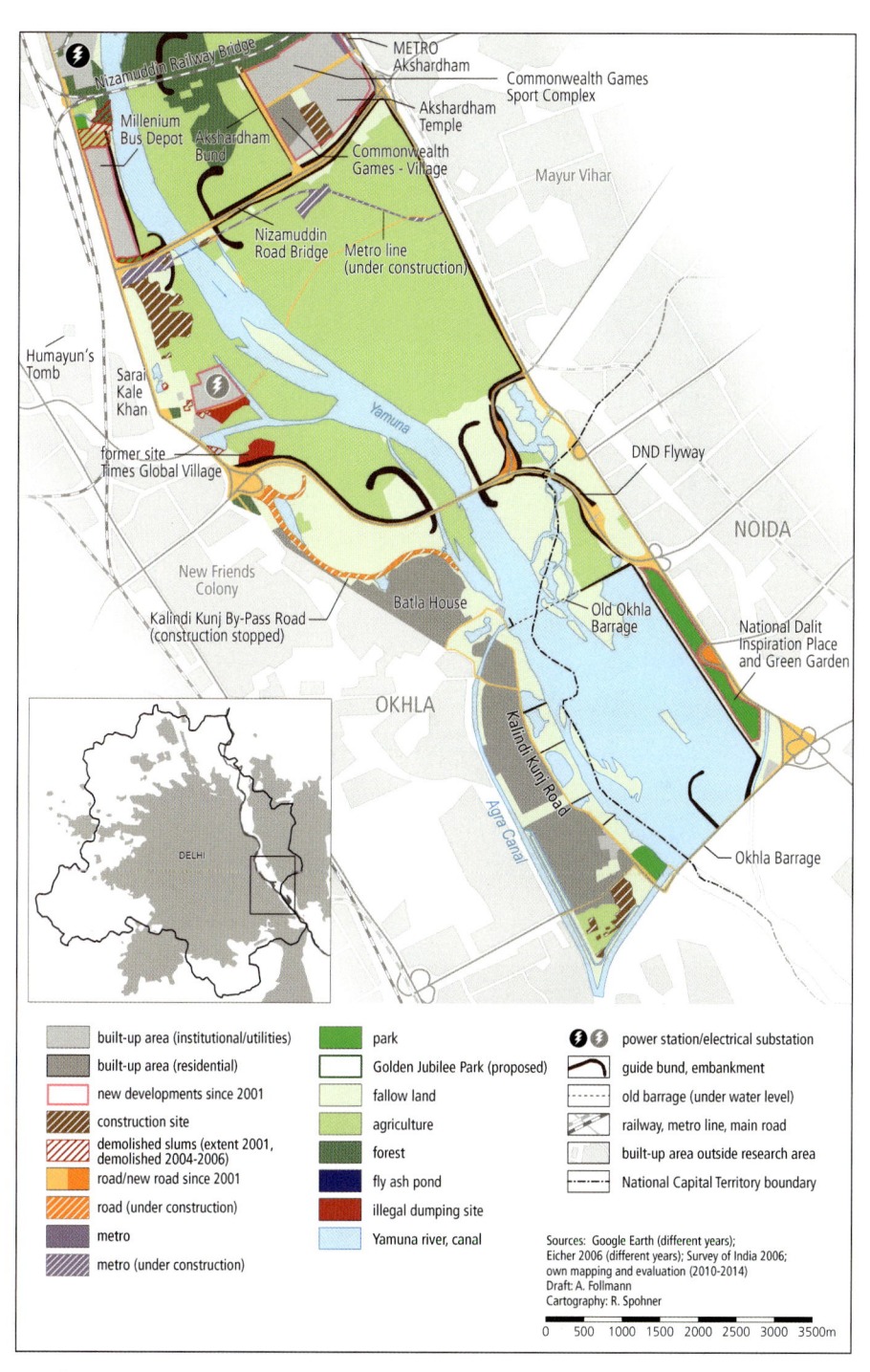

Map 3: Land-use change in the research area (south) 2001–2014

388　　　　　　　　　　　　　　　　　　　Maps

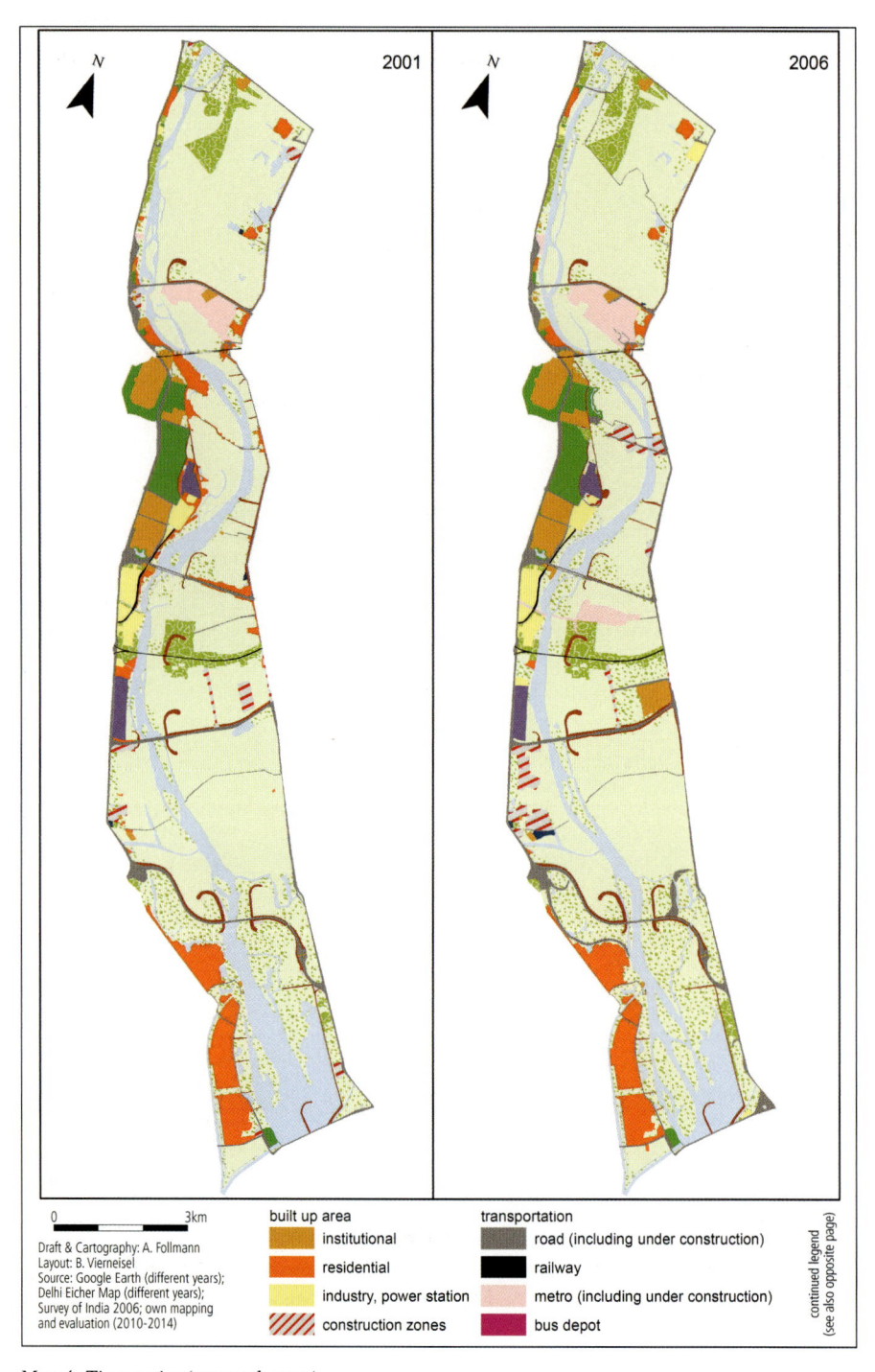

Map 4: Time series (research area)

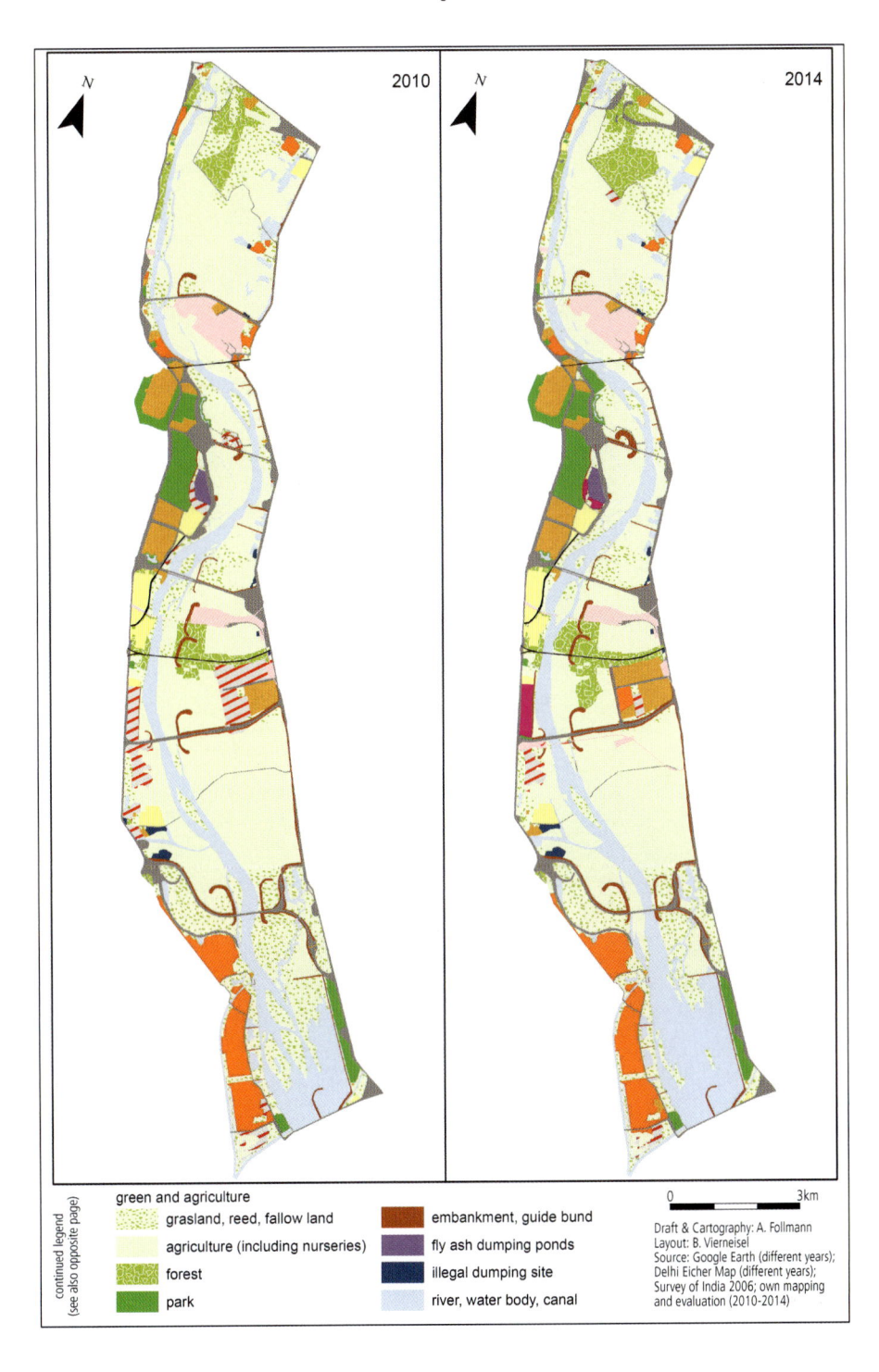

Maps

389

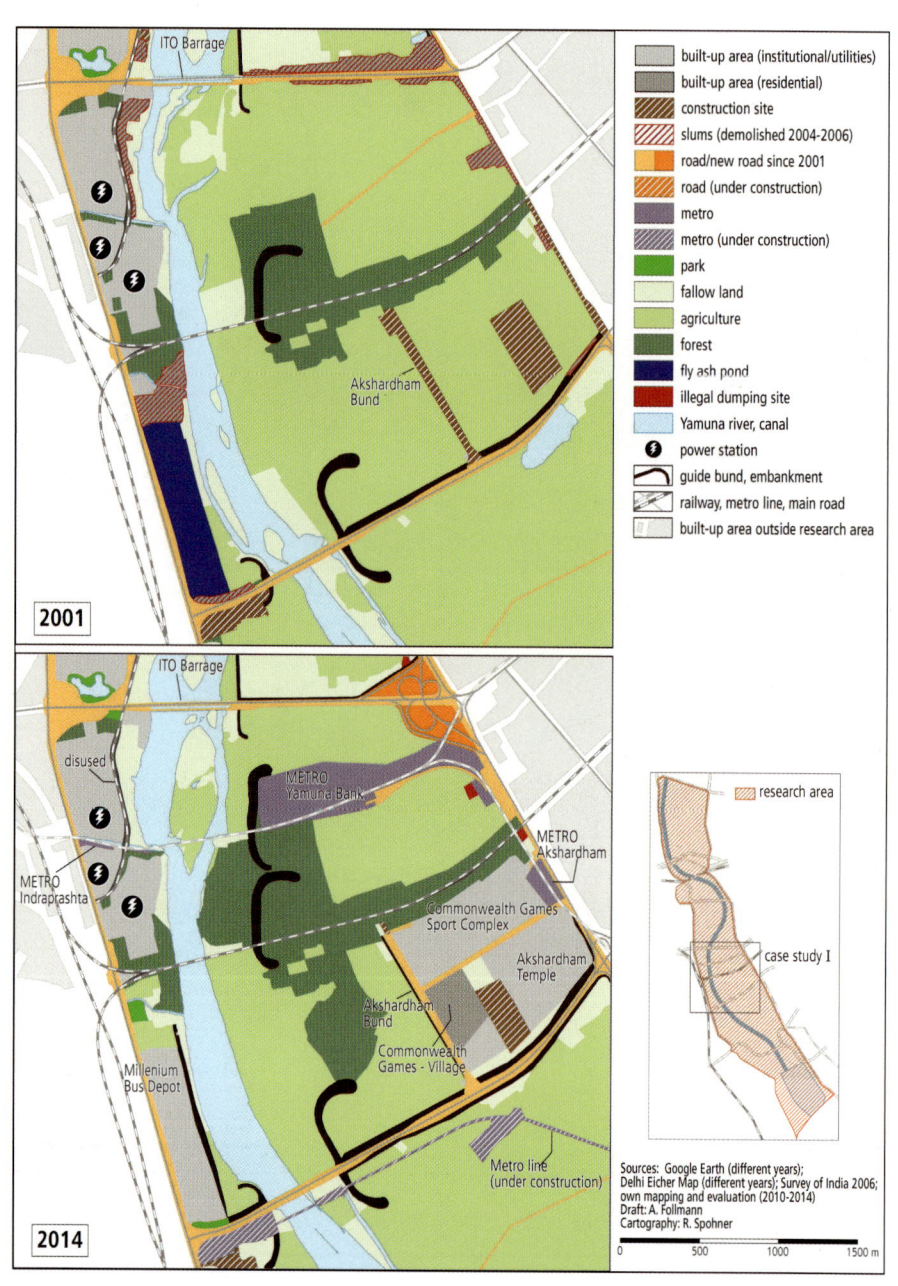

Map 5: The development of the interrelated mega-projects (focus area I)

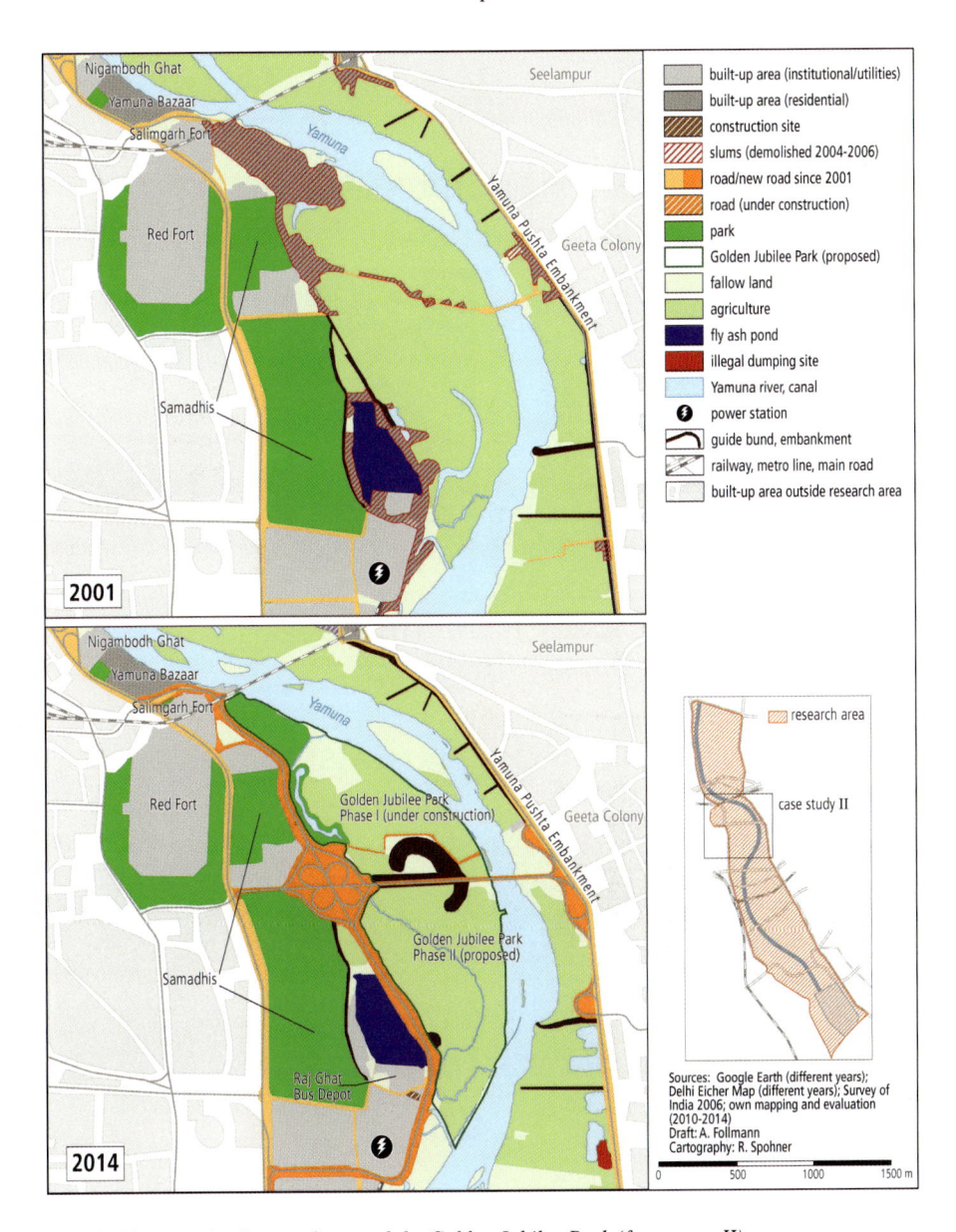

Map 6: *Changing land uses: slums and the Golden Jubilee Park (focus area II)*

COLOR PHOTOGRAPHS

Figure 50: Pollution of the river Yamuna, eastern bank, near CWGs-Village
Source: Follmann, November 2011

Figure 51: Pollution river Yamuna near ghats, Yamuna Bazar area
Source: Follmann, February 2011

Figure 52: Riverfront promenade along the river Musi in Hyderabad, Telangana
Source: Follmann, September 2014

Figure 53: Golden Jubilee Park along the Yamuna
Source: Follmann, February 2011

Color photographs 393

Figure 54: Agriculture and temporary settlements of farmers along the Yamuna
Source: Follmann, March 2013

Figure 55: Plant nursery, Yamuna Bank area
Source: Follmann, February 2011

Figure 56: Fencing off the river, Kalindi Kunj, South Delhi
Source: Follmann, November 2011

Figure 57: Okhla Bird Sanctuary, Kalindi Kunj, South Delhi
Source: Follmann, November 2011

Figure 58: Unauthorized Colony, Jogabai Extension, Kalindi Kunj area
Source: Follmann, November 2011
Note: In the foreground slum cluster and informal waste recycling

Figure 59: Unauthorized Colony, Majnu-Ka-Tila (New Aruna Nagar)
Source: Follmann, September 2014

Figure 60: Akshardham Temple
Source: Follmann, September 2014

Figure 61: CWGs-Village Entry Gate
Source: Follmann, September 2010

Figure 62: Construction work on the Akshardham Bund towards the river
Source: Follmann, December 2009

Figure 63: Akshardham Bund and CWGs-Village
Source: Follmann, February 2011

Color photographs 395

Figure 64: Farming Bela Estate near Golden Jubilee Park
Source: Follmann, March 2013

Figure 65: Former construction workers camp near CWGs-Village
Source: Follmann, November 2011

Figure 66: Yamuna Biodiversity Park Phase I
Source: Follmann, February 2011

Figure 67: Yamuna Biodiversity Park Phase II
Source: Follmann, November 2011

Figure 68: Yamuna Biodiversity Park Phase I Amphitheatre
Source: Follmann, February 2011

Figure 69: Yamuna Biodiversity Park Phase I Butterfly Park
Source: Follmann, February 2011

Figure 70: Golden Jubilee Park during flood of 2010
Source: Follmann, September 2010

Figure 71: Golden Jubilee Park
Source: Follmann, February 2011

Figure 72: Model for the development of Golden Jubilee Park (phase I)
Source: Follmann, November 2011, picture from model prepared by DDA and displayed during Yamuna-Elbe exhibition at the Golden Jubilee Park